FOUNDATIONS OF
GENETIC
ALGORITHMS·2

FOUNDATIONS OF
GENETIC
ALGORITHMS·2

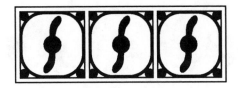

EDITED BY
L. DARRELL WHITLEY

MORGAN KAUFMANN PUBLISHERS
SAN MATEO, CALIFORNIA

Senior Editor: Bruce M. Spatz
Production Manager: Yonie Overton
Production Artist/Cover Design:
Susan M. Sheldrake

Morgan Kaufmann Publishers, Inc.

Editorial Office:
2929 Campus Drive, Suite 260
San Mateo, CA 94403

96 95 94 93   5 4 3 2 1

Library of Congress Cataloging in Publication Data is available for this book.
ISBN 1-55860-263-1

Dedicated to
Gunar E. Liepins

FOGA–92

THE PROGRAM COMMITTEE

Lashon Booker, *MITRE Corporation*

Lawrence Davis, *TICA Associates*

Stephanie Forrest, *University of New Mexico*

John Grefenstette, *Naval Research Laboratory*

Gregory J.E. Rawlins, *Indiana University*

Michael Vose, *University of Tennessee*

Kenneth A. DeJong, *George Mason University*

Larry Eshelman, *Philips Laboratories*

David Goldberg, *University of Illinois*

Gunar Liepins, *Oak Ridge National Laboratory*

Stephen F. Smith, *Carnegie Mellon University*

Darrell Whitley, *Colorado State University*

Contents

Part 5: Genetic Operators and Their Analysis

Part 6: Machine Learning

Author Index

Key Word Index

Introduction

The second workshop on Foundations of Genetic Algorithms (FOGA) was held July 26-29, 1992 in Vail, Colorado. Like the first FOGA workshop held in 1990 (Rawlins, 1991), the purpose of this workshop is to create a forum for theoretical work on genetic algorithms. It has often been difficult for researchers interested in the theoretical foundation of genetic algorithms to keep up with recent work in this area. Several important publications dealing with more theoretical aspects of genetic algorithms are scattered across different journals. One of the goals of FOGA is to not only provide a snapshot of current theoretical work in genetic algorithms, but also to help define some of the issues that could shape the direction of future research.

Since the purpose of FOGA is to provide a more visible forum for theoretical work on genetic algorithms, an effort has been made to make *Foundations of Genetic Algorithms -2-* both a high quality as well as a detailed collection of papers. The papers are much longer than the typical conference paper. In addition, papers went through a first round of reviewing to select the papers that appear in the current volume. A second round of reviewing and editing was done to improve the presentation and clarity of the papers.

There already exists a good general introductory textbook on genetic algorithms (Goldberg, 1989) and John Holland's classic work *Adaptation in Natural and Artificial Systems* (1975) has just been republished by MIT Press in an expanded, somewhat revised form (1992). Therefore, there is little need to present a basic introduction to genetic algorithms. The paper in the current volume by Ken DeJong, *Genetic Algorithms are NOT Function Optimizers*, provides a general overview of a "canonical" genetic algorithm and in many ways serves as a good starting point from which to approach many of the other papers in the book. The theme of DeJong's paper also appears to have the same intent as statements made by Holland in the new introduction to his classic text. Genetic algorithms were not originally designed to be function optimizers. Rather function optimization is an emergent property of genetic algorithms; the theoretical results that have been derived with respect to the canonical genetic algorithm also suggest that the canonical genetic algorithm is good at maintaining a population of robust individuals, especially in situations of competition and environmental change. In this regard, Holland makes the following statements concerning adaptive systems in the preface to the new edition of his book:

> I would ... put more emphasis on improvement and less on optimization. Work on the more complex adaptive systems–ecologies, for example–has convinced me that their behavior is not well described by the trajectories around global optima. Even when a relevant global optimum can be defined, the system is typically so "far away" from that optimum that basins of attraction, fixed points, and the other apparatus used in studying optima tell little about the system's behavior. Instead, competition between components of the system, aimed at "getting an edge" over neighboring competitors, determines the aggregate behavior (Holland, 1992, p. x).

This view of genetic algorithms is quite different from the view that has become part of the folklore, in which genetic algorithms are assumed to converge to the global optimum of a function space. This is not to say that genetic algorithms do not display a different, more global sampling of the search space than more traditional types of search algorithms. But at the same time a more global sampling of the search space does not automatically imply convergence to a global optimum. It also suggests that the term genetic algorithm is being used in two very different ways by researchers who may have very different goals. First, it refers to a specific adaptive plan, the canonical genetic algorithm around which most of the theory has been developed (Holland, 1975). Second, it is often used to refer to various members of a class of algorithms which utilize populations of strings and recombination operators. Several arguments about computational behavior and "how genetic algorithm really work" can be traced back to this confusion. It is time that we paid more attention to this distinction.

Researchers specifically interested in function optimization often modify the genetic algorithm, or develop new variants of "genetic algorithms." While experimentally oriented researchers point to the results of the schema theorem as a theoretical anchor, these new algorithms typically do not process strings or schemata in the same way as the canonical genetic algorithm and it is unclear what part of the schema theorem, if any, applies to these new algorithms. *This should not be interpreted* as a statement that genetic algorithms in their various forms should not be used as function optimizers. Rather, it is merely an acknowledgement that experimentation is currently leading the way in this area of research and that the existing theory provides only guiding principles which are largely based on indirect references back to Holland's canonical genetic algorithm.

This represents a gap in our theory. The models which are introduced in the second section of this book are also based on the canonical genetic algorithm, as are many of the analyses found in other papers in this collection. But it should be possible to extend these models to other variants of genetic algorithms, especially those that more closely relate to the canonical genetic algorithm. This is an area where future research could broaden our understanding of a more general class of genetic based algorithms.

Even less well understood from a theoretical point of view are algorithms that are evolutionary in nature, but which are quite different from genetic algorithms. Such algorithms are typically developed as tools for function optimization. But what kind of search results from an algorithm that uses only mutation as an operator acting on a population of strings? The search is likely to be quite different from that of a canonical genetic algorithm using crossover. Assumptions about processing schemata representing binary subpartitions of hyperspace are unlikely to apply. Perhaps such a search has some connection to simulated annealing or other stochastic hill-climbing algorithms. For that matter, what happens to a canonical genetic algorithm when high mutation rates are used or local hill-climbing is added as an operator? From a schema processing point of view this would appear to be a bad idea, since these operators disrupt schemata and therefore disrupt the statistical information implicitly contained in the population about the relative fitness of strings in different subpartitions of hyperspace. In practice, however, such local operators are often effective for function optimization. This inconsistency has lead to the view that theory and practice are in conflict. In part, this view is correct. But the job of theory is to bridge this gap, and that will only happen if we continue to make theoretical advances in our understanding of genetic algorithms and the use of evolutionary paradigms as tools for function optimization.

To make the picture even more complex, even if one is interested in optimization the results of any empirical test designed to demonstrate a "more effective optimizer" may say as much about the nature of the functions being optimized as it does about any two algorithms which might be compared. This observation reinforces the notion that empirical comparisons of algorithms is not an altogether satisfactory way of evaluating algorithms unless one has very specific computational goals. It also highlights the danger in the popular practice of declaring one algorithm "superior" to another based on a few empirical tests. Obviously experimental work will continue to play a leading role in the development of new strains of genetic algorithms, but the only general solution to this problem is a better theoretical foundation.

Another area of work that is largely lacking in the current volume is work on parallel genetic algorithms. From a pragmatic point of view, parallel and distributed genetic algorithms are likely to become increasingly important computational tools. Not only do they lend themselves to massively parallel implementation on a variety of hardware platforms, they have been numerous researchers who have reported better optimization results with less computational effort. These algorithms also appear to have interesting emergent properties. But here again, there has been little theoretical work. The paper on *Executable Models of Simple Genetic Algorithms* presents some preliminary work on parallel island models, but this is only a modest first step. This paper also tells us little about the computational behavior of more fine-grain, massively parallel types of models that impose locality considerations on selection and mating.

Research on the theoretical foundations of genetic algorithms represents only a fraction of the total work currently being done in the field of genetic algorithms. Given that this is the case, some of the papers in the current volume are more theoretical than others. While application papers were not considered for this volume, papers on new implementations and operators were considered, especially when these papers touched on more theoretical issues in one way or another.

There was a general consensus among the participants of the 1992 FOGA workshop that real advances are being made with respect to theoretical work in genetic algorithms. There was also a general feeling that both the quantity and quality of work on the theoretical foundations of genetic algorithms is on the rise. The new journal *Evolutionary Computation* will appear in 1993 and has as one of its goals the publication of theoretical work on genetic algorithms and related computational paradigms which are evolutionary in nature.

I wish to thank several people for helping to make the FOGA workshop and the publication of *FOGA-2-* possible. First, I wish to thank the program committee for helping to select the papers contained in this volume. Also, Michael Vose and Kalyanmoy Deb helped with additional reviewing tasks that, for one reason or another, could not be handled by the regular program committee. I also wish to thank all of the authors. Not only did they contribute excellent papers, in addition to editing each paper myself I also asked most of the authors to edit one other paper appearing in *FOGA-2-* in order to widen both the technical and editorial scope of the second set of editorial reviews. My thanks to Ken DeJong for offering both advise and support from both a technical and administration point of view. I doubt that few people realize how much time and effort Ken has devoted over the years to making sure the various conferences and workshops on genetic algorithms are successful and well-organized. A special thank also goes to Denise Hallman, who took care of local arrangements.

Finally, this book is delicate to Gunar Liepins, who died this spring. Gunar made some very significant and lasting contributions to the theoretical foundations of genetic algorithms. Several papers in this book build on ideas that orginially were developed by Gunar. Gunar was also a friend; it is extremely sad to know that he is gone.

Darrell Whitley
Fort Collins, Colorado
whitley@cs.colostate.edu

References

Goldberg, D. (1989) *Genetic Algorithms in Search, Optimization and Machine Learning.* Reading, MA: Addison-Wesley.

Holland, John (1975) *Adaptation In Natural and Artificial Systems.* University of Michigan Press.

Holland, John (1992) *Adaptation In Natural and Artificial Systems.* MIT Press Edition.

Rawlins, Gregory J.E., ed., (1991) *Foundations of Genetic Algorithms.* Morgan Kaufmann.

PART 1

FOUNDATION ISSUES REVISITED

Genetic Algorithms Are NOT Function Optimizers

Kenneth A. De Jong
Computer Science Department
George Mason University
Fairfax, VA 22030, USA
kdejong@aic.gmu.edu

Abstract

Genetic Algorithms (GAs) have received a great deal of attention regarding their potential as optimization techniques for complex functions. The level of interest and success in this area has led to a number of improvements to GA-based function optimizers and a good deal of progress in characterizing the kinds of functions that are easy/hard for GAs to optimize. With all this activity, there has been a natural tendency to equate GAs with function optimization. However, the motivating context of Holland's initial GA work was the design and implementation of robust adaptive systems. In this paper we argue that a proper understanding of GAs in this broader adaptive systems context is a necessary prerequisite for understanding their potential application to any problem domain. We then use these insights to better understand the strengths and limitations of GAs as function optimizers.

1 INTRODUCTION

The need to solve optimization problems arises in one form or other in almost every field and, in particular, is a dominant theme in the engineering world. As a consequence, an enormous amount of effort has gone into developing both analytic and numerical optimization techniques. Although there are now many such techniques, there are still large classes of functions which are beyond analytical methods and present significant difficulties for numerical techniques. Unfortunately, such functions are not bizarre, theoretical constructs; rather, they seem to be quite commonplace and show up as functions which are not continuous or differentiable everywhere, functions which are non-convex, multi-modal (multiple peaks), and functions which contain noise.

As a consequence, there is a continuing search for new and more robust optimization techniques capable of handling such problems. In the past decade we have seen an increasing interest in biologically motivated approaches to solving optimization problems, including neural networks (NNs), genetic algorithms (GAs), and evolution strategies (ESs). The initial success of these approaches has led to a number of improvements in the techniques and a good deal of progress in understanding the kinds of functions for which such techniques are well-suited.

In the case of GAs, there are now standard GA-based optimization packages for practical applications and, on the theoretical side, continuing efforts to better understand and characterize functions that are "GA-easy" or "GA-hard". However, with all this activity, there is a tendency to *equate* GAs with function optimization. There is a subtle but important difference between "GAs *as* function optimizers" and "GAs *are* function optimizers".

The motivating context that led to Holland's initial GA work was the design and implementation of robust *adaptive* systems. In this paper we argue that a proper understanding of GAs in this broader adaptive systems context is a necessary prerequisite for understanding their potential application to any problem domain. We then use these insights to better understand the strengths and limitations of GAs *as* function optimizers.

2   WHAT IS A GENETIC ALGORITHM?

Figure 1 provides a flow diagram of a fairly generic version of a GA. If we ask what this algorithm is intended to do, we find ourselves in an awkward "cart before the horse" situation. The algorithm in question wasn't designed to solve any particular problem (like sorting, tree traversal, or even function optimization). Rather, it is a high level simulation of a biologically motivated adaptive system, namely evolution. As such, the kinds of questions one asks have a different slant to them. We would like to know what kind of emergent behavior arises from such a simple set of rules, and how changes in the algorithm affect such behavior. As we understand better the kind of adaptive behavior exhibited by GAs, we can begin to identify potential application areas that might be able to exploit such algorithms.

Such discussions about evolutionary systems are not new, of course, and certainly predate the emergence of GAs. Although a precise and agreed upon statement of the role of evolution is difficult to find, I think it is fair to say that there is general agreement that its role is not function optimization in the usual sense of the term. Rather, one thinks of evolution in terms of a strategy to explore and adapt to complex and time-varying fitness landscapes.

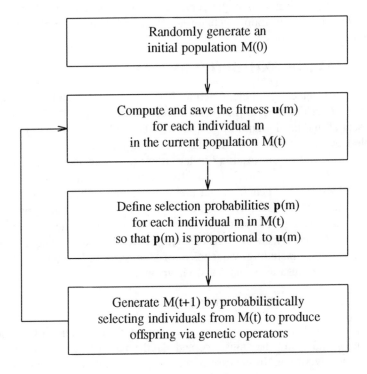

Figure 1. A Canonical Genetic Algorithm

Analyzing this evolutionary adaptive behavior is surprisingly difficult even for high level abstractions such as GAs. Holland's analysis [Holland75] in terms of near-optimal allocation of trials in the face of uncertainty is still today one of the few analytical characterizations we have of global GA behavior. Much of the early GA research was an attempt to gain additional insight via empirical studies in which the fitness landscape was defined by carefully chosen time-invariant memoryless functions whose surfaces were well understood and by observing how GA populations evolved and adapted to these landscapes (see, for example, [DeJong75] or [Bethke81]).

Several important observations came out of these early studies. First, the actual behavior exhibited by such a "simulation" varied widely as a function of many things including the population size, the genetic representation chosen, the genetic operators used, and the characteristics of the fitness landscape. Second, it became very clear that there was no universal definition of what it meant for one simulation run to exhibit "better" adaptive behavior than another. Such measures were more easily motivated by particular task domains and, as such, frequently emphasized different aspects of adaptive behavior. Finally, the robust adaptive behavior observed on these artificial fitness functions gave rise to the belief that GAs might serve as a *key element* in the design of more robust global function optimization techniques. We explore these observations in more detail in the next few sections.

3 BEHAVIOR EXHIBITED BY GAs

Suppose we select a simple function like $f(x)=x^2$ to define an (unknown) fitness landscape over the interval [0,4] with a precision for x of 10^{-4}. Suppose further that we represent all the legal values in the interval [0,4] as binary strings of fixed length in the simplest possible manner by mapping the value 0.0 onto the string 00...00, $0.0 + 10^{-4}$ onto 00...01, and so on, resulting in an order-preserving mapping of the interval [0,4] onto the set {00...00,...,11...11}. Finally, suppose we run the simulation algorithm in Figure 1 by randomly generating an initial population of binary strings, invoking f(x) to compute their fitness, and then use fitness-proportional selection along with the standard binary string versions of mutation and crossover to produce new generations of strings. What sort of behavior do we observe? The answer, of course, is not simple and depends on many things including the choice of things to be measured. In the following subsections we illustrate 3 traditional viewpoints.

3.1   A GENOTYPIC VIEWPOINT

The traditional biological viewpoint is to consider the contents of the population as a gene pool and study the transitory and equilibrium properties of gene value (allele) proportions over time. In the case of a canonical GA, it is fairly easy to get an intuitive feeling for such dynamics. If fitness-proportional selection is the only active mechanism, an initial randomly generated population of N individuals fairly rapidly evolves to a population containing only N duplicate copies of the best individual in the initial population. Genetic operators like crossover and mutation counterbalance this selective pressure toward uniformity by providing diversity in the form of new alleles and new combinations of alleles. When they are applied at fixed rates, the result is a sequence of populations evolving to a point of dynamic equilibrium at a particular diversity level. Figure 2 illustrates how one can observe these effects by monitoring the contents of the population over time.

Note that, for the fitness landscape described above, after a relatively few number of generations almost all individuals are of the form 1..., after a few more generations, the pattern 11... dominates, and so forth. After several hundred generations we see that, although more than 50% of the gene pool has converged, the selective pressures on the remaining alleles are not sufficient to overcome the continual diversity introduced by mutation and crossover, resulting in a state of dynamic equilibrium at a moderate level of genetic diversity.

Gen 0	Gen 5	Gen 50	Gen 200	Gen 500
01111100010011	11000001011100	11111101010101	11111111010011	11111111011110
00010101011100	11100001000010	11111110011101	11110111011110	11111111110011
01001000110010	11110100010100	11111011001001	11111011010110	11111110010001
00110001110010	11110011100011	11111101000001	11111110101011	11111111011010
01010001110010	11111110100001	11111100110111	11111100001011	11111111110000
00010000010001	11111100011001	11111101000101	11111111011111	11111111111011
11001100000110	11010100000110	11111101001001	11111110100001	11111011010101
10110001110010	11111101000001	11111001011001	11111111110001	11111111110110
10100001111100	11111100111001	11010111110001	11111111111111	11111111110000
00100001101001	11110111011001	11111101011101	11110110110101	11111111000010
01010100100011	11110001111111	11101111100111	11111101110011	11111111001010
01000101100011	11110111100011	11111100110111	11111100010001	11111111111000
00010001111110	11110111100001	11111111010011	11111111011001	11111111110000
10101111100111	11110111011100	11111111010011	11101111111001	11111111010001
01111101010100	11101010100100	11110100000111	11111111100011	11111111011000
11000011100110	11010100011000	11111101011001	11111111110011	11111111010001
11001101100001	11100001110001	11111100101001	11111111111011	11111111110110
11110111011100	11110110011011	11111011100101	11111001001010	11111111110010
11001010100000	11010101000001	11111100000001	11111110111010	11111111110100
00101000101111	11010001010110	11111111100101	11111110011011	11111111110011
01000011111110	11110100010101	11111111001001	11111110110111	11111111111000
11100101000001	11110101100001	11111001010011	11111111101001	11111111101111
01001001100110	11110111100001	11111100101011	11111101010011	11111111001111
01000101100011	11111101000001	11111011011000	10111100111010	11111111011000
00111001100000	11110010011100	11110110001101	11111111010011	11111111011000
01110100100100	11111100111001	11111100111011	11111101111010	11111111010110

Figure 2: A Genotypic View a GA Population

3.2   A PHENOTYPIC VIEWPOINT

An alternative to the genotypic point of view is to focus on the "phenotypic" characteristics of the evolving populations produced by GAs. The phenotype of an individual is the physical expression of the genes from which fitness is determined. In the example introduced above, the phenotype is simply the real number x defined by a particular binary gene value. Although not true in general, in this particular case there is a one-to-one mapping between genotypes and phenotypes.

Since fitness is defined over this phenotype space, one gains additional insight into the adaptive behavior of GAs by plotting over time where the individuals in the population "live" on the fitness landscape. Figure 3 illustrates this behavior for the fitness landscape defined above. As one might expect, the population is observed to shift quite rapidly to inhabiting a very small niche at the rightmost edge of the phenotype space.

Suppose one creates more interesting landscapes with multiple peaks. Does a GA population evolve to steady states in which there are subpopulations inhabiting each of the peaks? For the typical landscapes used which are memoryless and time-invariant,

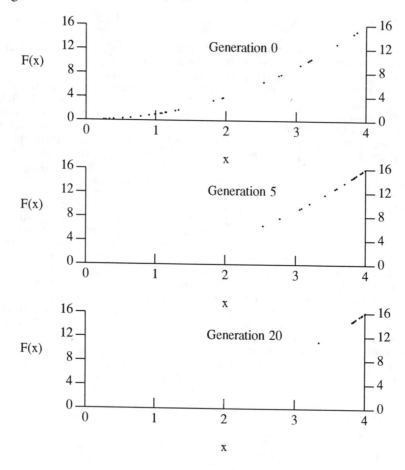

Figure 3: A Phenotypic View of a GA Population

there is no negative feedback for overcrowding. The net result for canonical GAs is a population crowding into one of the higher fitness peaks (see, for example, [DeJong75] or [Goldberg89]). Frequently, the equilibrium point involves the highest fitness peak, but not always.

3.3   AN OPTIMIZATION VIEWPOINT

There are of course other ways of monitoring the behavior of a GA. We might, for example, compute and plot the average fitness of the current population over time, or the average fitness of all of the individuals produced. Alternatively, we might plot the best individual in the current population or the best individual seen so far regardless of whether it is in the current population. In general such graphs look like the ones in Figure 4 and exhibit the properties of "optimization graphs" and "learning curves" seen in many other contexts.

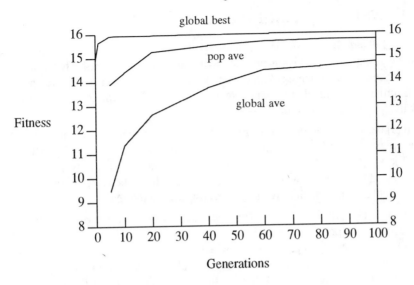

Figure 4: GA Fitness Measures

Measurements of these sorts have encouraged the view of GAs as function optimizers. However, one must be careful not to adopt too simplistic a view of their behavior. The genotypic and phenotypic viewpoints provide ample evidence that populations do not in general converge to dynamic steady states containing multiple copies of the global optimum. In fact it is quite possible for the best individual to never appear or appear and then disappear from the population forever. With finite and relatively small populations, key alleles can be lost along the way due to sampling errors and other random effects such as genetic drift. In theory, as long as mutation is active, one can prove that every point in the space has a non-zero probability of being generated. However, if the population evolves toward subspaces not containing the optimum, the likelihood of generating optimal points via mutations of the current population becomes so small (e.g., $< 10^{-10}$) that such an event is unlikely to be observed in practice.

With a canonical GA every point generated (good or bad) has a relatively short life span since the population is constantly being replaced by new generations of points. Hence, even if a global optimum is generated, it can easily disappear in a generation or two for long periods of time.

These observations suggest that a function optimization interpretation of these canonical GAs is artificial at best and raise the question as to whether there might be a more natural behavioral interpretation.

4 ANALYSIS OF GA BEHAVIOR

Holland [Holland75] has suggested that a better way to view GAs is in terms of optimizing a sequential decision process involving uncertainty in the form of lack of *a priori* knowledge, noisy feedback, and time-varying payoff function. More specifically, given a limited number of trials, how should one allocate them so as to maximize the *cumulative* payoff in the face of all of this uncertainty?

Holland argues that an optimal strategy must maintain a balance between exploitation of the best regions found so far and continued exploration for potentially better payoff areas. Too much exploitation results in local hill climbing, sensitivity to noise, and an inability to adapt to time-varying fitness functions. Too much exploration ignores the valuable feedback information available already early in the search process.

Holland has shown via his Schema Theorem that, given a number of assumptions, GAs are quite robust in producing near-optimal sequences of trials for problems with high levels of uncertainty. Regardless of whether or not one believes that real GA implementations, as opposed to mathematic models of GAs, achieve these near-optimal sequences of trials, I think that Holland's formulation in terms of trying to optimize a sequential decision process in the face of uncertainty is the best behavioral description we currently have for canonical GAs. [1]

5 GAs AS FUNCTION OPTIMIZERS

Once we view a canonical GA as attempting to maximize the cumulative payoff of a sequence of trials, we gain a better understanding of their usefulness as a more traditional function optimization technique. While maximizing cumulative payoff is of interest in some function optimization contexts involving online control of an expensive process, it is much more frequently the case that optimization techniques are compared in terms of the "bottom line" (the best value found), or in terms of the best value found as a function of the amount of effort involved (usually measured in terms of the number of trials).

The standard way of using GAs as function optimizers in this context is to keep track of the (globally) best individual produced so far regardless of whether that individual exists in the current population. That is, the population is viewed as a database (an accumulating world model) of samples from which potentially better individuals are to be generated. Hence, converged alleles represent a "focus of attention" in the sense that most new trials will share these values (i.e., live in the same subspace), while the residual diversity in other genes provides the building blocks for creating new individuals. The genetic operators in conjunction with selection use this focussed diversity to generate new individuals with potentially higher fitness.

If we apply GAs in this way as function optimizers, how well do they perform relative to other techniques? An accumulating body of experience has yielded performances

---

[1] Recent work by Vose [Vose92] and Whitley [Whitley92] presented elsewhere in this book provide the beginnings of additional behavior insights.

ranging from spectacular to lousy. As a consequence, considerable effort has gone into studying and improving GA-based function optimizers. We summarize a few of these efforts in the following subsections.

5.1   SCALING ISSUES

From a function optimization point of view, GAs frequently don't exhibit a "killer instinct" in the sense that, although they rapidly locate the region in which a global optimum exists, they don't locate the optimum with similar speed. If one looks at this behavior from a "maximizing cumulative returns" point of view, it makes perfect sense. If the range of payoff values is, for example, [0,100], the population evolves quite rapidly to one in which most individuals are in the range [99,100], and selective differential between 99.988 and 100.000 provides little incentive for the GA to prefer one over the other. However, if such differentials are important to the goal of function optimization, then the GA must be provided with a payoff incentive to exploit such differentials.

A typical solution involves providing feedback to the GA in the form of a dynamically scaled fitness function in order to maintain sufficient selective pressure between competing individuals in the current population (see, for example, [DeJong75] or [Goldberg89]). Unfortunately, there is currently no theory to suggest how to achieve an optimal scaling strategy.

5.2   RANKING APPROACHES

An alternative approach is to change the canonical GA itself, substituting rank-proportional selection for the traditional fitness-proportional selection. By keeping the population sorted by fitness and performing selection of the basis of rank, a constant selection differential is maintained between the best and worst individuals in the population (see, for example, [Whitley89]). The effect is to slow down initial convergence to promising subspaces, but to increase the killer instinct in the final stages.

Such approaches can frequently improve the performance of GA-based function optimizers, but depend on appropriate choices of rank-based selection functions (linear, quadratic, exponential, etc.). Unfortunately, currently theory is not strong enough to indicate how general such observations are or to provide guidance as to when to use a particular ranking scheme.

5.3   ELITIST STRATEGIES

Frequently, performance improvements can be obtained by singling out the best and/or worst individuals in the current population for special treatment. Examples of such tactics include always keeping the best individual found so far in the population (it only gets replaced by globally better individuals), or conversely, systematically replacing the worst members of the population with newly generated individuals (e.g., [DeJong75] or [Whitley89]).

The effect is to shift the balance toward more exploitation and less exploration which, for some classes of functions, works quite well. However, if no *a priori* information is available about the functions to be optimized, this can result in suboptimal hill-climbing behavior on multi-peaked functions.

5.4   REPRESENTATIONAL ISSUES

The binary representation chosen in the earlier example is just one of many ways to represent the space to be searched. It was noted very early that the choice of representation can itself affect the performance of a GA-based function optimizer. Suppose, for example the global optimum just happened to be represented by the string 0111111, but from a payoff point of view the subspace represented by 1... was slightly better. In such circumstances, a GA-based optimizer can get caught on a "Hamming cliff" by evolving a population dominated by individuals of the form 1000... which are very close to the optimum in the phenotype space, but far apart in terms of Hamming distance.

Such observations led to alternative representations such as gray codes in an attempt to avoid such problems. Unfortunately, without knowledge about the function to be optimized, it is always possible to have picked a poor representation. Adaptive representation strategies have been suggested as an alternative, but are difficult to efficiently implement.

In the case in which the function arguments are real-valued, one might view a floating point representation as more natural. Again, we have little in the way of theoretical guidance as to whether this will help or hinder the performance of a GA-based function optimizer.

5.5   DOMAIN-SPECIFIC ALTERATIONS

A canonical GA achieves its robustness in part by making no strong assumptions about the properties of the fitness landscape to be explored. If we then compare the performance of a GA-based optimizer with another optimizer on a class of functions for which that optimizer was specifically designed, the GA-based optimizer is frequently outperformed. There are two typical responses to such comparisons. One is to illustrate the robustness of the GA-based optimizer and the brittleness of the specialized one by making minor modifications to the test suite (e.g., add some noise and/or additional peaks).

The other response is to build a specialized GA-based optimizer which takes advantage of this *a priori* knowledge in the form of specialized representations, specialized genetic operators, etc. (e.g., traveling salesperson problems or continuous real-valued functions). This can lead to significant improvements in performance on the selected problems class, but the development process can be time consuming since current theory does not provide strong guidance for making such alterations.

6 SOME FUNDAMENTAL MISCONCEPTIONS

The previous sections have attempted to provide evidence for the following strong claims. First, the canonical GA given in Figure 1 is not a function optimizer in the traditional sense. Rather, it is better viewed as solving sequential decision processes. Second, by modifying a canonical GA in a variety of ways (such as those noted in the previous section) one can build very powerful and effective GA-based optimizers. Third, the strongest theoretical results currently available are for canonical GAs.

Fourth, and finally, there is a tendency to assume that the theoretical and behavioral characteristics of canonical GAs are identical to those of GA-based function optimizers. This is simply not true in general. As we have seen, an effective strategy for solving sequential decision problems does not necessarily also result in an effective function optimization strategy. A GA that satisfies the Schema Theorem does not necessarily make it a good optimizer. Conversely, when we add scaling, introduce ranking and/or elitist strategies, change the way crossover and mutation work, etc., we have significantly changed the behavior of a canonical GA so that there is no *a priori* reason to believe that it continues to solve sequential decision problems effectively. In fact, many of the strong sampling biases introduced into GA-based optimizers are known to negatively affect cumulative (online) payoff.

This last point leads to some fairly fundamental misconceptions about the meaning and implications of some of the current work in the field regarding understanding what makes problems hard (easy) for GAs to solve. [2] Although not clearly stated, the motivating context for characterizing GA-hard and GA-easy problems is function optimization. So to be precise we should be talking about classes of functions which are hard (easy) for GA-based optimizers to solve. The important point here is that GA-hard optimization problems may or may not represent hard sequential decision problems. In fact the notion of GA hard (easy) is much less well defined for canonical GAs solving sequential decision problems. Let me illustrate these points with two examples.

6.1 DECEPTIVE PROBLEMS

Since canonical GAs use sampled averages to bias the direction in which future individuals are generated, it is easy to construct fitness landscapes which "deceive" a GA-based optimizer by hiding the global optimum in a part of the landscape with low average payoff, thus making it highly unlikely to be found (see, for example, [Goldberg89] or [Whitley91]). What is important to understand here is that there is no strong sense of deception from a canonical GA's point of view. It is attempting to maximize *cumulative* payoff from arbitrary landscapes, deceptive or otherwise. In general, this is achieved by not investing too much effort in finding cleverly hidden peaks (the risk/reward ratio is too high).

On the other hand, since most interesting optimization problems are NP-hard, we might reasonably drop our demand (expectation) that a GA-based optimizer find the global

---

[2] See [Grefenstette92] elsewhere in this book for a related discussion of misconceptions.

optimum and think more in terms of using it as a heuristic that finds good values quickly. Notice that in this case deception becomes less of a concern since GA-based optimizers frequently turn out to be excellent heuristics for most deceptive problems.

This is not to say that the work on deception is irrelevant or incorrect. Rather, it is a step in a direction badly needed: theoretical results for GA-based optimizers (as opposed to canonical GAs). Without such work we are left in the uncomfortable position of assuming "what's good for the goose (canonical GAs) is good for the gander (GA-based optimizers)".

6.2   BUILDING BLOCK HYPOTHESES

The theoretical analysis that we have for canonical GAs suggests a view in which selection, mutation, and crossover work synergistically to construct new individuals from an evolving set of useful building blocks represented by the genetic makeup of the individuals in the population. It is easy to show that deception, finite populations and sampling errors can result in the loss of important low level building blocks which can, in turn, negatively affect future "construction projects".

Suppose, however, we select landscapes which have all the right properties with respect to encouraging the formation of useful building blocks. Are we correct in assuming that these landscapes are necessarily GA-easy from a function optimization point of view in that the global optimum is easily found? In general, the answer is no. Rather than view such results as "anomalous" (e.g., [Forrest91]), we should expect this once we understand that canonical GAs are not optimizers. Moreover, the addition of scaling, elitism, and other optimization-motivated features to a canonical GA can result in improved optimization behavior, but can also sufficiently distort and bias the sampling so as to make some of these "easy" landscapes appear hard.

7   SUMMARY AND CONCLUSIONS

The intent of this paper is not to question the usefulness of GAs as function optimization techniques. In fact, the intent is quite the opposite. GA-based function optimizers have already demonstrated their usefulness over a wide range of difficult problems. At the same time, there are obvious opportunities for improvements and extensions to new function classes. In order to achieve this we need to understand better how GA-based optimizers work.

The concern expressed here is that there is a tendency to think of GAs as function optimizers, and thus blur the distinction between a canonical GA and a GA which has been modified for use as an optimizer. By blurring that distinction, it is easy to mistakenly assume that what we know and understand about canonical GAs carries over to the GA-based optimizers. It is certainly the case that both canonical GAs and GA-based optimizers share many properties. However, it is also quite clear that they are quite different in some important respects. In order to improve our GA-based optimizers we must continue to clarify both the similarities and the differences.

REFERENCES

Bethke, Albert D. (1981). *Genetic Algorithms as Function Optimizers,* Doctoral Thesis, Department of Computer and Communication Sciences, University of Michigan, Ann Arbor.

De Jong, Kenneth A. (1975). *An Analysis of the Behavior of a Class of Genetic Adaptive Systems,* Doctoral Thesis, Department of Computer and Communication Sciences, University of Michigan, Ann Arbor.

Forrest, S. and Mitchell, M. (1991). The Performance of Genetic Algorithms on Walsh Polynomials: Some Anomalous Results and their Explanation. *Proc. 4th Intl. Conf. on Genetic Algorithms,* La Jolla, CA: Morgan Kaufmann.

Goldberg, D. (1989). *Genetic algorithms in search, optimization, and machine learning.* New York: Addison-Wesley.

Grefenstette, J. (1992). Deception Considered Harmful. In *Foundations of Genetic Algorithms 2,,* D. Whitley (ed.), Vail, CO: Morgan Kaufmann.

Holland, John H. (1975). *Adaptation in Natural and Artificial Systems,* The University of Michigan Press.

Vose, M. (1992). Modeling Simple Genetic Algorithms. In *Foundations of Genetic Algorithms 2,,* D. Whitley (ed.), Vail, CO: Morgan Kaufmann.

Whitley, D. (1989). The Genitor algorithm and selection pressure: Why rank-based allocation of reproductive trials is best, *Proc. 3rd Intl. Conf. on Genetic Algorithms,* Fairfax, VA: Morgan Kaufmann.

Whitley, D. (1991). Fundamental principles of deception in genetic search. In *Foundations of Genetic Algorithms,* G. Rawlins (ed.), Bloomington, IN: Morgan Kaufmann.

Whitley, D. (1992). An Executable Model of a Simple Genetic Algorithm. In *Foundations of Genetic Algorithms 2,,* D. Whitley (ed.), Vail, CO: Morgan Kaufmann.

Generation Gaps Revisited

Kenneth A. De Jong
George Mason University
Fairfax, VA 22030 USA
kdejong@aic.gmu.edu

Jayshree Sarma
George Mason University
Fairfax, VA 22030 USA
jsarma@cs.gmu.edu

Abstract

There has been a lot of recent interest in so-called "steady state" genetic algorithms (GAs) which, among other things, replace only a few individuals (typically 1 or 2) each generation from a fixed size population of size N. Understanding the advantages and/or disadvantages of replacing only a fraction of the population each generation (rather than the entire population) was a goal of some of the earliest GA research. In spite of considerable progress in our understanding of GAs since then, the pros/cons of overlapping generations remains a somewhat cloudy issue. However, recent theoretical and empirical results provide the background for a much clearer understanding of this issue. In this paper we review, combine, and extend these results in a way that significantly sharpens our insight.

1 INTRODUCTION

In Holland's book *Adaptation in Natural and Artificial Systems* [Holland75], he introduces and analyzes two forms of reproductive plans which have served as the basis for the field of Genetic Algorithms. The first plan, R_1, maintained a fixed size population and at each time step a single individual was selected probabilistically using payoff-proportional selection to produce a single offspring. To make room for this new offspring, one individual from the current population was selected for deletion via a uniform random distribution.

Since the *expected* lifetime of an individual in R_1 under uniform deletion in a population of size N is simply N, Holland was able to show that, if the average fitness of the population, f_t, didn't change much over the lifetime of an individual, then the expected number of offspring produced by individual i is $f(i)/f_t$, which is the basis for Holland's now famous schema theorem.

Holland also introduced a second plan, R_d, in which at each time step all individuals were deterministically selected to produce their expected number of offspring $f(i)/f_t$ in a temporary storage location and, when that process was completed, the offspring produced replaced the entire current population. Thus in R_d individuals were guaranteed to produce their expected number of offspring (within probabilistic roundoff).

At the time, the two plans were viewed as generally equivalent from a theoretical point of view. However, because of practical considerations relating to severe genetic drift in small populations and the overhead of recalculating selection probabilities, most early researchers favored the R_d approach. However, the emergence of classifier systems (e.g., [Holland78]) in which only a subset of the population is replaced each time step, and recent interest in "steady state" GAs (e.g., [Whitley88]) in which only 1 or 2 individuals are replaced each time step has again raised the issue of the advantages/disadvantages of overlapping generations. In spite of considerable progress in our understanding of GAs, the pros/cons of overlapping generations remains a somewhat cloudy issue, partly due the fact that other changes such as non-uniform deletion and ranking selection are frequently also incorporated making comparisons difficult. However, recent theoretical and empirical results provide the background for a much clearer understanding of this issue. In this paper we review, combine, and extend these results in a way that significantly sharpens our insight.

2 BACKGROUND

The earliest attempt at evaluating the properties of overlapping generations was a set of empirical studies in [DeJong75] in which a parameter G, called the generation gap, was defined to control the fraction of the population to be replaced in each generation. Thus, $G = 1.0$ corresponded to R_d and $G = 1/N$ represented R_1. These early studies suggested that any advantages that overlapping populations might have was dominated by the negative effects of genetic drift (allele loss) due to the high variance in expected lifetimes and expected number of offspring given the generally modest population sizes in use at the time (N <= 100). These effects were shown to increase in severity as G was reduced. These studies also suggested the advantages of an *implicit* generation overlap in the sense that, even with $G = 1.0$, the optimal crossover rate of 0.6 and optimal mutation rate of 0.001 identified for the test suite used meant that approximately 40% of the offspring were clones of their parents.

Early experience with classifier systems yielded quite the opposite behavior in that replacing a smaller number of classifiers was generally more beneficial than replacing a large number or possibly all of them. In this case, however, the performance drop off observed as G increased was attributed to the fact that the population as a whole represented a single solution and could not tolerate large changes in content.

In recent years with the increased capacity of easily available computing equipment and the desire to solve increasing larger problems, a variety of researchers have reported positive results with so called "steady state" systems involving larger population sizes in which at most 1 or 2 individuals are replaced each generation. Whitley's GENITOR system [Whitley88] is a good example of this approach. Unfortunately, other simultaneous changes such as the use of ranking selection and deletion of the worst individual makes it difficult to understand precisely where the improved performance is coming from.

This, in turn, has led to renewed attempts to analyze the behavior of these alternatives in an attempt to resolve some of the conflicting empirical results, and ultimately build better GA-based systems. Several recent results [Goldberg90, Syswerda90] have shed light on these issues and set the stage for the work presented here.

3   GENERATION GAP ANALYSIS

A recurring theme in Holland's work is the importance of a proper balance between exploration and exploitation when adaptively searching an unknown space for high performance solutions [Holland75]. Many aspects of a GA affect this balance including selection, reproduction and deletion mechanisms, genetic operators, and population size. One way of understanding the effects of overlapping generations is to focus on the changes it makes in the exploitation/exploration balance. This is can be done by imagining a system in which there are no genetic operators. Without the ability to generate any novel individuals, selection, reproduction and deletion drive the system to a state of equilibrium in which one of members of the initial population takes over the entire population. In the following sections we analyze the effects that varying G from a value of $1/N$ (replacing a single individual) to $G = 1.0$ (replacing the entire population) will have on the actual equilibrium points reached and the rates at which equilibrium is reached.

3.1   VARIANCE EFFECTS

If we let $g = G * N$ (the number of offspring produced each generation) and let $m_i(t)$ represent the number of individuals of type i in the population at time t, then we can easily express the growth equations of type i individuals as:

$$m_i(t+1) = m_i(t) + m_i(t) * p_s(i,t) * g - m_i(t) * p_d(i,t) * g \qquad (1)$$

$$= m_i(t)[\, 1 + g * (p_s(i,t) - p_d(i,t))\,]$$

where $p_s(i,t)$ is the probability of selecting a type i individual at time t for cloning, and $p_d(i,t)$ the probability of deleting a type i individual at time t.

The second line of equation 1 highlights quite clearly one immediate effect of changing g (or G), namely the amount of variance expected among individual growth curves. The "steady state" value of $g = 1$ represents the case with the highest variance. As we increase g (or G), the variance decreases since the average of g samples from the probability distribution $(p_s(i,t) - p_d(i,t))$ is more likely to approximate the mean of that distribution than a single sample. In the extreme case of $g = N$, one can deterministically replace each individual by its expected offspring (as in R_d), driving the variance to its minimal value (which is always > 0 for finite populations).

This analysis immediately provides an explanation for two known phenomena. The first is that small values of g (or G) result in actual growth curves of individuals on a single GA run (without genetic operators) which vary much more widely from the expected curves than for larger g (or G) values, including cases where the best individuals disappear and the population converges to a suboptimal equilibrium point. Figures 1 and 2 illustrate this effect by plotting the averages and variances of the growth curves obtained from 100 independent simulation runs. Using a population of size 50 in which the best individual occupies 10% of the initial population and using only payoff proportional selection, reproduction and uniform deletion (no crossover or mutation), notice that for g=1 the best individuals take over the population only about 80% of the time and the growth curves exhibit much higher variance than the case of g=N.

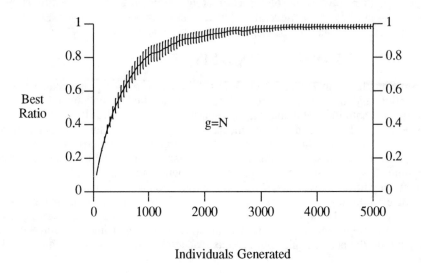

Individuals Generated

Figure 1: Mean and variance of the growth of the best with POP=50 and g=50.

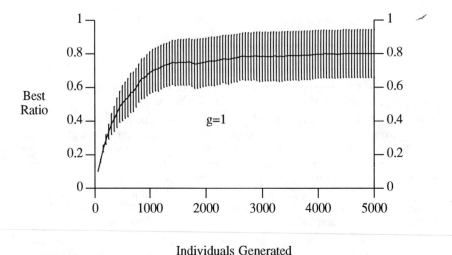

Individuals Generated

Figure 2: Mean and variance of the growth of the best with POP=50 and g=1.

A more detailed analysis of these effects can be found in [Syswerda90].

This high variance for small g values gives rise to a second phenomena: more genetic drift. With smaller population sizes the higher variance makes it is easier for alleles to disappear. Figure 3 illustrates the typical differential in allele loss rate one can see on almost any test function and smaller population sizes.

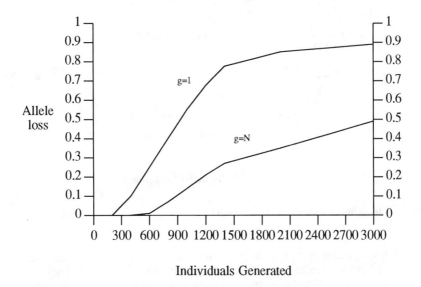

Figure 3: Allele loss for g=1 and g=N for N=50 on F1

Looked at in this way one can gain considerable insight as to why "steady state" systems (g=1) are generally run with larger population (offsetting allele loss), and prefer deletion biased towards poorer individuals (thus increasing the life expectancy of the better individuals).

3.2   LOWER VARIANCE DELETION STRATEGIES

The previous section raises the interesting question as to whether one could identify alternate implementations of steady state systems (g=1) which have lower variance, and thus reduce the need for larger population sizes or the need to introduce a deletion bias towards less fit individuals. The analysis above suggests two alternative deletion strategies which have the right theoretical properties and deserve some experimental evaluation: FIFO deletion, and excess offspring deletion.

With FIFO deletion the population is simply a first-in-first-out queue with new individuals added to one end and deleted individuals removed from the other end. The variance reduction effect is obvious in that the variance in individual lifetimes is reduced to zero. Figure 4 illustrates this by showing the effects of rerunning the Figure 2 experiments with FIFO deletion rather than uniform random deletion.

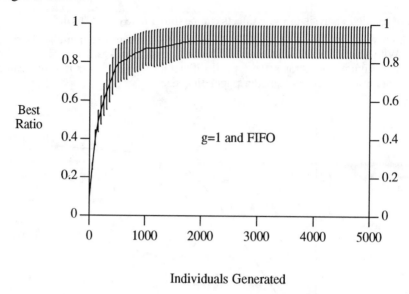

Figure 4: Mean and variance of the growth of the best with g=1 and FIFO deletion.

The variance remaining is due to the differences between the actual and expected number of offspring generated. This suggests an alternate deletion strategy in which one monitors the "excess offspring", defined as the actual number of offspring produced minus the expected number of offspring. If we bias deletion towards those with excess offspring, the effect will be to reduce the variance in the actual number of offspring produced. Care has to be taken in the implementation of this strategy in much the same spirit as Baker's analysis of selection schemes in order not to introduce unintended additional biases.

3.3   COMPOUNDING EFFECTS

One of the strong intuitive motivations for small g (or G) values is that getting good individuals back in the selection pool earlier leads to the possibility of a compounding effect due to earlier exploitation. The analysis goes something like this. If we assume payoff-proportional selection and uniform deletion, then $p_s(i,t) = f(i) / \sum f(j)$ and $p_d(i,t) = 1/N$. The change in the number of individuals of type i in one generation as specified in equation (1) above becomes:

$$m_i(t+1) = m_i(t)[1 + g * (\frac{f(i)}{\sum f(j)} - \frac{1}{N})]$$

$$= m_i(t)[1 + \frac{g}{N} * (\frac{f(i)}{\overline{f_t}} - 1)]$$

After N such iterations (generations) we have:

$$m_i(t+N) = m_i(t) * \prod_{j=0}^{N-1}[1 + \frac{g}{N} * (\frac{f(i)}{\overline{f_{t+j}}} - 1)]$$

If we assume for the time being that $f(i) / \bar{f_t}$ is relatively constant, we can simplify this to:

$$m_i(t+N) = m_i(t) * [1 + \frac{g}{N} * (\frac{f(i)}{\bar{f}} - 1)]^N$$

which one can argue grows faster than the case for g=N which is simply:

$$m_i(t+N) = m_i(t) * \frac{f(i)}{\bar{f_t}}$$

Unfortunately, the assumption that $f(i) / \bar{f_t}$ is relatively constant is a bad one and leads incorrectly to the "early compounding" intuition that many of us have held in the past. In fact, the growth curves for both cases (g=1 and g=N) turn out to be theoretically identical. The effect of increasing the number of copies of an above average individual leads to a corresponding decrease in its fitness relative to the population average, resulting in an ever decreasing compounding factor rather than a constant one. This is nicely illustrated in [Syswerda90] for a population of size 100 by computing and plotting the recurrence relations for g=1 and g=100.

We can obtain more formal insight into this phenomena by noting that the analysis in [Goldberg90] extends to the case when $g < N$. Goldberg and Deb suggest that we imagine individuals of type i reproducing at an unconstrained rate given by:

$$m_i(t+1) = m_i(t) * f(i)$$

Then the proportion of the unbounded total population taken up by individuals of type i is:

$$p_i(t+1) = \frac{m_i(t+1)}{\sum m_j(t+1)}$$

$$= \frac{m_i(t) * f(i)}{\sum m_j(t) * f(j)} \qquad (2)$$

Note that we could control the number of individuals produced each time step by uniformly changing the growth rates to

$$m_i(t+1) = m_i(t) * \frac{f(i)}{k_t}$$

of each subpopulation. However, such a change has no effect on proportion equation (2) above since the k_t appearing in the numerator and denominator cancel out. Hence, the theoretical growth curves (expected values) are identical for any setting of g (or G), and in particular for the cases of g=1 and g=N.

4 WHAT ABOUT REAL GAs?

The growth curve analysis of the previous section describes the behavior of a simplified environment in which crossover and mutation are inactive and quantities like $f(i)$ are exact and constant. In real GAs, of course, crossover and mutation play a critical role in the exploration/exploitation balance, and $f(i)$ represent changing estimates of hyperplane averages. Currently, our mathematical tools are inadequate to formally extend the results of the previous section to this more realistic environment.

It is not difficult, however, to generate empirical evidence to support the hypothesis that the results do indeed extend to real GAs. If this hypothesis is true, the effects of decreasing the generation gap G on actual GA systems should produce no significant increase in selection pressure (due to compounding effects). One should see only an increase in variance on individual runs and the side effects of the resulting increased allele loss.

Figures 5 and 6 show typical performance curves for the two extreme values (g=1 and g=N) for population sizes of 50 and 200. All curves are averages of 50 runs. Note how the behavioral differences diminish when the effects due to variance are reduced by increasing population size and by averaging over a sufficient number of trials.

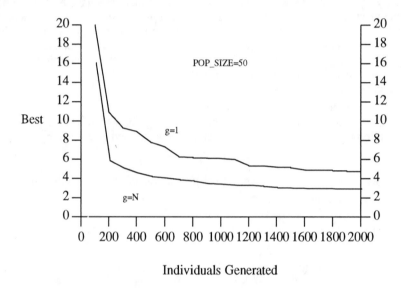

Figure 5: Performance Curves for g=1 and g=N for POP=50 on F5

5   CONCLUSIONS AND FURTHER WORK

These observations leave us with two important insights into the role of generation gaps. First, as we have seen, increasing G reduces the variance seen on individual GA runs which can be an important issue for an end user of GA-based systems. It suggests avoiding small G values in conjunction with small populations if the usual form of uniform random deletion is used. The analysis also suggests two alternate deletion strategies which could reduce variance and permit the use of steady-state systems with smaller populations.

Second, the analysis emphasizes that the important behavioral differences between current steady-state and generational systems have little if anything to do with their choice of generation gap size. In that sense, the terms "generational" and "steady-state" don't appear to be all that significant as semantic categories. Rather, the important behavioral changes are due to the changes in the exploration/exploitation balance

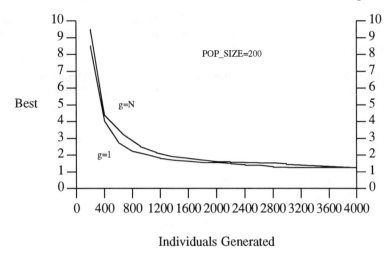

Figure 6: Performance Curves for g=1 and g=N for POP=200 on F5

resulting from the different selection and deletion strategies used. This is where we should continue to press our analysis efforts.

Note that, while a steady-state implementation of rank-based selection and worst-biased deletion is in some sense natural, one could imagine a system with $g > 1$ in which groups of individuals are generated before they are inserted back into the ranked population and the worst deleted. This raises the interesting question as to whether the effects of increasing g in such systems would also result in variance reduction without changing the selection pressure.

References

De Jong, Kenneth A. (1975). *An Analysis of the Behavior of a Class of Genetic Adaptive Systems,* Doctoral Thesis, Department of Computer and Communication Sciences, University of Michigan, Ann Arbor.

Goldberg, D. & Deb, K. (1990). A Comparative Analysis of Selection Schemes Used in Genetic Algorithms, *Proceedings of the Foundations of Genetic Algorithms Workshop,* Indiana, July 1990.

Holland, John H. (1975). *Adaptation in Natural and Artificial Systems,* The University of Michigan Press.

Holland, John H. & Reitman, Judith S. (1978). Cognitive systems based on adaptive algorithms. In D. A. Waterman & F. Hayes-Roth (Eds.), *Pattern-directed inference systems.* New York: Academic Press.

Syswerda, Gilbert. (1990). A Study of Reproduction in Generational and Steady-State Genetic Algorithms, *Proceedings of the Foundations of Genetic Algorithms Workshop*, Indiana, July 1990.

Whitley, D. & Kauth, J., (1988). Genitor: A Different Genetic Algorithm, *Proceedings 4th Rocky Mountain Conference on Artificial Intelligence*, Denver, 1988.

PART 2

MODELING GENETIC ALGORITHMS

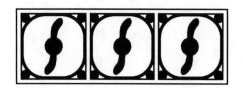

Recombination Distributions
for Genetic Algorithms

Lashon B. Booker
The MITRE Corporation
7525 Colshire Drive
McLean, VA 22102-3481
booker@mitre.org

Abstract

Though genetic algorithms are loosely based on the principles of genetic variation and natural selection, the theory of mathematical genetics has not played a large role in most analyses of genetic algorithms. This paper reviews some well known results in mathematical genetics that use probability distributions to characterize the effects of recombination on multiple loci in the absence of selection. The relevance of this characterization to genetic algorithm research is illustrated by using it to quantify certain inductive biases associated with crossover operators. The potential significance of this work for the theory of genetic algorithms is discussed.

1 Introduction

Holland's (1975) original analysis of genetic algorithms (GAs) placed considerable emphasis on the role of recombination in genetic search. In many ways, recombination is the most distinguishing feature of a genetic algorithm. It is therefore not surprising that analyses of recombination have always been an important topic in the theory of genetic algorithms. Early work extended Holland's analysis and focused on the disruptive properties of crossover operators (De Jong, 1975), trying to quantify the way recombination interferes with the transmission of schemata from parents to offspring. Because the effects of crossover have proven to be difficult to characterize precisely, analyses of crossover disruption have focused on finding good bounds for the disruptive effect. One notable exception is the work of Bridges and Goldberg (1987), which analyzes the special case of one-point crossover and provides an exact expression for the probability of disruption. Crossover has also been studied in terms of other properties besides disruption. Eshelman, Caruana, and Schaffer (1989) characterize recombination operators in terms of the inductive biases they impose on genetic search. Liepins and Vose (1992) have characterized crossover operators mathematically as binary operators that preserve schemata and commute with projection.

Studies such as these have greatly increased our understanding of how crossover works in genetic algorithms, but many questions remain unanswered. Recent work, for example, has challenged traditional views about the role of crossover in genetic search. There is strong empirical evidence that highly disruptive crossover operators can sometimes search more effectively than less disruptive ones (Syswerda, 1989; Eshelman, 1991). There is also a growing appreciation of the role of mutation as an important genetic search operator (Schaffer and Eshelman, 1991). One response to these results has been a more careful analysis of disruption for crossover operators (Spears and De Jong, 1991) to help clarify the role of recombination in genetic algorithms. Another line of research (Mitchell, Forrest, and Holland, 1991) is investigating a class of fitness functions in which the effects of crossover operators can be isolated and understood in greater detail.

A surprising aspect of all this theoretical work on recombination is that, since Holland's original analysis, no theoretical developments in mathematical genetics have been brought to bear to help answer these questions. There seems to be a presumption among GA researchers that mathematical genetics has few theoretical results that are relevant to the questions most often asked about genetic algorithms. Perhaps this is because much of the work in mathematical genetics assumes single-locus models, restrictive conditions on fitness, or large populations at equilibrium. Even though mathematical genetics might not address the same set of questions that arise in the theory of genetic algorithms, however, the analytical tools and insights of mathematical genetics can be usefully applied to genetic algorithms. For instance, there is a well developed line of research in mathematical genetics — the theory of **recombination distributions** — that characterizes the effects of recombination on multiple loci in the absence of selection. This research can be significant for genetic algorithm theory because it provides tools that may facilitate a much deeper analysis of the role of recombination as a search operator.

The landmark result on recombination distributions was published almost fifty years ago (Geiringer, 1944). Clearly, the application of this work to genetic algorithms is long overdue! This paper reviews Geiringer's results and more recent work that builds on it. We then derive expressions for the recombination distributions associated with the crossover operators most commonly used in genetic algorithms. The utility of this characterization is illustrated by showing how it helps us quantify certain inductive biases associated with crossover operators. We conclude with a discussion of the potential significance of recombination distributions for GA theory.

2   The Theory of Recombination Distributions

Geiringer's (1944) paper analyzed the limiting distribution of genotypes for arbitrarily linked loci having multiple alleles per locus. The assumptions used in this analysis were fairly standard: a large, diploid, random mating population with no sex differences that reproduces in non-overlapping generations. She also assumed that recombination acts without selection or mutation to produce the next generation. In this section we summarize Geiringer's analysis in a way that (hopefully) makes it accessible to readers familiar with genetic algorithms but unfamiliar with classical genetics models and terminology.

Let $S_\ell = \{1, \ldots, \ell\}$ be the set of ℓ numbers designating the loci in strings of length ℓ. The number of alleles allowed at each locus can be any arbitrary integer. Geiringer describes the transmission of alleles during recombination in terms of a distribution $\mathcal{R}$, called the linkage distribution or **recombination distribution**, defined as follows. Each subset $A \subseteq S_\ell$ denotes a recombination event in which an offspring receives its alleles at the loci in A from one parent and alleles at the loci in $A' = S_\ell \setminus A$ from the other parent[1]. $\mathcal{R}(A)$ denotes the probability of that event. For example, suppose two parent strings mate to produce two offspring strings and a crossover occurs between loci 1 and 2. Two recombination events occur during this instance of crossover. The set $A = \{1\}$ denotes the recombination event in which one offspring receives the allele at the first locus from one parent and the alleles at loci $2, \ldots, \ell$ from the other parent. The set A' denotes the complementary event describing the transmission of alleles to the other offspring. Clearly, under Mendelian segregation, $\mathcal{R}(A) = \mathcal{R}(A')$ since all alleles will be transmitted to one offspring or the other. It is also clear that $\sum_{A \subseteq S_\ell} \mathcal{R}(A) = 1$. We can therefore view recombination distributions as probability distributions over the power set 2^{S_ℓ} (Schnell, 1961). Geiringer also defines marginal distributions $\mathcal{R}_A$ that describe the transmission of the loci in A. The marginal recombination distribution $\mathcal{R}_A$ is given by the probabilities

$$\mathcal{R}_A(B) = \sum_{C \subseteq A'} \mathcal{R}(B \cup C) \quad B \subseteq A$$

$\mathcal{R}_A(B)$ is the marginal probability of the recombination event in which one parent transmits the loci in $B \subseteq A$ and the other parent transmits the loci in $A \setminus B$.

Given these definitions, Geiringer proceeds to examine how recombination without selection modifies the proportions of individuals in a population over time. Assume that each individual $x \in \{1, 2, \ldots, m\}^\ell$ is a string of length ℓ in a finite alphabet of m characters. We also assume in the following that $B \subseteq A \subseteq S_\ell$. Let $p^{(t)}(x)$ be the frequency of individual x in a population at generation t, and $p_A^{(t)}(x)$ denote the marginal frequency of individuals that are identical to x at the loci in A. That is,

$$p_A^{(t)}(x) = \sum_y p^{(t)}(y) \text{ , for each } y \text{ satisfying } y_i = x_i \ \forall i \in A$$

Geiringer derives the following important recurrence relations.

$$p^{(t+1)}(z) = \sum_{A,x,y} \mathcal{R}(A)p^{(t)}(x)p^{(t)}(y) \quad \text{where} \quad \begin{cases} A \subseteq S_\ell \text{ is arbitrary} \\ y_i = z_i \ \forall i \in A \\ x_i = z_i \ \forall i \in A' \end{cases} \quad (1)$$

---

[1] We use the operator $\setminus$ to denote set difference.

$$p^{(t+1)}(z) = \sum_{A,B,x,y} \mathcal{R}_A(B)p^{(t)}(x)p^{(t)}(y) \quad \text{where} \quad \begin{cases} B \subseteq A \subseteq S_\ell \text{ are arbitrary subsets} \\ x_i \neq z_i, y_i = z_i \ \forall i \in B \\ x_j = z_j, y_j \neq z_j \ \forall j \in A \setminus B \\ x_k = y_k = z_k \ \forall k \in A' \end{cases}$$

$$(2)$$

$$p^{(t+1)}(z) = \sum_{A \subseteq S_\ell} \mathcal{R}(A)p_A^{(t)}(z)p_{A'}^{(t)}(z) \tag{3}$$

These recurrence relations are equivalent, complete characterizations of how recombination changes the proportion of individuals from one generation to the next. Equation 1 has the straightforward interpretation that alleles appear in offspring if and only if they appear in the parents and are transmitted by a recombination event. Each term on the right hand side of (1) is the probability of a recombination event between parents having the desired alleles at the loci that are transmitted together. A string z is the result of a recombination event A whenever the alleles of z at loci A come from one parent and the alleles at loci A' come from the other parent. The change in frequency of an individual string is therefore given by the total probability of all these favorable occurrences. Equation 2 is derived from (1) by collecting terms based on marginal recombination probabilities. Equation 3 is derived from (1) by collecting terms based on marginal frequencies of individuals.

The last equation is perhaps the most significant, since it leads directly to a theorem characterizing the expected distribution of individuals in the limit.

Theorem (Geiringer's Theorem II) *If ℓ loci are arbitrarily linked, with the one exception of "complete linkage", the distribution of transmitted alleles "converges toward independence". The limit distribution is given by*

$$\lim_{t \to \infty} p^{(t)}(z) = \prod_{i=1}^{\ell} p_{\{i\}}^{(0)}(z)$$

which is the product of the ℓ marginal distributions of alleles from the initial population.

This theorem tells us that, in the limit, random mating and recombination without selection lead to chromosome frequencies corresponding to the simple product of initial allele frequencies. A population in this state is said to be in **linkage equilibrium** or Robbin's equilibrium (Robbins, 1918). This result holds for all recombination operators that allow any two loci to be separated by recombination[2]. Note that Holland (1975) sketched a proof of a similar result for schema frequencies and one-point crossover. Geiringer's theorem applied to schemata gives us a much more general result. Together with the recurrence equations, this work paints a picture of "search pressure" from recombination acting to reduce departures from linkage equilibrium for all schemata. The convergence rate for any particular schema is given by the probability of the recombination event specified by the schema's defining loci. In this view the only difference between recombination operators is the rate at

---

[2]We have "complete linkage" whenever a set of loci cannot be separated by recombination.

which, undisturbed by selective pressures, they drive schemata to their equilibrium proportions.

Analyses using Geiringer's results can be simplified for binary strings $x \in \{0,1\}^\ell$ if we represent individual strings using index sets (Christiansen, 1989). Each x can be represented uniquely by the subset $A \subseteq S_\ell$ using the convention that A designates the loci where $x_i = 1$ and A' designates the loci where $x_i = 0$. In this notation S_ℓ represents the string $11 \dots 11$, $\emptyset$ represents the string $00 \dots 00$, and A' represents the binary complement of the string represented by A. Index sets can greatly simplify expressions involving individual strings. Consider, for example, the marginal frequency $p_A(x)$ of individuals that are identical to x at the loci in A. The index set expression

$$p_A(B) = \sum_{C \subseteq A'} p(B \cup C) \quad B \subseteq A$$

makes it clear that $p_A(B)$ involves strings having the allele values given by B at the loci designated by A. Note that $p_\emptyset(B) = 1$ and $p_{S_\ell}(B) = p(B)$.

With this notation we can also succinctly relate recombination distributions and schemata. If A designates the defining loci of a schema s and $B \subseteq A$ specifies the alleles at those loci, then the frequency of s is given by $p_A(B)$ and the marginal distribution $\mathcal{R}_A$ describes the transmission of the defining loci of s. These definitions provide us with an important perspective on previous analyses of crossover disruption in genetic algorithms. Disruption (or survival) probabilities are given by the marginal probabilities $\mathcal{R}_A(\emptyset) = \mathcal{R}_A(A)$. There are $2^{|A|-1} - 2$ other probabilities to compute in the marginal distribution $\mathcal{R}_A$.

3   Recombination Distributions for Crossover Operators

The first step in applying the theory of recombination distributions to genetic algorithms is to derive marginal distributions for the standard crossover operators. In what follows we will assume, without loss of generality, that the elements of the index set A for a schema s are in increasing order so that the kth element $A_{(k)}$ is the locus of the kth defining position of s. This means, in particular, that the outermost defining loci of s are given by the elements $A_{(1)}$ and $A_{(o(s))}$ where $o(s)$ is the order of s. It will be convenient to define the following property relating the order of a schema to its defining length $\delta(s)$.

Definition   The kth *component of defining length* for schema s, $\delta_k(s)$, is the distance between the kth and $k+1$st defining loci, $1 \le k < o(s)$, with the convention that $\delta_0(s) \equiv \ell - \delta(s)$

Note that the defining length of a schema is equal to the sum of its defining length components:

$$\delta(s) = \sum_{k=1}^{o(s)-1} \delta_k(s) = A_{(o(s))} - A_{(1)}$$

3.1   One-point Crossover

In deriving the marginal recombination distribution for one-point crossover we will follow the analysis given by Feldman and Holland (in preparation) for the restricted class of index sets $S_{[i,j]} = \{i, i+1, \ldots, j-1, j\}$, $1 \le i < j \le \ell$. However, the formulation given here will be in terms of arbitrary index sets to facilitate the generalization to multi-point crossover.

Assume exactly one crossover point in a string of length ℓ, chosen between loci i and $i+1$ with probability $\frac{1}{\ell-1}$ for $i = 1, 2, \ldots, \ell-1$. The only recombination events with non-zero probability are $S_x = [1, x]$ and $S'_x = [x+1, \ell-1]$ for $x = 1, 2, \ldots, \ell-1$. The probability of each event is

$$\mathcal{R}^1(S_x) = \mathcal{R}^1(S'_x) = \frac{1}{2(\ell-1)}$$

since each parent is equally likely to transmit the indicated loci. The marginal distribution $\mathcal{R}^1_A$ for an arbitrary index set A can be expressed solely in terms of these recombination events. We will refer to these events as the primary recombination events.

Now for any arbitrary event $B \subseteq A$ there are two cases to consider:

1. $B = \emptyset$
   This corresponds to the primary recombination events S_x, $x < A_{(1)}$ and S'_x, $x \ge A_{(o(s))}$. There are $\ell - 1 - \delta(s)$ such events.

2. $B \ne \emptyset$
   These situations involve the primary events S_x, $A_{(1)} \le x < A_{(o(s))}$. The events B having non-zero probability are given by $B_i = \{A_{(1)}, \ldots, A_{(i)}\}$, $1 \le i < o(s)$. For each i, there are $\delta_i(s)$ corresponding primary events.

The complete marginal distribution is therefore given by

$$\mathcal{R}^1_A(B) = \begin{cases} \frac{\ell-1-\delta(s)}{2(\ell-1)} & \text{if } B = \emptyset \text{ or } B = A \\ \frac{\delta_i(s)}{2(\ell-1)} & \text{if } B = B_i, \ 1 \le i < o(s) \\ 0 & \text{otherwise} \end{cases}$$

Note that if we restrict our attention to disruptive events, we obtain the familiar result

$$1 - (\mathcal{R}^1_A(\emptyset) + \mathcal{R}^1_A(A)) = 1 - 2\left(\frac{\ell-1-\delta(s)}{2(\ell-1)}\right) = 1 - \left(1 - \frac{\delta(s)}{\ell-1}\right) = \frac{\delta(s)}{\ell-1}$$

3.2   Multi-Point Crossover

The generalization to n crossover points in a string of length ℓ uses the standard convention (De Jong, 1975) that when the number of crossover points is odd, a final crossover point is defined at position zero. We also assume that all the crossover points are distinct, which corresponds to the way multi-point crossover is usually implemented. Given these assumptions, there are $2 \begin{pmatrix} \ell \\ n \end{pmatrix}$ non-zero recombination events if n is even or $n = \ell$, and $2 \begin{pmatrix} \ell - 1 \\ n \end{pmatrix}$ such events if n is odd. Since the n points are randomly selected, these events are equally likely to occur.

We derive an expression for the marginal distributions in the same way we proceeded for one-point crossover. First we identify the relevant recombination events, then we count them up and multiply by the probability of a single event. Identification of the appropriate recombination events begins with the observation (De Jong, 1975) that crossover does not disrupt a schema whenever an even number of crossover points (including zero) fall between successive defining positions. We can use this to identify the configurations of crossover points that transmit all the loci in $B \subseteq A$ and none of the loci in $A \backslash B$. Given any two consecutive elements of A, there should be an even number of crossover points between them if they both belong to B or $A \backslash B$. Otherwise there should be an odd number of crossover points between them. This can be formalized as a predicate $\mathcal{X}_A$ that tests these conditions for a marginal distribution $\mathcal{R}_A$

$$\mathcal{X}_A(B, n, i) = \begin{cases} 1 & \text{if } n \text{ is even and } \{A_{(i)}, A_{(i-1)}\} \cap B = \emptyset \text{ or } \{A_{(i)}, A_{(i-1)}\} \\ 1 & \text{if } n \text{ is odd and } \{A_{(i)}, A_{(i-1)}\} \cap B \neq \emptyset \text{ or } \{A_{(i)}, A_{(i-1)}\} \\ & \text{where } 2 \leq i \leq o(s) \\ 0 & \text{otherwise} \end{cases}$$

The recombination events can be counted by simply enumerating all possible configurations of crossover points and discarding those not associated with the marginal distribution. The following function $\mathcal{N}_A$ computes this count recursively (as suggested by the disruption analysis of Spears and De Jong (1991)).

$$\mathcal{N}_A(B, n, i) = \begin{cases} \displaystyle\sum_{j=0}^{n} \begin{pmatrix} \delta_{i-1}(s) \\ j \end{pmatrix} \mathcal{X}_A(B, j, i) \mathcal{N}_A(B, n - j, i - 1) & 2 < i \leq o(s) \\ \\ \begin{pmatrix} \delta_1(s) \\ n \end{pmatrix} \mathcal{X}_A(B, n, 2) & i = 2 \end{cases}$$

Putting all the pieces together, we can now give an expression for the complete marginal distribution.

$$\mathcal{R}_A^n(B) = \begin{cases} \dfrac{\displaystyle\sum_{j=0}^{n} \begin{pmatrix} \delta_0(s) \\ j \end{pmatrix} \mathcal{N}_A(B, n - j, o(s))}{2 \begin{pmatrix} \ell \\ n \end{pmatrix}} & \text{if } n \text{ is even or } n = \ell \\ \\ \dfrac{\displaystyle\sum_{j=0}^{n} \begin{pmatrix} \delta_0(s) - 1 \\ j \end{pmatrix} \mathcal{N}_A(B, n - j, o(s))}{2 \begin{pmatrix} \ell - 1 \\ n \end{pmatrix}} & \text{otherwise} \end{cases}$$

Figure 1 shows how the marginal probability of transmission for 2nd order schemata — $2 \mathcal{R}_A^n(A), |A| = 2$ — varies as a function of defining length. The shape of the

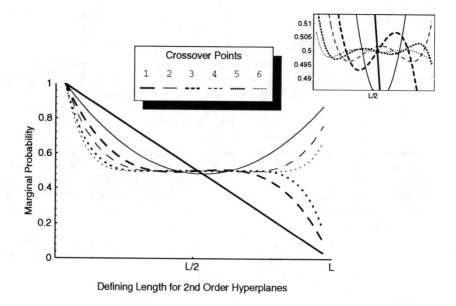

Figure 1: Transmission probabilities for 2nd order schemata. The inset shows the behavior of these curves in the vicinity of the point $L/2$.

curves depends on whether n is odd or even. This is similar to the family of curves for non-disruptive crossovers given by Spears and De Jong (1991). The difference is that Spears and De Jong assume crossover points are selected randomly with replacement. This means that their measure $P_{2,even}$ is a polynomial function of the defining length having degree n, with n identical solutions to the equation $P_{2,even} = 1/2$ at the point $\ell/2$. The function $\mathcal{R}_A^n(A)$, on the other hand, has n distinct solutions to the equation $2\,\mathcal{R}_A^n(A) = 1/2$ as shown in the upper right hand corner of Figure 1. This property stems from our assumption that crossover points are distinct and hence selected without replacement.

3.3   Parameterized Uniform Crossover

The marginal distribution $\mathcal{R}_A^{u(p)}$ for parameterized uniform crossover with parameter p is easily derived from previous analyses (Spears and De Jong, 1991). It is given by

$$\mathcal{R}_A^{u(p)}(B) = p^{|B|}(1-p)^{|A \setminus B|}$$

4   Quantifying Bias in Crossover Operators

There are many ways recombination distributions like those given above can be used to facilitate the analysis of genetic algorithms. Once we have a true probability distribution, we are in a position to treat various measures characterizing crossover operators as random variables and compute descriptive statistics for them. In this section we show how to use these distributions to derive quantitative measures of certain biases in crossover operators.

Bias in search operators is a fundamental characteristic of all symbolic approaches to induction. It has long been recognized that inductive bias is *necessary* in any generalization scheme that makes inductive leaps beyond the data that has been observed (Mitchell, 1980). Furthermore, there is no "universal" bias that facilitates induction across all problem domains (Dietterich, 1989). In order to devise effective inductive procedures, we need to be able to characterize the biases in the search operators, recognize when the bias is "correct" or "incorrect" for a given problem, and recover from incorrect biases whenever possible.

Two types of bias have been attributed to crossover operators in genetic search: *distributional* bias and *positional* bias (Eshelman, Caruana, and Schaffer, 1989). Distributional bias refers to the number of loci transmitted during a recombination event and the extent to which some values might be more likely than others. The marginal distribution $\mathcal{R}_A^1$ for one-point crossover assigns equal probability to each of the primary events, so it clearly has no distributional bias. Positional bias refers to how much the probability that a set of loci will be transmitted together during a recombination event depends on the relative positions of those loci on the chromosome. In terms of recombination distributions, this is simply the extent to which the marginal distribution depends on defining length and the components of defining length. The marginal distribution $\mathcal{R}_A^{u(p)}$ for parameterized uniform crossover clearly has no positional bias since it depends only on the cardinality of the index sets, not their contents.

One benefit of recombination distributions is that they give us a way to characterize these biases using a single scalar quantity. For example, distributional bias can be interpreted as the dissimilarity between two probability distributions imposed over the set $\{1, \ldots, \ell - 1\}$: the uniform distribution, and the marginal probabilities

$$q(x) = \sum_{|A|=x} \mathcal{R}(A), x \in \{1, \ldots, \ell - 1\}$$

derived from a recombination distribution $\mathcal{R}$ by collecting terms based on the number of bits transmitted by one parent during crossover. By measuring the extent to which these probabilities are not uniform, we characterize distributional bias in exactly the same way Eshelman et al. (1989) defined it. In a similar manner, positional bias can be interpreted as the dissimilarity between $\mathcal{R}$ and the distribution having the same marginal probabilities $q(x)$ as $\mathcal{R}$ but equal probabilities

$$q'(x) = \frac{q(x)}{\binom{\ell}{x}}$$

for all recombination events A having the same size x. This "piecewise uniformity" criterion captures the intuition that an operator has no positional bias if we can change the relative locations of the elements of A without affecting the probability that A is transmitted.

A quantitative measure for each kind of bias requires some way to measure the difference between two probability distributions. There are many ways to quantify the dissimilarity between two distributions. The class of information-theoretic *divergence measures* (Lin, 1991) is attractive because these measures have been extensively studied and widely used. For our purposes the L divergence measure given by

$$L(P, P') = \sum_x \left(P(x) \log \frac{2P(x)}{P(x) + P'(x)} + P'(x) \log \frac{2P'(x)}{P(x) + P'(x)} \right)$$

is a reasonable choice[3]. It is a symmetric, non-negative, bounded quantity that approaches zero as the "resemblance" between P and P' increases. The L measure equals zero exactly when the two distributions are identical, and equals 2 when the distributions are as dissimilar as possible.

We can use this to define the following measure of distributional bias for a recombination distribution $\mathcal{R}$:

$$B_D(\mathcal{R}) = q(0) + q(\ell) + \sum_{x=1}^{\ell-1} \left(q(x) \log \frac{2q(x)}{q(x) + \frac{1}{\ell-1}} + \frac{1}{\ell-1} \log \frac{2\frac{1}{\ell-1}}{q(x) + \frac{1}{\ell-1}} \right)$$

Note that this definition accommodates recombination distributions like those for parameterized uniform crossover that can transmit 0 or ℓ bits. In order to maintain the intuitive characterization of one-point crossover as having no distributional bias, we extended the uniform distribution over $\{1, \ldots, \ell-1\}$ to the set $\{0, \ldots, \ell\}$ by assigning zero probability to the points 0 and ℓ.

The distribution of the marginal probabilities $q(x)$ for multi-point crossover and parameterized uniform crossover are shown in Figure 2 for strings of length 15. In the case of multi-point crossover we see that the probabilities are uniformly distributed for one-point and two-point crossover. As the number of crossover points increases, we obtain an increasingly narrow symmetric distribution centered at $\ell/2$. For parameterized uniform crossover we see a symmetric distribution around $\ell/2$ bits when the crossover parameter is 1/2. As the crossover parameter decreases, the mode of the distribution approaches zero. The effects of this intuitively obvious behavior on distributional bias is illustrated in Figure 3. Distributional bias changes monotonically for both multi-point crossover and parameterized uniform crossover in a way that directly reflects the changes observed in $q(x)$.

Things are somewhat more interesting when we quantify positional bias. We define the following measure of positional bias for a recombination distribution $\mathcal{R}$:

$$B_P(\mathcal{R}) = \sum_{A \subseteq S_\ell} \left(q'(|A|) \log \frac{2q'(|A|)}{q'(|A|) + \mathcal{R}(A)} + \mathcal{R}(A) \log \frac{2\mathcal{R}(A)}{q'(|A|) + \mathcal{R}(A)} \right)$$

---

[3] We use the logarithmic base 2 in this definition.

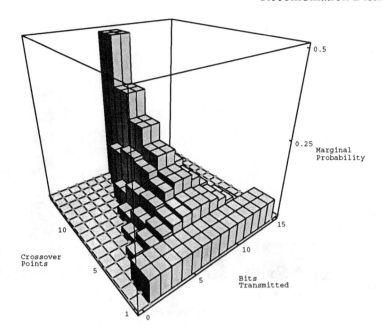

(a) Multi-Point crossover

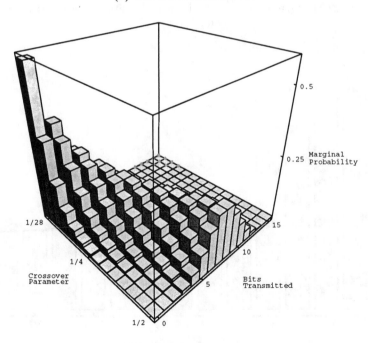

(b) Parameterized uniform crossover

Figure 2: Distribution of the marginal probabilities $q(x)$

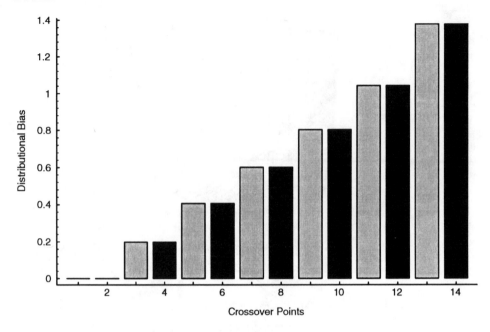

(a) Multi-Point crossover.

(Gray (black) histogram bars are used when the number of crossover points is odd (even).)

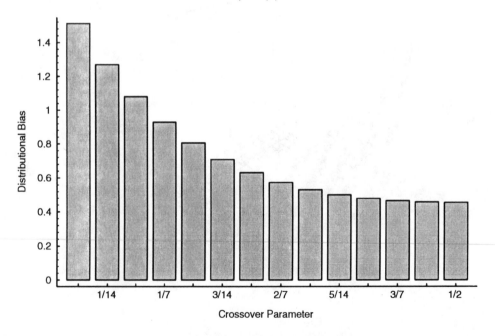

(b) Parameterized uniform crossover

Figure 3: The distributional bias measure B_D

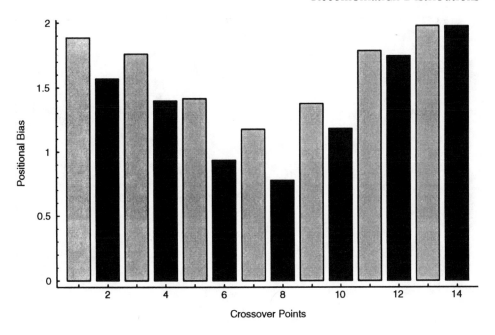

Figure 4: The positional bias B_P for multi-point crossover. Gray (black) histogram bars are used when the number of crossover points is odd (even).

This measure assigns a positional bias of zero to parameterized uniform crossover for all values of the crossover parameter. To see this, simply observe that

$$q'(|A|) = \frac{q(|A|)}{\left(\begin{array}{c} \ell \\ |A| \end{array} \right)} = \frac{\sum_{|B|=|A|} \mathcal{R}(B)}{\left(\begin{array}{c} \ell \\ |A| \end{array} \right)} = \frac{\left(\begin{array}{c} \ell \\ |A| \end{array} \right) p^{|A|}(1-p)^{\ell-|A|}}{\left(\begin{array}{c} \ell \\ |A| \end{array} \right)} = \mathcal{R}^{u(p)}(A)$$

which implies that each term in the sum defining $B_P(\mathcal{R}^{u(p)})$ must be zero. The values for multi-point crossover are shown in Figure 4. This data can be understood by recalling that there are $2 \left(\begin{array}{c} \ell \\ n \end{array} \right)$ non-zero recombination events if n is even or $n = \ell$, and $2 \left(\begin{array}{c} \ell-1 \\ n \end{array} \right)$ such events if n is odd. This means that the more n differs from $\ell/2$, the more events there are having zero probability. It is clear that as the number of events having zero probability increases, the departure of a probability distribution from uniformity becomes more pronounced. Consequently, we should expect to see some symmetry in the shape of $B_P(\mathcal{R}^n)$. It should be pointed out that this is not the way Eshelman et al. (1989) expected positional bias to behave.

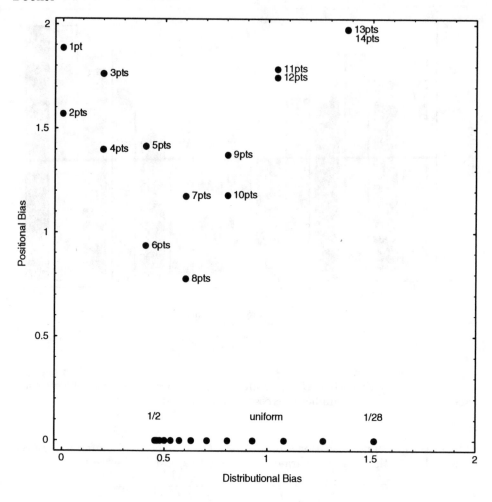

Figure 5: Overall view of crossover bias

They predicted that positional bias would continue to increase as n gets larger. Finally, note that the difference in bias between consecutive odd/even crossover points $2k-1$ and $2k$ is explained by the fact that $\begin{pmatrix} \ell - 1 \\ 2k - 1 \end{pmatrix} < \begin{pmatrix} \ell \\ 2k \end{pmatrix}$ whenever $k < \ell/2$.

Quantifying bias is an important step toward understanding how to use bias to make learning and generalization more effective. The overall bias "landscape" for crossover operators is summarized in Figure 5. Since each bias measure is derived from the recombination distribution $\mathcal{R}$, perhaps $\mathcal{R}$ itself is the best quantitative summary of bias in crossover. We might therefore be able to to use $\mathcal{R}$ to help understand when crossover is helpful in genetic search. For example, we could think of selection in terms of the probability distribution it imposes over 2^{S_ℓ}. The way this distribution aligns or fails to align with $\mathcal{R}$ may be correlated with the effectiveness of crossover.

5   Discussion

The quantification of bias in crossover operators is just one example of how recombination distributions can be used to analyze genetic algorithms. Geiringer's convergence results suggest that the most important difference among recombination operators is the rate at which they converge to equilibrium in the absence of selection. A useful quantity for studying this property is the coefficient of **linkage disequilibrium** which measures the deviation of current chromosome frequencies from their equilibrium levels. The convergence to linkage equilibrium is well understood analytically (Christiansen, 1989). GA theory may benefit from an effort to apply what is known about the dynamics of linkage disequilibrium to characterize the search behavior of GA crossover operators.

Another, perhaps ultimately more important, use of recombination distributions is to facilitate the analysis of schema dynamics. The dynamics of simple genetic algorithms are usually studied with models that explicitly represent populations of binary strings (e.g., Vose and Liepins (1991)). Such models are exact and complete, but it is not clear if they will help us easily answer questions about the dynamics of schemata and how they compete over time. Marginal recombination distributions tell us directly how schema and all their components are transmitted by recombination from parents to offspring. Feldman and Holland (in preparation) have developed a model of genetic algorithms using recombination distributions that includes selection. This line of research may help us analyze and understand the transient behavior of schemata.

† Acknowledgements

This work was supported by MITRE Sponsored Research Project No. 96430. Marcus Feldman pointed out to the author that Geiringer's paper is a significant result in mathematical genetics that is relevant to any study of recombination without selection. I would also like to thank Darrell Whitley for several helpful comments on an earlier draft of this paper.

References

Bridges, C. L. and Goldberg, D. E. (1987). An analysis of reproduction and crossover in a binary-coded genetic algorithm. *Proceedings of the Second International Conference on Genetic Algorithms* (pp. 9–13). Cambridge, MA: Lawrence Erlbaum.

Christiansen, F. B. (1989). The effect of population subdivision on multiple loci without selection. In M. W. Feldman (Ed.) *Mathematical Evolutionary Theory*. Princeton, NJ: Princeton University Press.

De Jong, K. A. (1975). *An analysis of the behavior of a class of genetic adaptive systems*. Doctoral dissertation, Department of Computer and Communication Sciences, University of Michigan, Ann Arbor, MI.

Dietterich, T. G. (1989). Limits of inductive learning. *Proceedings of the Sixth International Conference on Machine Learning* (pp. 124–128). Ithaca, NY: Morgan Kaufmann.

Eshelman, L. J. (1991). The CHC adaptive search algorithm: How to have safe search when engaging in nontraditional genetic recombination. In G. Rawlins (Ed.) *Foundations of Genetic Algorithms*. San Mateo, CA: Morgan Kaufmann.

Eshelman, L. J., Caruana, R. A. and Schaffer, J. D. (1989). Biases in the crossover landscape. *Proceedings of the Third International Conference on Genetic Algorithms* (pp. 10–19). Fairfax, VA: Morgan Kaufmann.

Feldman, M. W. and Holland, J. H. (in preparation). Unpublished manuscript.

Geiringer, H. (1944). On the probability theory of linkage in Mendelian heredity. *Annals of Mathematical Statistics, 15*, 25–57.

Liepins, G. and Vose, M. (1992). Characterizing crossover in genetic algorithms. *Annals of Mathematics and Artificial Intelligence, 5(1)*.

Lin, J. (1991). Divergence measures based on the Shannon entropy. *IEEE Transactions on Information Theory, 37(1)*, 145–151.

Mitchell, M., Forrest, S., and Holland, J. H. (1991). The royal road for genetic algorithms: Fitness landscapes and GA performance. *Proceedings of the First European Conference on Artificial Life*. Cambridge, MA: MIT Press.

Mitchell, T. M. (1980). The need for biases in learning generalizations. Technical report CBM-TR-117, Department of Computer Science, Rutgers University, New Brunswick, NJ. (Reprinted in J. Shavlik and T. Dietterich (Eds.) *Readings in Machine Learning*. San Mateo, CA: Morgan Kaufmann 1990).

Robbins, R. B. (1918). Some applications of mathematics to breeding problems, III. *Genetics, 3*, 375–389.

Schaffer, J. D. and Eshelman, L. J. (1991). On crossover as an evolutionarily viable strategy. *Proceedings of the Fourth International Conference on Genetic Algorithms* (pp. 61–68). La Jolla, CA: Morgan Kaufmann.

Schnell, F. W. (1961). Some general formulations of linkage effects in inbreeding. *Genetics, 46*, 947–957.

Spears, W. M. and De Jong, K. A. (1991). An analysis of multi-point crossover. In G. Rawlins (Ed.) *Foundations of Genetic Algorithms*. San Mateo, CA: Morgan Kaufmann.

Syswerda, G. (1989). Uniform crossover in genetic algorithms. *Proceedings of the Third International Conference on Genetic Algorithms* (pp. 2–9). Fairfax, VA: Morgan Kaufmann.

Vose, M. and Liepins, G. (1991). Punctuated equilibria in genetic search. *Complex Systems, 5*, 31–44.

An Executable Model of a Simple Genetic Algorithm

Darrell Whitley
Computer Science Department
Colorado State University
Fort Collins, CO 80523
whitley@cs.colostate.edu

Abstract

A set of executable equations are defined which model the ideal behavior of a simple genetic algorithm. The equations assume an infinitely large population and require the enumeration of all points in the search space. When implemented as a computer program, the equations can be used to study problems of up to approximately 15 bits. These equations represent an extension of earlier work by Bridges and Goldberg. At the same time these equations are a special case of a model introduced by Vose and Liepins. The various models are reviewed and the predictive behavior of the executable equations is examined. Then the executable equations are extended by quantifying the behavior of a reduced surrogate operator and a uniform crossover operator. In addition, these equations are used to study the computational behavior of a parallel island model implementation of the simple genetic algorithm.

1 Introduction

The original schema theorem (Holland, 1975) provides a lower bound on changes in the sampling rate for a hyperplane from generation t to generation $t + 1$. The following version of the schema theorem is based on Schaffer's work (1987).

$$P(h, t+1) \geq P(h, t) \frac{f(h)}{\bar{f}} \left\{ 1 - P_c \frac{l(h)}{L-1} (1 - P(h, t) \frac{f(h)}{\bar{f}}) \right\}$$

where h represents a particular hyperplane and $P(h, t)$ indicates the fraction of the population that are members of a hyperplane h at time t, where t is an index to a particular generation of strings. The value $f(h)/\bar{f}$ indicates the average fitness of the members of h relative to average fitness of all strings in the population. The defining length of the schema, $l(h)$, includes only the significant part of the defining schema for h. In other words, the defining length of ****10*1*0***** is 5 since there are 5 potential break points between the bits in the significant part of the schema. The length of the string is represented by L and P_c is the probability that crossover will be applied. Note that no mutation is included in this particular formulation of the schema theorem.

This formula captures the following ideas. Together, P_c and $\{l(h)/L - 1\}$ define the probability that crossover will cut the schema representing the hyperplane h. However, this need not always be disruptive if the other parent string also samples the same hyperplane h. The probability that a random string in the population contains a copy of the schema representing h is $P(h, t)$ so $1 - P(h, t)$ is the probability that the mate *does not* sample the schema representing h. Therefore, the probability of disruption is (1) the probability of crossover multiplied by (2) the probability that crossover will occur within the defining length of the schema multiplied by (3) the probability the other parent is not a member of $P(h, t)$. The probability that a schema survives intact is 1 minus the probability of disruption. Note that the equation is an inequality because it ignores string gains and overestimates the sources of disruption. The sources of disruption are over estimated because recombination with strings containing "similar schemata" (i.e., strings that have many of the relevant bits in common with h, but not all the bits are identical) do not necessarily produce disruption. *String gains* occur when new schema samples are generated by recombination of strings that do not contain copies of the schema h.

The following equations have recently been used to study 3 and 4 bit problems (Whitley, 1992). These equations are based on similar equations for 2 bit problems used by Goldberg (1987) to study the minimally deceptive problem.

$$P(Z, t+1) = (1 - P_c)P(Z, t)\frac{f(Z)}{\bar{f}} + P_c \left\{ P(Z, t)\frac{f(Z)}{\bar{f}}(1 - \text{losses}) + \text{gains.} \right\}$$

which reduces to $\quad P(Z, t+1) = P(Z, t)\frac{f(Z)}{\bar{f}}(1 - \{P_c \text{ losses}\}) + \{P_c \text{ gains.}\}$

This is an idealized version of a simple genetic algorithm (Goldberg, 1989a). In the current formulation, Z might refer to either a string or a schema representing a hyperplane. P(Z,t) is the proportion of the population that samples the string (schema) Z at time t. The expression $f(Z)/\bar{f}$ indicates the change in representation of Z in the population due to fitness, where f(Z) is the fitness associated with the string (schema) Z and $\bar{f}$ is the average fitness of the population. Given a specification of Z, one can calculate string losses and gains. Losses occur when a string (schema) crosses with another string and the resulting offspring fails to preserve the original string. Gains occur when two different strings cross and independently create a new copy of some string. For example, if Z = 000 then recombining with 100 or 001 will always produce a new copy of 000. Assuming 1-point crossover is used as an operator, the probability of "losses" and "gains" for the string Z = 000 are calculated as follows:

$$\text{losses} = P_{I0}\frac{f(111)}{\bar{f}}P(111,t) + P_{I0}\frac{f(101)}{\bar{f}}P(101,t)$$

$$+P_{I1}\frac{f(110)}{\bar{f}}P(110,t) + P_{I2}\frac{f(011,t)}{\bar{f}}P(011,t).$$

$$\text{gains} = P_{I0}\frac{f(001)}{\bar{f}}P(001,t)\frac{f(100)}{\bar{f}}P(100,t) + P_{I1}\frac{f(010)}{\bar{f}}P(010,t)\frac{f(100)}{\bar{f}}P(100,t)$$

$$+P_{I1}\frac{f(011)}{\bar{f}}P(011,t)\frac{f(100)}{\bar{f}}P(100,t) + P_{I2}\frac{f(001)}{\bar{f}}P(001,t)\frac{f(110)}{\bar{f}}P(110,t)$$

$$+P_{I2}\frac{f(001)}{\bar{f}}P(001,t)\frac{f(010)}{\bar{f}}P(010,t).$$

When Z is a string $P_{I0} = 1$ and represents the probably of crossover somewhere on the string. (If Z were a schema, then this term would be the same as $i(h)/(L-1)$) where $l(s)$ is the defining length of the schema representing hyperplane h and L is the length of the string.) For Z = 000 this can be further broken into two order-2 schemata that are of interest. The first order-2 schema of interest, h_1, spans from the first to the second bit and the second order-2 schema of interest, h_2, spans from the second to the third bit. These are of interest because the defining lengths of these order-2 schemata provide probability information about the disruption of the order-3 schema. The probability that one-point crossover will fall between the first and second bit is merely $l(h_1)/(L-1)$, which will be referred to as P_{I1}. Likewise, P_{I2} will denote the probability that one-point crossover will fall between the second and third bit which is given by $l(h_2)/(L-1)$.

The equations can be generalized to cover the remaining 7 strings in the space by using $(S_i \oplus Z)$ to transform each bit string, S_i, contained in the formula to the appropriate corresponding strings for computing the formula for any and all terms of the form P(Z,t+1). The translation is accomplished using bitwise addition modulo 2 (i.e., a bitwise exclusive-or denoted by $\oplus$) (c.f., Liepins and Vose, 1990).

The equations already presented are applicable to 3 bit problems corresponding to either schemata or strings of 3 bits; of course, if schemata are modeled (as opposed to complete strings) the equations would not include information about higher order schemata competitions or the influence of "co-lateral" schemata competitions on the search process. It would also be difficult to obtain fitness information about competing sets of hyperplanes. Therefore in the remainder of this paper, the equations are developed only for strings.

2   A Generalized Form Based on Equation Generators

The 3 bit equations were generated by hand; however, a general pattern exists which allows the automatic generation of equations for arbitrary problems (Whitley, Das and Crabb, 1992). The number of terms in the equations is greater than the number of strings in the search space. Therefore it is only practical to develop exact equations for problems with approximately 15 bits in the encoding.

The development of a general form for these equations is illustrated by generating the loss and gain terms in a systematic fashion. Note that the equations need only be defined once for one string in the space; the *standard form* of the equation is always defined for the string composed of all zero bits. Let S represent the set of binary strings of length L, indexed by i. In general, the string composed of all zero bits is denoted S_0.

All the information needed to define the equations comes from the schema theorem except for the calculation of exact gains and losses. In the process of developing the general pattern for these gains and losses it was realized that the "generators" developed for these executable equations are equivalent to those used by Bridges and Goldberg (1987). The "generators" of Bridges and Goldberg will be reviewed and a general graph structure is introduced to allow better visualization of the relationship between generators.

2.1 Generating String Losses for 1-point crossover

Bridges and Goldberg (1987) present a "middle function" which is used to calculate losses dues to crossover disruption. Consider two strings 00000000000 and 00010000100. Using 1-point crossover, if the crossover occurs before the first "1" bit or after the last "1" bit, no disruption will occur. Any crossover between the 1 bits, however, will produce disruption: neither parent will survive crossover. Also note that recombining 00000000000 with any string of the form 0001####100 will produce the same pattern of disruption (Bridge and Goldberg denoted this as 0001****100; # is used here instead of * to better distinguish between a generator and the corresponding hyperplane.) Bridges and Goldberg formalize the notion of a generator in the following way. Consider strings B and B' where the first x bits are equal, the middle $(\delta + 1)$ bits have the pattern $b\#\#...\#b$ for B and $\bar{b}\#\#...\#\bar{b}$ for B'. Given that the strings are of length L, the last $(L - \delta - x - 1)$ bits are equivalent. The $\bar{b}$ bits are referred to as "sentry bits" and they are used to define the probability of disruption. The full generator for a middle function M[B,δ,x] is defined as follows:

$$M[B, \delta, x] = b_0 \ ... \ b_{x-1}\bar{b}_x \ ... \ \bar{b}_{x+\delta}b_{x+\delta+1} \ ... \ b_{l-1}$$

or, for the standard form of the executable equations,

$$M[S_0, \delta, x] = 0_0 \ ... \ 0_{x-1}1_x \ ... \ 1_{x+\delta}0_{x+\delta+1} \ ... \ 0_{l-1}$$

since $B = S_0$ and therefore the sentry bits must be 1.

The middle function generators are then used to count all strings that are capable of disruption as well as the probability of disruption which is derived from the δ length (analogous to the schema defining length) of each particular generator. The probability of disruption for B given by Bridges and Goldberg (1987) is:

$$P_d(B) = \left[\sum_{\delta=1}^{L-1} \frac{\delta}{L-1} \cdot \sum_{x=1}^{L-\delta-1} \sum_{\{j \mid S_j \epsilon M[B,\delta,x]\}} R_j^t \right]$$

where $R_j{}^t = P_b{}^t(f_b/\bar{f})$ and $P_b{}^t$ is the representation of B in the population at time t and $f_b/\bar{f}$ is its change in representation due to fitness.

The following directed acyclic graph illustrates all generators for "string losses" for the standard form of a 5 bit equation for S_0.

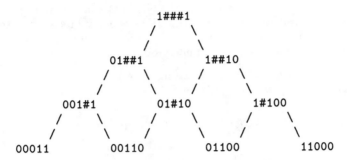

The graph structure allows one to visualize the set of all generators for string losses. In general, the root of this graph is defined by a string with a sentry bit in the first and last bit positions, and the generator token "#" in all other intermediate positions. A move down and to the left in the graph causes the leftmost sentry bit to be shifted right; a move down and to the right causes the rightmost sentry bit to be shifted left. Since the standard form of the equation is written for S_0, the sentry bits are always "1" bits. All bits outside the sentry positions are "0" bits. Summing over the graph, one can see that there are $\sum_{j=1}^{L-1} j \cdot 2^{L-j-1}$ or $(2^L - L - 1)$ strings generated as potential sources of string loss.

Note that there are $(2^L - L - 1)$ strings that cause disruption and that the term $\delta/(L-1)$ used in the Bridges and Goldberg equations corresponds to the disruption probability for each string. This term is the same as the continuous "interiors" used to define the 3 bit equations. These generators are used to define all string loss source terms for the executable equations and the "interior" disruption probabilities. For each string S_i produced by one of the "middle" generators in the above graph structure, a term of the following form is added to the *losses* equations:

$$\delta(S_i)\frac{f(S_i)}{\bar{f}} \, P(S_i, t)$$

where $\delta(S_i)$ is a function that calculates the number of crossover points between sentry bits in string S_i, thus defining the critical "interior" region for this particular string. This function does not explicitly require S_0 as an input, since its form is implicitly given by any S_i. This notation attempts to be consistent with Bridges and Goldberg's use of δ. Note, however, that $\delta(S_i)$ does not refer to the *defining length* of a schema (as is often the case in the genetic algorithm literature).

2.2   Generating String Gains for 1-point crossover

Bridges and Goldberg note that string gains for a string B are produced from two strings Q and R which have the following relationship to B.

Region ->	beginning	middle	end
Length ->	a	r	w
Q Characteristics	$\#\#...\#\bar{b}$	=	=
R Characteristics	=	=	$\bar{b}\#...\#$

The "=" symbol denotes regions where the bits in Q and R match those in B; again $B = S_0$ for the *standard form* of the equations. Sentry bits are located such that 1-point crossover between sentry bits produces a new copy of B, while crossover of Q and R outside the sentry bits will not produce a new copy of B.

Bridges and Goldberg define a beginning function $A[B,\alpha]$ and ending function $\Omega[B,\omega]$, assuming $L - \omega > \alpha - 1$, where

$$A[B,\alpha] = \#\#...\#\#\bar{b}_{\alpha-1}b_\alpha...b_{L-1}$$

and $$\Omega[B,\omega] = b_0...b_{L-\omega-1}\bar{b}_{L-\omega}\#\#...\#\#$$

or, for the standard form of the equations:

$$A[S_0,\alpha] = \#\#...\#\#1_{\alpha-1}0_\alpha...0_{l-1}$$

and $$\Omega[S_0,\omega] = 0_0...0_{L-\omega-1}1_{L-\omega}\#\#...\#\#.$$

The length of the "middle" overlapping region between sentry bits in generators A and Ω (as well as any pair of resulting strings Q and R) is denoted by σ. Bridges and Goldberg use the generators to calculate the probability of string gains for some string B:

$$P_g(B) = \frac{1}{2}\begin{pmatrix} 2 \\ 1 \end{pmatrix} P_c \cdot \sum_{\alpha=1}^{L-1}\sum_{\omega=1}^{L-\alpha}\frac{\sigma+1}{L-1}\left[\sum_{\{j|S_j\epsilon A[B,\alpha]\}} R_j{}^t \left[\sum_{\{k|S_k\epsilon\Omega[B,\omega]\}} R_k{}^t\right]\right]$$

These generators can again be presented as a directed acyclic graph structure composed of paired templates which will be referred to as the upper A-generator and lower Ω-generator. The following are the generators in a 5 bit problem.

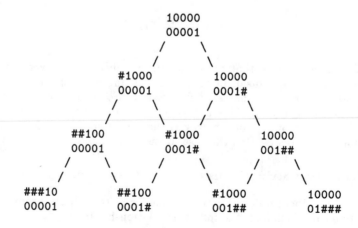

In this case, the root of the directed acyclic graph is defined by starting with the most specific generator pair. The A-generator of the root has a "1" bit as the sentry bit in the first position, and all other bits are "0." The Ω-generator of the root has a "1" bit as the sentry bit in the last position, and all other bits are "0." A move down and left in the graph is produced by shifting the left sentry bit of the current upper A-generator to the right. A move down and right is produced by shifting the right sentry bit of the current lower Ω-generator to the left. Each vacant bit position outside of the sentry bits which results from a shift operation is filled using the # symbol.

This graph structure allows one to observe that the number of generator pairs at each level of the tree is equal to the depth of the level at which the pair is located (where the root is level 1). Also, for any level k, the number of string pairs generated at that level is 2^{k-1} for each pair of generators. Therefore, the total number of string pairs that must be included in the equations to calculate string gains is:

$$\sum_{k=1}^{L-1} k \cdot 2^{k-1}$$

where L is the length of the string. When expressed in graph form all generators in a particular level in the graph have the same critical defining length between sentry bits *when modeling binary strings*; if the equations are used to model schemata however, the different "interiors" between sentry bits associated with generators at any one level in the tree may have different defining lengths.

Bridges and Goldberg use the equations for $P_d(B)$ and $P_g(B)$ to directly write a formula for the expected proportion of string B in generation t+1 under recombination and crossover with the generators for losses and gains included in the formula. The main difference in the approach offered in this paper is that the generators are used as part of a two stage computation where the generators are first used to create an exact equation in standard form instead of being included directly in the equation for some arbitrary string H. This allows the resulting equations to be implemented in an executable form with a simple transformation function for all other strings in the space.

The probability of a string gain is calculated using σ; since σ is based on the location of sentry bits, each generator pair has its own σ value which is associated with all string pairs produced by that particular generator. All possible string pairs are produced using the generators. Let $S_{\alpha+x}$ and $S_{\omega+y}$ be two strings produced by a generator pair, such that $S_{\alpha+x}$ was produced by the A-generator and has a sentry bit at location $\alpha - 1$ and $S_{\omega+y}$ was produced by the Ω-generator with a sentry bit at $L - \omega$. (The x and y terms are simply correction factors added to α and ω in order to uniquely index a string in S.) Let the σ associated with $S_{\alpha+x}$ and $S_{\omega+y}$ be computed by a corresponding function $\sigma(S_{\alpha+x}, S_{\omega+y}) = L - \omega - \alpha$. For each string pair $S_{\alpha+x}$ and $S_{\omega+y}$ a term of the following form is added to the *gains* equations:

$$\frac{\sigma(S_{\alpha+x}, S_{\omega+y}) + 1}{L - 1} \frac{f(S_{\alpha+x})}{\bar{f}} P(S_{\alpha+x}, t) \frac{f(S_{\omega+y})}{\bar{f}} P(S_{\omega+y}, t)$$

where $(\sigma(S_{\alpha+x}, S_{\omega+y}) + 1)/(L - 1)$ is the probability that 1-point crossover will fall within the "interior" region defined by the sentry bits located at $\alpha - 1$ and $L - \omega$.

3 The Vose and Liepins Models

The executable equations represent a special case of the model of a simple genetic algorithm introduced by Vose and Liepins (1991). In the following discussion the relationship between the standard form of the executable equations presented here and the Vose and Liepins model is examined.

In the Vose and Liepins model, the vector $s^t \, \epsilon \, \Re$ represents the t th generation of the genetic algorithm and the i th component of s^t is the probability that the string i is selected for the gene pool. Using i to refer to a string in s can sometimes be confusing. The symbol S has already been used to denote the set of binary strings, also indexed by i (in this case, the numeric value of i); this notation will be used where appropriate to avoid confusion. Note that s^t corresponds to the expected distribution of strings in what is sometimes referred to as the "intermediate population" in the generational reproduction process (after selection has occurred, but before recombination).

In the Vose and Liepins formulation,

$$s_i^t \sim P(S_i, t) f(S_i)$$

where $\sim$ is the equivalence relation such that $x \sim y$ if and only if $\exists \lambda > 0 \mid x = \lambda y$. In this formulation, the term $1/\bar{f}$, which would represent the average population fitness normally associated with fitness proportional reproduction, can be absorbed by the λ term.

The expected number of strings is also represented as a vector $p^t \, \epsilon \, \Re$ where the k th component of the vector is equal to the proportional representation of k at generation t **before** selection occurs. This would be the same as $P(S_k, t)$ in the notation more commonly associated with the schema theorem. Finally let $r_{i,j}(k)$ be the probability that k results from the recombination of strings i and j. Now, using $\mathcal{E}$ to denote expectation,

$$\mathcal{E} \, p_k^{t+1} = \sum_{i,j} s_i^t \, s_j^t \, r_{i,k}(k).$$

To further generalize this model, the function $r_{i,j}(k)$ is used to construct a mixing matrix M where the i, jth entry $m_{i,j} = r_{i,j}(0)$. Note that this matrix gives the probabilities that crossing strings i and j will produce the string S_0. For current purposes, assume *no mutation* is used and 1-point crossover is used as the recombination operator. Then matrix M is obviously symmetric and is zero everywhere on the diagonal except for entry $m_{0,0}$ which is 1.0. Note that M is expressed entirely in terms of string gain information. Therefore, the first row/column of the matrix is the *inverse* of that portion of the string *losses* probabilities that corresponds to $\delta(S_i)$, where each string in the set S is crossed with S_0. For completeness, the $\delta(S_i)$ for strings not produced by the string generators is 0.0 and, thus, the probability of obtaining S_0 during reproduction is 1.0. The remainder of the matrix can be calculated using string gain generators and

$$\frac{\sigma(S_{\alpha+x}, S_{\omega+y}) + 1}{L - 1}.$$

For each pair of strings produced by the string gains generators determine their index and enter the value returned by the function into the corresponding location in M. For completeness, $\sigma(S_j, S_k) = -1$ for all arbitrary pairs of strings not generated by the string gains generators, which implies entry $m_{j,k} = 0$.

Once defined M does not change since it is not affected by variations in fitness or proportional representation in the population. Thus, given the assumption of no mutations, that λ is updated each generation to correct for changes in the population average, and that 1-point crossover is used, then the standard form of the executable equations corresponds to the following portion of the Liepins and Vose model:

$$s^T M s$$

where T denotes transpose.

While the fitness terms are defined explicitly in the executable equation, so far it has been assumed that s includes fitness information without explicitly indicating how this is calculated. Fitness information is later added to the Vose and Liepins model. For now, however, this expression is first generalized to cover all strings in the search space. Vose and Liepins formalize the notion that bitwise exclusive-or can be used to remap all the strings in the search space, in this case represented by the vector s. They show that if recombination is a combination of crossover and mutation then

$$r_{i,j}(k \oplus q) = r_{i \oplus k, j \oplus k}(q).$$

This allows one to reorder the elements in s with respect to any particular point in the space. This reordering is equivalent to remapping the variables in the executable equations. A permutation function, ρ, (denoted by σ in Vose and Liepins' (1991) original work) is defined as follows:

$$\rho_j < s_0, ..., s_{N-1} >^T = < s_{j \oplus 0}, ..., s_{j \oplus (N-1)} >^T$$

where the vectors are treated as columns and N is the size of the search space. A general operator $\mathcal{M}$ can now be defined over s which remaps $s^T M s$ to cover all strings in the search space.

$$\mathcal{M}(s) = < (\rho_0 \; s)^T M \rho_0 \; s, ..., (\rho_{N-1} \; s)^T M \rho_{N-1} \; s >^T$$

Recall that s carries fitness information such that it corresponds to the intermediate phase of the population (after selection, but before recombination) as the genetic algorithm goes from generation t to $t + 1$. Thus, to complete the cycle and reach a point at which the Vose and Liepins models can be executed in an iterative fashion, fitness information is now explicitly introduced to transform the population at the beginning of iteration $t + 1$ to the next intermediate population. A fitness matrix F is defined such that fitness information is stored along the diagonal; the i, i th element is given by $f(i)$ where f is the fitness function. (In the simplest case the fitness function may directly be the objective function, but it could also be some function of the value returned by the objective function.)

The transformation from the vector p^{t+1} to the next intermediate population represented by s^{t+1} is given as follows:

$$s^{t+1} \sim F p^{t+1}$$

or, in terms of $\mathcal{M}$,

$$s^{t+1} \sim F \mathcal{M}(s^t).$$

Vose and Liepins give equations for calculating the mixing matrix M which not only includes probabilities for 1-point crossover, but also mutation. The executable model does not extend their work, but the introduction of the directed acyclic graphs does help in the visionalization of string losses and gains and the executable model has been related to earlier work by Bridges and Goldberg. The executable equations are also expressed in terms more commonly used in the genetic algorithm literature. Because the matrices F and M are sparse, the Vose and Liepins models also have the same general complexity as the executable equations.

In the following sections are some simple extensions of the executable models to include other recombination operators as well as preliminary work using these equations to study parallel island model genetic algorithms. More complex extension of the Vose and Liepins model include finite population models using Markov Models (Nix and Vose, 1992). Vose (1992) surveys the current state of this research.

4   Implementation Complexity and Preliminary Results

A computer program has been written which requires access to a copy of an evaluation function, the length of the strings to be processed and the crossover operator to be employed. The program then generates the standard equation for the string S_0 of appropriate length. The strings that are referenced in the standard equation are generated as variables and are redefined at each iteration using $(S_i \oplus Z)$ to remap all strings S_i. One generation is produced by iterating through all strings in the space and calculating P(Z,t+1) for each string. String distributions are updated after each generation.

The following complexity results pertain to the executable equations using 1-point crossover. Note that the complexity of each version of these equations is dominated by the string gains: $\sum_{j=1}^{L-1} j \cdot 2^{j-1}$. The total number of string loss terms $= 2^L - L - 1$. Also, for each of the 2^L strings representing variables the equations must be redefined at a cost of L per variable. Thus the computation required to remap and execute one equation must be repeated 2^L times per generation yielding a total computation cost on the order of:

$$2^L \left[\sum_{j=1}^{L-1} j \cdot 2^{j-1} + (2^L - L - 1) + (2^L \cdot L) \right]$$

which reduces to

$$2^L \cdot L \left[2^L + 2^{L-1} - L \right].$$

This expression grows extremely fast. For 10 bit problems the equations can be executed for 15 generations in approximately 5 minutes on a 25 MIPS workstation.

5   Other Operators and Computational Behavior

5.1   Reduced Surrogate 1-point Crossover

A reduced surrogate crossover operator reduces parent strings to a skeletal form in which only those bits that differ in the two parents are represented. Thus the strings have corresponding reduced surrogates as follows:

```
001011101101001 ==>   -0--11---1--0-1
011000101001100 ==>   -1--00---0--1-0
```

Recombination is now limited so that only one possible recombination point exists between the bits in the reduced surrogates. Thus, in the strings illustrated here, there are only 5 possible points of recombination for 1-point crossover; there are 6 possible points of recombination if the strings are treated as structures that wrap-around, as is the case in 2-point crossover.

There are several attributes of the reduced surrogate form that makes this an interesting operator for recombination. First, if at least 1 crossover point occurs between the first and last bits in the reduced surrogate, then the offspring will never duplicate the parents. Second, it causes the recombination process to equally weight the probability of generating each offspring which can potentially be produced by a particular operator. Note that 1-point crossover in any continuous region of matching bits in the parents produces exactly the same offspring. Thus, some offspring are more likely to be produced by crossover than others when recombination operators are applied to the non-reduced forms. The reduced surrogate removes this potential source of bias. The calculation for string losses and gains for the reduced surrogate form of 1-point crossover is also simple compared to other operators.

The same sets of strings and paired strings that can produce string losses and string gains for regular 1-point crossover are also sources for the reduced surrogate 1-point crossover. However, the probabilities associated with the losses and gains do change. Assume crossover must occur between the first and last bits in the reduced surrogates unless there is only 1 bit different between parents. This means that *all potential sources of loss will produce a loss with a probably of 1.0*, since recombination on the reduced surrogate forms forces crossover to occur between the sentry bits of strings produced by the *string losses* generators. In general, if only 1 crossover point is allowed at each possible crossover site in the reduced surrogate forms, then all potential sources of loss will produce a loss with probability of 1.0 regardless of whether 1-point, 2-point or n-point crossover is used.

The probabilities for gains are also simple when the reduced surrogate form is used, although the probabilities do not automatically generalize to an arbitrary number of crossover points. The probabilities of a string gain from a pair of strings produced by the 1-point crossover string generators is $1/(H-1)$, where H is the Hamming distance between the two parents. The sentry bits in the string gain generators are effectively adjacent bits in the reduced surrogate forms, so there is only 1 crossover point between the sentry bits. Since there are H bits different in the reduced surrogate form, there are $H-1$ possible crossover points, only 1 of which results in a string gain.

5.2   String Gains and Losses for Uniform Crossover

One can also calculate string gains and losses for the uniform crossover operator. For string losses, one must enumerate all strings in the space that contain 2 or more 1 bits (i.e., all strings that are more than 2 bits different from S_0). The Hamming distance, H, between parents is again critical, since one offspring must inherit all zeros if a loss is to be avoided. The probability of losses is $(1 - 0.5^{H-1})$. The exponent is $H - 1$ instead of H since it does not matter which offspring inherits the first 0 bit, but the next $H - 1$ bits are critical.

The probability of gains is more complicated and it is again useful to introduce a set of generators. Clearly a zero bit must occur in every position in one of the two parents before it is possible to generate S_0 as an offspring. Thus, for every position in the string, either a zero occurs in both parents or complementary bits occur in both parents. This suggests a general mechanism for calculating string gains.

First, there are $\begin{pmatrix} L \\ k \end{pmatrix}$ ways to select positions in which both strings have k zeros. All of the cases for $k = 0$ to $L - 2$ must be examined. If $k = L$, both parents are S_0 and if $k = L - 1$, then exactly one of the parents are S_0; in both of these cases, computing string *gains* is not relevant.

It has already been noted that for those positions where both bits are not zero, the bits must be complements. Consider the case where L = 4 and k = 0. A parent string and the complementary parent string are represented as follows:

$$B = b_1 b_2 b_3 b_4$$
$$C = c_1 c_2 c_3 c_4$$

This provides a generator, except B cannot $= S_0$, so we start enumerating B at 1. Furthermore, only half of this space must be enumerated since this is sufficient to enumerate all possible pairs of strings. (While B is enumerated starting at 1 and counting up, the complement is simultaneously enumerated starting at $2^L - 2$ and is decremented.) Fixing the value of the first bit in B which has a complement in C cuts the space in half. Thus, the generator for k=0 is as follows:

$$B = 0_b b_2 b_3 b_4$$
$$C = 1_c c_2 c_3 c_4$$

The bits 0_b and 1_c are marked for purposes of illustration to indicate they are part of the *bit and complement* portion of the string generator.

In general then, for $k = 0$ to $L - 2$ the following process is repeated for each possible assignment of the $\begin{pmatrix} L \\ k \end{pmatrix}$ ways to select positions in which both strings have zero bits. First the k zero bits are assigned a position, then the remaining bits positions are assigned the symbol "b" (or "c" in the complement) as a place holder. Next the rightmost "b" and "c" is converted to 0 and 1 respectively, which is denoted here by 0_b and 1_c. This produces the set of generators, each indexed by k, which indicates the number of common zero bits. Note that k is directly related to Hamming distance $(L - k = H)$ and all pairs of strings produced by a particular generator are at exactly the same Hamming distance from one another. From each generator the strings are produced by enumerating over the $L - k - 1$ "b" bits for B from 1 to $2^{L-k-1} - 1$ while generating the corresponding complements bits for C. The complete set of generators for a 4 bit problem is as follows:

For $\begin{pmatrix} 4 \\ 0 \end{pmatrix}$ $\begin{array}{l} 0_b b_2 b_3 b_4 \\ 1_c c_2 c_3 c_4 \end{array}$

For $\begin{pmatrix} 4 \\ 1 \end{pmatrix}$ $\begin{array}{llll} 00_b b_3 b_4 & 0_b 0 b_3 b_4 & 0_b b_2 0 b_4 & 0_b b_2 b_3 0 \\ 01_c c_3 c_4 & 1_c 0 c_3 c_4 & 1_c c_2 0 c_4 & 1_c c_2 c_4 0 \end{array}$

For $\begin{pmatrix} 4 \\ 2 \end{pmatrix}$ $\begin{array}{llllll} 000_b b_4 & 00_b 0 b_4 & 00_b b_3 0 & 0_b 00 b_4 & 0_b 0 b_3 0 & 0_b b_2 00 \\ 001_c c_4 & 01_c 0 c_4 & 01_c c_3 0 & 1_c 00 c_4 & 1_c 0 c_3 0 & 1_c c_2 00 \end{array}$

Having generated all possible sources of string gains for uniform crossover, for any string pairs produced by a generator indexed by k the probability of generating the string S_0 is 0.5^{L-k-1}, or 0.5^{H-1} since $L - k = H$.

5.3   Crossover Probability Versus Schema Length and Fully Deceptive Functions

Recall that the crossover probability affects only the term for losses and gains in the following equation:

$$P(Z, t+1) = P(Z, t) \frac{f(Z)}{\bar{f}}(1 - \{P_c \text{ losses}\}) + \{P_c \text{ gains.}\}$$

Examination of the executable equations reveals that lowering the crossover probability from 1.0 to 0.5 is exactly the same in these equations as reducing the length of the crossover interiors by exactly half. The assumption of an infinitely large population is critical here. In actual finite populations the effect is not identical, since lowering P_c to 0.5 produces increased duplicates and acts to lower genetic diversity. It also exaggerates biases due to sampling error, since it tends to reinforce any sampling bias that exists in the current population sample. However, given the relationship between P_c and the crossover interiors, it is reasonable to infer that if P_c is sufficiently low, the defining length of critical schemata will *appear* to be shorter than they actually are. Figure 1 illustrate the execution of these equations on an embedded 3 bit fully deceptive problem (c.f., Goldberg, 1989b; Whitley, 1991) as well as the results of executing an actual simple genetic algorithm using a population size of 625. The three bits corresponding to the deceptive function are located at positions 1, 5 and 9; the remaining bits contribute nothing to the evaluation function. When P_c is 0.4 or above, the equations show convergence to the deceptive attractor 000 (Whitley, 1991), while the global optimal solution is at 111. However, when P_c is lowered to 0.35 or below, the equations show that a simple genetic algorithm should correctly converge to the global solution 111 (Whitley, Das and Crabb, 1992). The actual behavior *for a single run* of the simple genetic algorithm is closely predicted by the equations. Note that in order to overcome deception in this way the population must be sufficiently large to prevent the loss of genetic diversity which will result from a low value of P_c.

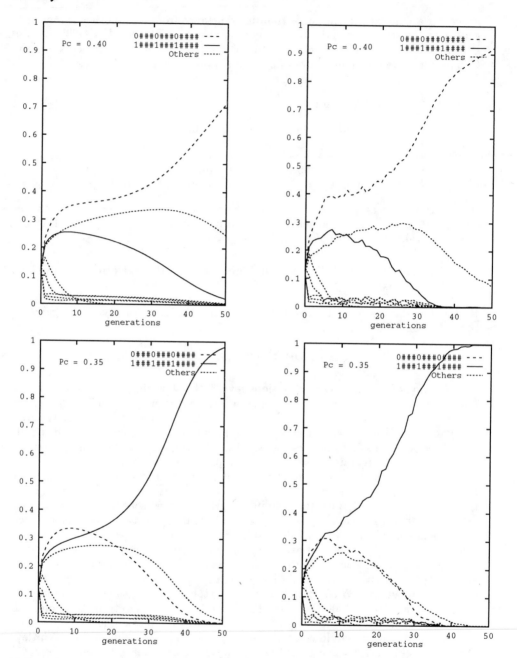

Figure 1: The left hand column shows the results of executing the equations for a fully deceptive three bit subproblem while the right hand column shows the results of executing an actual genetic algorithm (population size 625.) The effects of lowering the probability of crossover, P_c, to 0.40 and 0.35 are clearly illustrated.

5.4   Parallel Island Models

The equations presented in this paper are used to better understand the computational behavior of a parallel island model implementation of a simple genetic algorithm. The island model assumes that the total population used by the genetic algorithm is broken down into a number of subpopulations referred to as islands. Each island is in effective executing some version of the genetic algorithm on the reduced subpopulation. In the equations used here, each island is executing a separate version of the equations for the standard genetic algorithms using 1-point crossover. *Migration* occurs between subpopulations, usually in some restrictive fashion based on temporal and spatial considerations.

In the models implemented here, 4 subpopulations are modeled. The effects of population size are only partially modeled. The distribution of strings in the subpopulations are initialized by generating distinct finite subpopulations. Infinite population sizes are still assumed, however, after the first initial generation. Thus, the equations only model the effects of having different starting points in the different subpopulations based on finite populations. In the equations, migration occurs every X generations and copies of the individuals that make up the top 5% of the island population are allowed to migrate. Note that this top 5% may all be copies of the same string, or it may be some combination of strings. The island receiving these strings deletes the bottom 5% of its own population to absorb the new strings.

The migration scheme used in these equations assumes that the islands are arranged in a ring, and on the first migration strings move from their current island to their immediate neighbor to the left. Migrations occur between all island simultaneously. On the second migration the islands send copies of the top 5% of their current population to the island which is 2 moves to the left in the ring. In general, the migration destination address is incremented by 1 and moves around the ring. This continues with migrations occurring every X generations until each island has sent one set of strings to every other island (not including itself); then the process is repeated.

One thing that is immediately obvious is that migration of the best strings, or of any set of strings that are above average, acts to increase selection due to fitness. Those strings which migrate from one island to another automatically double their representation. Thus, one effect of the parallel island model is to allocate additional reproductive opportunities to those strings which have a high enough fitness to migrate. This could potentially cause premature convergence in a single population genetic algorithm. However, the equations also suggest that the island model may be especially well suited to handling the additional selective pressure. Although the island model equations are very much restricted by the assumption of infinite populations *after the first generation*, the equations nevertheless show that having different starting points for the individual islands may be sufficient to cause each island to follow a unique search trajectory, even though migration will tend to cause the set of equations representing the various islands to become more similar over time. Modeling finite populations and mutation should further distinguish the computational behavior of each of the individual subpopulations.

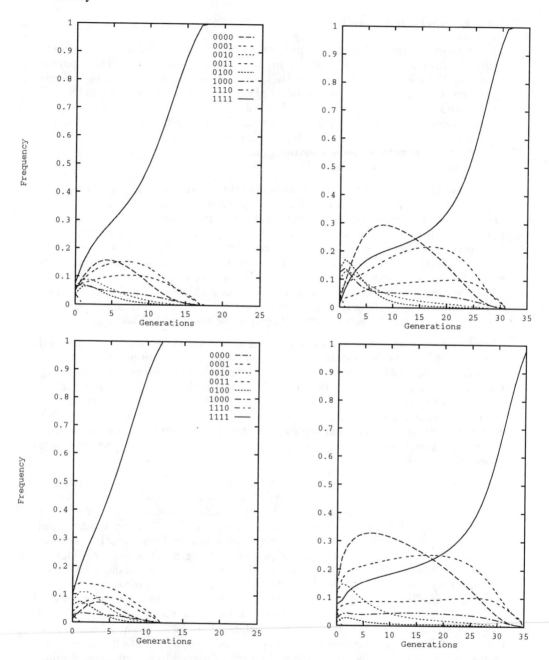

Figure 2: These graphs show the proportional representation of strings in four subpopulations of a Parallel Island Model, where each island executes a simple genetic algorithm. The search behavior of each island is different due to the different initial starting points associated with each island.

Results for a 4 bit fully deceptive problem (Whitley, 1991) are shown in Figure 2. In this figure, strings migrate every generation. The behavior of the models becomes more complex when migration occurs after two or more generations. In this case, every recombination breaks apart the critical order-4 level schemata (corresponding to strings in this case) which are needed to direct the population toward convergence to the global optimum. In the single population model, the genetic algorithm converges to the local optimum 0000; but the equations show that the parallel island model converges to 1111. This is clearly due in part to the higher selective pressure which results from migration but Figure 2 also suggests that the effects of the different search trajectories may be important. If any one subpopulation converges toward the global solution, migration (with its additional selective pressure) can sometimes help to pull the other subpopulations in the same direction.

6 Discussion

Several test problems have recently been constructed to test the optimization abilities of genetic algorithms. These problems are often characterized by the construction and concatenation of smaller functions of 15 bits or less. However, one criticism of these efforts is that the theoretical tools used to construct problems such as deceptive functions are based on static analyses. The models reviewed in this paper introduce a dynamic model of the behavior of a simple genetic algorithm.

The equations can be used to study the computational behavior of the genetic algorithm on small problems. The equations can also be used to study larger function which are constructed by concatenating small functions, since these subproblems typically have no higher-order or co-lateral interaction between hyperplanes (except those that occur due to sampling error and genetic hitch-hiking).

These equations do not include information about sampling error. However, they do offer the potential for obtaining precise information about other aspects of genetic search, such as hyperplane sampling behavior. These equations can also be extended to include various genetic operators and to model other type of genetic search.

Acknowledgements

This research was supported in part by NSF grant IRI-9010546 and by an equipment grant from IBM. An earlier version of the text in section 2 and subsection 5.2 appears in Whitley, Das and Crabb (1992). My thanks to Curtis Crabb for implementing the generalized form of the executable equations and Rajarshi Das for helping to develop the results presented in subsection 5.2.

References

Bridges, C. and Goldberg, D. (1987) An analysis of reproduction and crossover in a binary-coded genetic Algorithm. *Proc. 2nd International Conf. on Genetic Algorithms and Their Applications.*

Goldberg, D. (1987) Simple Genetic Algorithms and the Minimal, Deceptive Problem. In, *Genetic Algorithms and Simulated Annealing,* L. Davis, ed., Pitman.

Goldberg, D. (1989a) *Genetic Algorithms in Search, Optimization and Machine Learning.* Reading, MA: Addison-Wesley.

Goldberg, D. (1989b) Genetic Algorithms and Walsh Functions: Part I, A Gentle Introduction. *Complex Systems* 3:129-152.

Holland, John (1975) Adaptation In Natural and Artificial Systems. University of Michigan Press.

Liepins, G. and Vose, M. (1990) Representation Issues in Genetic Algorithms. Journal of Experimental and Theoretical Artificial Intelligence, 2:101-115.

Nix, A. and Vose, M. (1992) Modeling Genetic Algorithms with Markov Chains. *Annals of Mathematics and Artificial Intelligence.* 5:79-88.

Schaffer, D. (1987) Some Effects of Selection Procedures on Hyperplane Sampling by Genetic Algorithms. In, *Genetic Algorithms and Simulated Annealing,* L. Davis, ed. Pitman.

Vose, M. (1992) Modeling Simple Genetic Algorithms. *Foundations of Genetic Algorithms -2-,* D. Whitley, ed., Morgan Kaufmann.

Vose, M. and Liepins, G. (1991) Punctuated Equilibria in Genetic Search. *Complex Systems* 5:31-44.

Whitley, D. (1991) Fundamental Principles of Deception in Genetic Search. *Foundations of Genetic Algorithms.* G. Rawlins, ed. Morgan Kaufmann.

Whitley, D. (1992) Deception, Dominance and Implicit Parallelism in Genetic Search. *Annals of Mathematics and Artificial Intelligence.* 5:49-78.

Whitley, D., R. Das and C. Crabb (1992) Tracking Primary Hyperplane Competitors During Genetic Search. To Appear: *Annals of Mathematics and Artificial Intelligence.*

Modeling Simple Genetic Algorithms

Michael D. Vose
C.S. Dept., 107 Ayres Hall
The University of Tennessee
Knoxville, TN 37996-1301
(615) 974-5067 vose@cs.utk.edu

Abstract

Two models of the simple genetic algorithm are reviewed, extended, and unified. The result incorporates both short term (transient) and long term (asymptotic) GA behavior. This leads to a geometric interpretation of genetic search which explains population trajectories.

1 Introduction

Mathematical models have been used as analytical tools in the investigation of genetic algorithms. An early example is Goldberg's work on the minimal deceptive problem [3]. He used equations for the expected next generation to model the evolutionary trajectory of a two bit GA under crossover and proportional selection. The results were used to obtain the now familiar type-I and type-II classifications.

Vose and Liepins [6] simplified and extended these equations, incorporating mutation into the recombination of arbitrarily long binary strings. Like Goldberg, their equations are deterministic and compute what is equivalent to the evolutionary path taken by a GA with infinite population.

Holland questioned the relationship between GAs and infinite population models. Real GAs are based on finite populations and are stochastic, not deterministic. Moreover, the relationship between the trajectory of expectations (which is the path infinite population models follow) and the evolution of finite populations in real GAs was unclear.

Following up on Holland's constructive criticism, Nix explored these issues in his masters thesis under Vose's supervision; the result was twofold [4]. First, an exact model for real GAs was obtained in the form of a Markov chain. Second, the trajectory followed by finite populations was related to the evolutionary path predicted by the infinite population model.

Coincidentally, Davis independently modeled GAs with Markov chains [1]. However, his work focused on whether annealing the mutation rate would imply convergence to the global optimum (it doesn't).

This paper builds on the results of Vose and Nix to further tie real GAs to the infinite population model. The key idea dates back to a conference presentation where a geometric object was related to the infinite population model [7]. Roughly speaking, populations correspond to points on a smooth "*GA-surface*", and the progression from one generation to the next forms a path leading to a local minimum of the surface (downhill $\sim$ increasing fitness).

Therefore, at least in the infinite population case, genetic algorithms should not be thought of as global optimizers. Populations move quickly downhill to be trapped in some local basin. But what about *finite* GAs? Vose and Nix proved that for large populations, the evolutionary path of a real GA follows very closely, with large probability, and for a long period of time that path predicted by the infinite population model. Thus the local geometry of the GA-surface indicates the transient behavior of a real GA, i.e., that behavior which depends on the initial population.

When mutation is nonzero (which we assume throughout this paper) a real GA forms an ergodic Markov chain, visiting every state infinitely often. Hence, after some period of time, a GA will escape every local minimum of the surface. The interesting question is: *where is it likely to be?*

The answer to this question is a probability distribution which is converging to the steady state distribution of the Markov chain. Vose and Nix proved that this steady state distribution concentrates probability near fixed points of the infinite population model, i.e., near the local minima of the GA-surface. So a real GA will escape one local minimum only to be temporarily trapped at another.

The theme of this paper is that *these local minima are not equally likely*. A real GA will with large probability be asymptotically at that local minimum having largest basin of attraction. As population size grows, the probability of a GA being asymptotically anywhere else converges to zero. Therefore, the GA-surface provides a geometric interpretation of genetic search which explains population trajectories. Short term behavior is determined by which basin the initial population finds itself within, and long term behavior is determined by that minimum having largest basin.

We began this overview with "Roughly speaking ..." for two good reasons. First, some results have intricate mathematical form and are not so directly translated into common language. Second, some results are conditional on mathematical details working out as conjectured; the nonlinear mathematics of GAs are complex. The following sections will make all clear.

2   The Infinite Population Model

This section summarizes from [6], further details may be obtained from that source.

Let Ω be the set of all length ℓ binary strings, and let $N = 2^\ell$. Thinking of elements of Ω as binary numbers, we identify Ω with the interval of integers $[0, N-1]$. We also regard Ω as the product group

$$\mathcal{Z}_2 \times \ldots \times \mathcal{Z}_2$$

where $\mathcal{Z}_2$ denotes the additive group of integers modulo 2. The group operation $\oplus$ acts on integers in $[0, N-1]$ via these identifications, and we use $\otimes$ to represent componentwise multiplication.[1]

Given a vector x, let $|x|$ denote the sum of its coordinates. Hence $|\cdot|$ agrees with the ℓ_1 norm on nonnegative vectors.

The t th generation of the genetic algorithm can be modeled by a vector $s^t \, \epsilon \, \Re^N$, where the i th component of s^t is the probability that i is selected for the gene pool. Populations excluding members of Ω are modeled by vectors s^t having corresponding coordinates zero.

Let $p^t \, \epsilon \, \Re^N$ be a vector with i th component equal to the proportion of i in the t th generation. Since selection is by proportional fitness, we have $s^t = | \, Fp^t \, |^{-1} \, Fp^t$ where F is the diagonal matrix having ii th entry the fitness of string i. We assume fitness is positive hence F is invertible. The t th generation is therefore equivalently modeled by either the selection vector s^t or the population vector p^t.

Let $M_{i,j}$ be the probability that 0 results from the recombination process based on parents i and j. If recombination is 1-point crossover and mutation with crossover rate χ and mutation rate μ, then the i, j th entry of M is

$$\frac{(1-\mu)^\ell}{2} \left\{ \eta^{|i|} \left(1 - \chi + \frac{\chi}{\ell - 1} \sum_{k=1}^{\ell-1} \eta^{-\Delta_{i,j,k}} \right) + \eta^{|j|} \left(1 - \chi + \frac{\chi}{\ell - 1} \sum_{k=1}^{\ell-1} \eta^{\Delta_{i,j,k}} \right) \right\}$$

where $\eta = \mu/(1 - \mu)$, integers are to be regarded as bit vectors when occurring in $|\cdot|$, where division by zero at $\mu = 0$ and $\mu = 1$ is to be removed by continuity, and where

$$\Delta_{i,j,k} = |(2^k - 1) \otimes i| - |(2^k - 1) \otimes j|$$

Define permutations σ_j on $\Re^N$ by $\sigma_j <x_0, \ldots, x_{N-1}>^T = <x_{j \oplus 0}, \ldots, x_{j \oplus (N-1)}>^T$ where T denotes transpose, and define the operator $\mathcal{M}$ by

$$\mathcal{M}(x) = <(\sigma_0 \, x)^T M \sigma_0 \, x, \ldots, (\sigma_{N-1} \, x)^T M \sigma_{N-1} \, x>^T$$

The operators F and $\mathcal{M}$ form the basis of the model of Vose and Liepins. The utility of these operators follows from:

- Fp^t is a vector pointing in the same direction as s^t.

---

[1] Hence, $\oplus$ is *exclusive-or* on integers and $\otimes$ is *logical-and*.

- $\mathcal{M}(s^t) = p^{t+1}$

It follows that if $\mathcal{G}(x) = \mathcal{M}(|Fx|^{-1} Fx)$ then $\mathcal{G}(p^t) = p^{t+1}$ so that $\mathcal{G}$ is the transition operator from one population to the next ($\mathcal{G}$ was so defined in [4]). Alternatively, if $\mathcal{G}(x) = F\mathcal{M}(x)$ then $\mathcal{G}(s^t)$ points in the same direction as s^{t+1} making $\mathcal{G}$ the transition operator when generations are modeled by (directions of) selection vectors ($\mathcal{G}$ was so defined in [6]). Which definition for $\mathcal{G}$ is best depends on the application, and in this paper we use both; $\mathcal{G}_1$ will refer to the first and $\mathcal{G}_2$ to the second.

3   The Finite Population Model

This section summarizes from [4], further details may be obtained from that source.

Let Z be an $N \times R$ matrix whose columns represent the possible populations of size r. The ith column $\phi_i = <Z_{0,i}, \ldots, Z_{N-1,i}>^T$ of Z is the incidence vector

$$Z_{y,i} = \text{ the number of occurrences of string } y \text{ in the } i\text{th population}$$

Given an enumeration of the populations, Z is well defined. However, we do not specify an enumeration here.

An *exact model* of a simple GA is provided by the following Markov chain. The states are given by (indices of) the columns of the matrix Z, and the transition probabilities are given by the matrix Q where

$$Q_{i,j} = r! \prod_{y=0}^{N-1} \frac{\{\mathcal{G}_1(\phi_i)_y\}^{Z_{y,j}}}{Z_{y,j}!}$$

Note that the states of this Markov chain can be regarded as points on the simplex

$$\Lambda = \{x \in \Re^N : x \text{ is nonnegative and } |x| = 1\}$$

through the correspondence $j \longleftrightarrow \phi_j/r$, since a population incidence vector has coordinates which sum to r.

4   The GA-surface

This section summarizes from and extends [7].

Let $\|\cdot\|$ be a continuous function mapping vectors to nonnegative reals such that

- $\|x\| = 0$ if and only if $x = \vec{0}$
- $\|\alpha x\| = \alpha \|x\|$ for $\alpha > 0$

Since $\|\cdot\|$ is continuous and Λ is compact, it's extrema are attained. Hence there is a bounded continuous surface $\mathcal{B}$ in the nonnegative orthant on which $\|\cdot\|$ is 1,

$$\mathcal{B} = \left\{ \frac{x}{\|x\|} \; : \; x \in \Lambda \right\}$$

Thinking of $\| \cdot \|$ as like a norm, $\mathcal{B}$ is the surface of its unit ball. In fact, if $\| \cdot \|$ is $| \cdot |$, then $\mathcal{B} = \Lambda$.

Let $\varphi_{t+1} = \|\mathcal{G}_2(s^t)\|$ and $s^{t+1} = \varphi_{t+1}{}^{-1} \mathcal{G}_2(s^t)$. When $\| \cdot \|$ is $| \cdot |$, this inductive definition of s^t coincides with previous usage so that s^t is the selection vector for generation t. When alternate choices for $\| \cdot \|$ are made, the length but not the direction of s^t may change.

The *GA-surface* corresponding to the simple genetic algorithm is the set

$$\mathcal{S} = \left\{ s^0 \prod_{i=1}^{\infty} \varphi_i^{-2^{-i}} \; : \; s^0 \in \mathcal{B} \right\}$$

Theorem 1 $\mathcal{S}$ is smooth, independent of $\| \cdot \|$, and mapped by $\mathcal{G}_2$ into itself.

Proof: First suppose that $\| \cdot \|$ is $| \cdot |$. A generic point of $\mathcal{S}$ is

$$s^0 \exp \left\{ - \sum_{i=1}^{\infty} 2^{-i} \ln \varphi_i \right\}$$

Since $\mathcal{G}_2$ has quadratic components, $\varphi_{t+1} = |\mathcal{G}_2(s^t)|$ is infinitely differentiable if s^t is. Moreover, φ_{t+1} is both uniformly bounded from above and away from 0 from below because

$$|\mathcal{G}_2(s^t)| \in |F\mathcal{M}(\Lambda)| \subset |F\Lambda|$$

and F is invertible. Now $s^{t+1} = \varphi_{t+1}{}^{-1} \mathcal{G}_2(s^t)$ is infinitely differentiable if s^t is, and so by induction s^t and hence φ_{t+1} are infinitely differentiable. The uniform bounds on φ_{t+1} now imply the smoothness of

$$- \sum_{i=1}^{\infty} 2^{-i} \ln \varphi_i$$

since we may differentiate term by term. Now drop the assumption that $\| \cdot \|$ is $| \cdot |$ and note that because $\mathcal{G}_2(\alpha x) = \alpha^2 \mathcal{G}_2(x)$, we have

$$\mathcal{G}_2\left(s^0 \prod_{i=1}^{\infty} \varphi_i^{-2^{-i}}\right) = \mathcal{G}_2(s^0) \prod_{i=1}^{\infty} \varphi_i^{-2^{1-i}} = s^1 \varphi_1 \prod_{i=1}^{\infty} \varphi_i^{-2^{1-i}} = s^1 \prod_{i=1}^{\infty} \varphi_{i+1}^{-2^{-i}}$$

Since the last expression is the point on $\mathcal{S}$ corresponding to $s^1 \epsilon \mathcal{B}$, we have $\mathcal{G}_2(\mathcal{S}) \subset \mathcal{S}$. The proof is finished by noting that $\mathcal{S}$ does not depend on $\| \cdot \|$ since $\mathcal{S}$ is the invariant surface of $\mathcal{G}_2$ in the nonnegative orthant, and $\mathcal{G}_2$ does not depend on $\| \cdot \|$ ☐

Choosing $\|\cdot\|$ to be $|\cdot|$, *we may regard any probability vector $s^0 \in \Lambda$ as a point of S* through the correspondence

$$s^0 \quad\longleftrightarrow\quad s^0 \prod_{i=1}^{\infty} \varphi_i^{-2^{-i}}$$

In particular, the progression of selection vectors corresponding to successive generations traces out a path in S which defines the evolutionary trajectory taken by a GA (this holds for for finite as well as infinite populations, since the states of the Markov model may be viewed as belonging to Λ).

The rationale for defining S in terms of $\|\cdot\|$ is to provide a natural and flexible concept of "downhill" for the surface. If $x_1, x_2, x_3, \ldots$ is a sequence of points (a path) in S, we say the path moves downhill if

$$\|x_i\| \geq \|x_{i+1}\|$$

with strict inequality except when $x_i = x_{i+1}$. A path of particular interest is the evolutionary trajectory followed by the sequence of selection vectors

$$s, \quad G_2(s), \quad G_2^2(s), \quad G_2^3(s), \quad \ldots$$

for arbitrary initial vector $s \in S$.

Theorem 2   A necessary and sufficient condition for the sequence of selection vectors with initial vector s^0 to move downhill is that

$$\forall j \ . \ \ln \varphi_j \leq \sum_{k=1}^{\infty} 2^{-k} \ln \varphi_{k+j}$$

with strict inequality except when $G_2^j(s^0) = G_2^{j-1}(s^0)$.

Proof: By (the proof of) Theorem 1, we have

$$\frac{\|G_2^j(s^0)\|}{\|G_2^{j-1}(s^0)\|} \ = \ \frac{\| s^j \prod_{i=1}^{\infty} \varphi_{i+j}^{-2^{-i}} \|}{\| s^{j-1} \prod_{i=1}^{\infty} \varphi_{i+j-1}^{-2^{-i}} \|} \ = \ \exp\left\{ -\sum_{i=1}^{\infty} 2^{-i}(\ln \varphi_{i+j} - \ln \varphi_{i+j-1}) \right\}$$

which is less than 1 when $\ln \varphi_j - \sum 2^{-k} \ln \varphi_{k+j} < 0$ $\qquad\square$

We conjecture that G_2 is invertible for all string lengths ℓ. In fact, for $\ell < 3$, the inverse can be obtained by simply changing the matrices F and M (whether this holds in general is not known). If this conjecture is true, then the exceptional condition in Theorem 2 simplifies to "...except when s^0 is a fixed point of G_2".

It is enlightening to consider Theorem 2 when $\|\cdot\|$ is $|\cdot|$. For this choice,

$$\varphi_t \ = \ |G_2(s^{t-1})| \ = \ |FM(s^{t-1})| \ = \ |Fp^t|$$

so that φ_t is the average population fitness at generation t. In the infinite population model, stochastic fluctuations are averaged out ($\mathcal{G}$ is deterministic) and population fitness rarely decreases. Moreover, increasing population fitness implies $\ln \varphi_{j+1} < \sum 2^{-k} \ln \varphi_{k+j}$, hence Theorem 2 shows the evolutionary trajectory moves downhill on $\mathcal{S}$. In fact, if decreases in population fitness are slight and transient, then trajectories *still* move downhill; the sufficient condition in Theorem 2 is *weaker* than monotonicity of the φ_t.

We conjecture that every evolutionary path (sequence of iterates of $\mathcal{G}_2$) converges to a fixed point, and with appropriate choice of $\|\cdot\|$, evolutionary trajectories on $\mathcal{S}$ move downhill.

5   The Fixed Point Graph

The component equations of $\mathcal{G}_2(x) = x$ are equivalent to a system of quadratics,

$$(\sigma_k x)^T M \sigma_k x \;=\; x_k / F_{k,k} \qquad (0 \le k < N)$$

If x solves all equations except possibly the j th, then an arbitrarily small perturbation of $F_{j,j}$ will guarantee x does not satisfy them all. Therefore, the generic situation is that these N equations in N unknowns are independent and the fixed points are isolated and finite in number.

Assuming the generic case and our previous conjectures, the local minima $\{\omega_1, \ldots, \omega_w\}$ of $\mathcal{S}$ are therefore the stable attracters on the GA-surface. We define the basin of attraction U_j of ω_j to be the set of all $s \in \mathcal{S}$ which converge to ω_j under iterations of $\mathcal{G}_2$.

The *fixed point graph* is the complete directed graph $\Im$ on vertices $\{1, \ldots, w\}$ with edge $i \to j$ having weight

$$\inf_{x \in F^{-1}U_j} \left\{ \ln \frac{|F^{-1}\omega_i|}{|x|} \;+\; \frac{1}{|x|} \sum_{k=0}^{N-1} x_k \ln \frac{x_k}{(F^{-1}\omega_i)_k} \right\}$$

for $i \ne j$. This weight is the minimum relative entropy (or Kullback-Lieber information) of U_j relative to ω_i after mapping each by F^{-1} and scaling to probability vectors.

A Markov chain is represented by a complete directed graph over $\{1, \ldots, w\}$ if the $i \to j$ edge is labeled by a weight which encodes the i,j th entry of its transition matrix A.

We define a *tributary* of a complete directed graph as a spanning intree[2]. Let Tree_k be the set of tributaries rooted at k, and for $t \in \text{Tree}_k$ let its cost $|t|$ be the sum of its edge weights (this matches previous usage of $|\cdot|$ when thinking of a tributary as its vector of weights). We conjecture that in the generic case, the minimum cost tributary of $\Im$ is unique; let it be t' and be rooted at k'.

---

[2]i.e., a tree containing every vertex and such that all edges point towards the root.

There is a beautiful connection between a Markov chain's steady state distribution and the tributaries of its graphical representation. The following result can be established by adapting the methods in chapter 2 of [2].

Theorem 3   Let A be a positive transition matrix for a Markov chain with states $\{1, \ldots, w\}$, and let the corresponding graphical representation have edge $i \to j$ weighted by $-\ln A_{i,j}$ for $i \neq j$. A solution to the steady state equation $xA = x$ is

$$x_k = \sum_{t \in \text{Tree}_k} \exp\{-|t|\}$$

If we leave the graph (hence all weights and tributaries) referred to in Theorem 3 unchanged but modify the $i \to j$ transition probability to $(A_{i,j})^r$, we obtain a Markov chain parametrized by r and with steady state solution

$$< \sum_{t \in \text{Tree}_1} \exp\{-r\,|t|\}\, , \ldots,\ \sum_{t \in \text{Tree}_w} \exp\{-r\,|t|\} >$$

where the Tree_k are computed from the graph representing A. The minimum cost tributary corresponds to the dominant term, and as $r \to \infty$ the steady state distribution converges to point mass at that state at which it is rooted.

By equating $-\ln A_{i,j}$ to the weight of edge $i \to j$ in $\mathfrak{S}$, we obtain a Markov chain with parameter r whose steady state distribution converges to k'. The next section shows how this Markov chain captures the asymptotic behavior of the finite population model as the population size $r \to \infty$. We will conclude that for large populations a GA will with large probability be asymptotically near that local minimum $\omega_{k'}$ of $\mathcal{S}$ corresponding to the minimum tributary t' of $\mathfrak{S}$.[3]

6   Asymptotic Approximation

The main result of [4] is that as $r \to \infty$, the steady state distribution of the finite population model can give nonvanishing probability only to fixed points of $\mathcal{G}_1$. Since what time a GA spends away from a fixed point is negligible, steady state behavior for large populations is therefore captured by a Markov chain having the fixed points as states.

We conjecture that the basins of attraction for unstable fixed points have measure zero. By using a multinormal approximation to the multinomial transition probabilities of the finite population model, this would imply that a GA spends time near an unstable fixed point only with vanishing probability (the probability of remaining in an unstable basin converges to an integral over a set of measure zero). This is antipodal to the behavior near a stable fixed point; as $r \to \infty$ the probability of remaining in its basin converges to 1.

Since mapping by F^{-1} and scaling transforms selection vectors into population vectors, the stable fixed points of $\mathcal{G}_1$ are given by $v_i = |F^{-1}\omega_i|^{-1} F^{-1}\omega_i$.

---

[3]This is what was meant in the introduction concerning "... that local minimum having largest basin of attraction".

We will need the following result from large deviation theory [5].

Theorem 4 Let λ be a positive probability vector, and consider the multinomial distribution

$$P_r\left\{<\frac{z_0}{r},\ldots,\frac{z_{N-1}}{r}>\right\} = r!\prod_{y=0}^{N-1}\frac{\{\lambda_y\}^{z_y}}{z_y!}$$

If $U \subset \Lambda$ is an open set, then

$$\lim_{r\to\infty}\frac{1}{r}\ln P_r(U) = -\inf_{x\in U}\sum_{k=0}^{N-1}x_k\ln\frac{x_k}{\lambda_k}$$

Choosing $z = \phi_j$ makes $z_y = Z_{y,j}$ and so when $\lambda = \mathcal{G}_1(\phi_{i'})$ the first equation in Theorem 4 represents the probability $Q_{i',j}$ of a transition from state i' to state j in the finite population model.

Let U be the basin of attraction of v_j (i.e., the basin U_j of ω_j after mapping by F^{-1} and scaling). From the second equation in Theorem 4 we deduce that the logarithm of the probability that a GA moves from state i' into the basin of attraction of v_j is asymptotic to

$$-r\inf_{x\in F^{-1}U_j}\left\{\ln|x|^{-1} + \frac{1}{|x|}\sum_{k=0}^{N-1}x_k\ln\frac{x_k}{\mathcal{G}_1(\phi_{i'})_k}\right\} \qquad (1)$$

If $\phi_{i'}$ is chosen to depend on r such that $\phi_{i'}/r \to v_i$ as $r \to \infty$ (i.e., state i' is chosen near v_i), then

$$\mathcal{G}_1(\phi_{i'}) \longrightarrow \mathcal{G}_1(v_i) = v_i = |F^{-1}\omega_i|^{-1}F^{-1}\omega_i$$

so that expression (1) is asymptotic to $\ln(A_{i,j})^r$ (recall that $-\ln A_{i,j}$ is the weight of edge $i \to j$ in $\mathfrak{S}$) and represents the logarithm of the probability that a GA with population size r moves from near fixed point v_i into the basin of attraction of v_j.

It follows from our previous discussion that the Markov chain $\mathcal{C}$ having states $\{1,\ldots,w\}$ and transition $i \to j$ with probability that a GA of population size r moves from near fixed point v_i into the basin of attraction of fixed point v_j has steady state distribution converging to k'.

As $r \to \infty$, the evolutionary path of a GA is converging toward that path predicted by the infinite population model [4]. Hence the movement modeled by $\mathcal{C}$ from near v_i into the basin of attraction of v_j corresponds to a transition in GA behavior from spending time near v_i to spending time near v_j. As already noted, what time a GA spends away from a fixed point is negligible, hence the Markov chain $\mathcal{C}$ captures the asymptotic behavior of a large population GA.

In terms of selection vectors, this means that large population GAs will with large probability be asymptotically near that local minimum $\omega_{k'}$ of $\mathcal{S}$ corresponding to the minimum tributary t' of $\mathfrak{S}$. In terms of population vectors, the GA will be near the fixed point $v_{k'}$ of $\mathcal{G}_1$.

7   Conclusion

We have introduced the GA-surface $\mathcal{S}$ to provide a geometric interpretation of genetic search and to explain population trajectories.

The shape of $\mathcal{S}$ governs both short and long term behavior for large population GAs. Short term behavior is determined by which local basin the initial population finds itself within, and long term behavior is determined by that local minimum having largest basin (i.e., minimum spanning intree).

A reasonable next step in the analysis of large population GAs is to explore "mid term" behavior (the Markov chain $\mathcal{C}$ should play an important role).

This paper is based on positive fitness, nonzero mutation, and the conjectures that

1. $\mathcal{G}_2$ is invertible for all string lengths ℓ.

2. With appropriate choice of $\|\cdot\|$, evolutionary trajectories move downhill on $\mathcal{S}$.

3. For every starting point, iterates of $\mathcal{G}_2$ converge.

4. In the generic case, the minimum cost tributary in $\Im$ is unique.

5. The basins of attraction for unstable fixed points of $\mathcal{G}_2$ have measure zero.

To put these conjectures into perspective, it should be noted that by including all fixed points in $\Im$, conjecture 5 is obviated. Moreover, conjecture 4 is not necessary to comprehend the limit of the steady state distribution of $\mathcal{C}$ as $r \to \infty$; when there are several minimum tributaries, they contribute mass equally. Conjectures 1 and 2 are not crucial since choosing $\|\cdot\|$ to be $|\cdot|$ is a good approximation to desired behavior, and our asymptotic approximation requires only that evolutionary paths converge – thus conjecture 3 is critical. The reason we have made these conjectures is because they simplify results and we believe them to be true.

While these conjectures are interesting in themselves, the fact that our work has woven them into real (finite population) GA behavior will hopefully encourage their consideration and the further development of the interrelationships between the finite and infinite population models. The GA surface is particularly fascinating and is as yet largely unexplored.

Understanding large population GA behavior is prerequisite to the successful analysis and eventual comprehension of the small population case. The tools appropriate for both tasks will involve our models for finite and infinite populations for the reason that they are exact in capturing stochastic and expected behavior. Formalization is the first step of analysis, and as this paper demonstrates, these models form foundations of simple genetic algorithms from which further steps may be taken.

Acknowledgements

This research was supported by the National Science Foundation (IRI-8917545) and AFOSR (90-0135).

References

[1] T. Davis (1991), Toward An Extrapolation Of The Simulated Annealing Convergence Theory Onto The Simple Genetic Algorithm. Dissertation presented to the University of Florida.

[2] A. Gibbons (1987), *Algorithmic Graph Theory*. Cambridge University Press.

[3] D. E. Goldberg (1987), Simple genetic algorithms and the minimal, deceptive problem. In L. Davis (ed.), *Genetic algorithms and simulated annealing* (pp 74 –88). London: Pitman.

[4] A. Nix & M. D. Vose (1991), Modeling Genetic Algorithms With Markov Chains. Annals of Mathematics and Artificial Intelligence 5 (1992) 79-88.

[5] S. R. S. Varadhan (1984), *Large Deviations and Applications* Society for Industrial and Applied Mathematics.

[6] M. D. Vose & G. E. Liepins (1991), Punctuated Equilibria In Genetic Search. *Complex Systems* 5, 31-44.

[7] M. D. Vose (1991), Formalizing Genetic Algorithms. Technical Report (CS-91-127), Computer Science Department, The University of Tennessee.

PART 3

DECEPTION AND THE
BUILDING BLOCK HYPOTHESIS

Deception Considered Harmful [*]

John J. Grefenstette
Navy Center for Applied Research in Artificial Intelligence
Code 5514
Naval Research Laboratory
Washington, DC 20375-5000
E-mail: GREF@AIC.NRL.NAVY.MIL

Abstract

A central problem in the theory of genetic algorithms is the characterization of problems that are difficult for GAs to optimize. Many attempts to characterize such problems focus on the notion of *Deception*, defined in terms of the static average fitness of competing schemas. This article examines the Static Building Block Hypothesis (SBBH), the underlying assumption used to define Deception. Exploiting contradictions between the SBBH and the Schema Theorem, we show that Deception is neither necessary nor sufficient for problems to be difficult for GAs. This article argues that the characterization of hard problems must take into account the basic features of genetic algorithms, especially their dynamic, biased sampling strategy.

1 INTRODUCTION

Since Holland's early work on the analysis of genetic algorithms (GAs), the usual approach has been to focus on the allocation of search effort to subspaces described by schemas representing hyperplanes of the search space. The Schema Theorem (Holland, 1975) provides a description for the growth rate of schemas that depends on the observed relative fitness of the schemas represented in the population. Bethke (1981) initiated work on the formal characterization of problems that might be difficult for GAs to solve, and presented an analysis of problems in terms of the Walsh transformation of the fitness

---

[*] With apologies to Edsger Dijkstra (Dijkstra, 1968).

function. Goldberg (1987) introduced the notion of *Deception* in GAs, and defined a Minimal Deceptive Problem (MDP). Subsequently, Deception has come to be widely regarded as a central feature in the design of problems that are difficult for GAs (Das and Whitley, 1991; Homaifar, Qi and Fost, 1991). Goldberg and his colleagues (Goldberg, Korb, and Deb, 1989) have defined messy GAs (mGAs) specifically to handle Deceptive problems, and consider the use of Deceptive functions as test functions to be "critical to understanding the convergence of mGAs, traditional GAs, or any other similarity-based search technique". The literature on Deception in GAs is growing rapidly (Battle & Vose, 1991; Davidor, 1990; Deb and Goldberg, 1992; Goldberg 1989a, 1989b, 1989c, 1991, 1992; Goldberg, Deb and Korb, 1990, 1991; Goldberg, Deb and Clark, 1992; Liepins & Vose, 1990, 1991; Mason, 1991; Whitley, 1991, 1992), so this is clearly a topic that deserves careful scrutiny.

In previous papers (Grefenstette and Baker, 1989; Grefenstette, 1991) we have raised some questions about this approach to the analysis of GAs, and others have begun to express similar concerns (Forrest and Mitchell, 1993). This paper will try to clarify and expand on the argument that the current definitions of Deception appear to be based on faulty assumptions about the dynamics of GAs. This paper only addresses definitions of Deception that are based on the static analysis of hyperplanes. By static analysis, we mean the analysis based on the average fitness of hyperplanes, when the average is taken over the entire search space. Our fundamental point is that the dynamic behavior of genetic algorithms cannot be predicted in general on the basis of the static analysis of hyperplanes. This paper does *not* argue against the notion that there are classes cf problems that are "deceptive", in the sense that properties of the response surface lead the GA away from the optimal regions of the search space. It is clearly important to try to characterize such problems, but any useful characterization will have to be based on correct assumptions about genetic algorithms.

The remainder of the paper is organized as follows: Section 2 discusses the Static Building Block Hypothesis (SBBH) that appears to underlie much of the work on static hyperplane analysis. Section 3 shows that some functions that are highly Deceptive according to the SBBH are, in fact, very easy for GAs to optimize. Section 4 shows that some functions that have no Deception and therefore should be easy, according to the SBBH, are nearly impossible for GAs to optimize. These counterexamples show that Deception is neither necessary nor sufficient to make a problem difficult for GAs. More importantly, the analysis of these results highlights the shortcomings of the static analysis of hyperplanes. As an aside to our main point, Section 5 shows that, even if the SBBH were true, many Deceptive functions could be easily solved by simple changes to the basic GA. Section 6 summarizes the paper.

2   THE STATIC BUILDING BLOCK HYPOTHESIS

The fundamental description of the dynamics of genetic algorithms is the Schema Theorem (Holland, 1975), which describes the relationship between the expected growth in a hyperplane from one generation to the next as a function of the hyperplane's observed relative fitness:

Schema Theorem.

$$M(H,t+1) \geq M(H,t)(\frac{f(H,t)}{\overline{f}(t)})(1-p_d(H,t))$$

where $M(H,t)$ is the expected number of samples of a hyperplane H in the population at time t, $f(H,t) / \overline{f}(t)$ is the *observed* relative fitness of H at time t, and $p_d(H,t)$ is the probability that H will be disrupted by the genetic operators such as crossover and mutation. The Schema Theorem is only directly applicable to a single generational cycle, but one can get an intuitive feel for the dynamics of GAs by considering what happens to a hyperplane that consistently has an observed fitness that is higher than the population average. This case is often described as follows:

> "The usual interpretation of [the Schema Theorem] is that subspaces with higher than average payoffs will be allocated exponentially more trials over time, while those subspaces with below average payoffs will be allocated exponentially fewer trials. This assumes that ... the effects of crossover and mutation are not disruptive." (Spears and De Jong, 1991).

Strictly speaking, the word *observed* should be inserted before each occurrence of *payoff*, but the intended meaning is usually clear and the shorthand phrase is often more convenient to use. Many even more reckless statements appear in the introductory sections of many articles on genetic algorithms, for example:

> "[The Schema Theorem] says that the number of samples allocated to an above-average hyperplane H grows exponentially over time." (Grefenstette, 1990)

Sometimes, as in the case above, such generalizations are qualified at some later points in the article. Using the same form of shorthand, the overall dynamics of GAs are often expressed in the form of the so-called "Building Block Hypothesis":

> "[A] genetic algorithm seek[s] near optimal performance through the juxtaposition of short, low-order, high performance schemata, or building blocks." (Goldberg 1989a, p. 41)

The Building Block Hypothesis is a rough but serviceable first explanation of how GAs operate, and has been used by most of us when explaining GAs to newcomers. There is usually little harm in informal statements about the dynamics of GAs, but they can lead to rather more serious misinterpretations if they are taken as the basis for an operational theory for GAs. This paper will focus on the implications of one operational version of the Building Block Hypothesis, which we call the:

> *Static Building Block Hypothesis* (SBBH): Given any short, low-order hyperplane partition, a GA is expected to converge to the hyperplane with the best static average fitness (the "expected winner").

For example, consider the following 2nd-order hyperplane partition.

		$f(H)$
H_a:	00######	1
H_b:	01######	3
H_c:	10######	10
H_d:	11######	7

Here, and throughout this paper, the value $f(H)$ refers to the static average fitness for schema H, that is, the mean fitness value of every point described by that schema. This *static* average is independent of whatever points happen to be in the population at any time. According to the SBBH, the "expected winner" of the above schema competition is the hyperplane H_c.

The SBBH implies that functions for which the low-order schemas associated with the optimum have higher static average fitness than the competing schemas in their partitions ought to be easy for GAs. For example, suppose the global optimum is 00...0, and that

$$f(0\#...\#) > f(1\#...\#)$$
$$f(00\#...\#) > f(01\#...\#)$$
$$f(00\#...\#) > f(10\#...\#)$$
$$f(00\#...\#) > f(11\#...\#)$$

and so on for every hyperplane partition of the search space. According to the SBBH, a GA should find f simple to optimize. In fact, such functions are commonly called "GA-easy" (Wilson, 1991).[1]

Conversely, the SBBH implies that functions for which which the low-order schemas associated with the optimum have lower static average fitness than the competing schemas in their partitions ought to be difficult for GAs. For example, suppose the global optimum is 00...0, and that

---
[1] We are not asserting that Wilson subscribes to the SBBH.

$$f(0\#...\#) \; < \; f(1\#...\#)$$
$$f(00\#...\#) \; < \; f(01\#...\#)$$
$$f(00\#...\#) \; < \; f(10\#...\#)$$
$$f(00\#...\#) \; < \; f(11\#...\#)$$

and so on. Then this function would be called *Deceptive* (Goldberg, 1987).

The SBBH appears to underlie much of the recent published work in GA theory, especially work on Deception. The SBBH itself is rarely cited as an assumption in the Deception literature. Instead, the Schema Theorem is usually invoked, along with an informal statement of the building block hypothesis. For example,

> "Let's construct the simplest problem that should cause a GA to diverge from the global optimum ... To do this, we want to violate the building block hypothesis in the extreme." (Goldberg introducing the MDP, 1989a)

or

> "It follows from the Schema Theorem that the number of instances of a schema is expected to increase in the next generation if it is of above average utility and is not disrupted by crossover. Therefore, such schemata indicate the area within the search space that the GA explores, and hence it is important that, at some stage, these schema contain the object of search. Problems for which this is not true are called *deceptive*." (Liepins and Vose, 1991)

or

> "It has been shown that genetic algorithms work well when building blocks, short, low-order ... schemata with above average fitness values, combine to form optimal or near-optimal solutions." (Homaifar et. al, 1991)

When such statements are followed by the analysis of static hyperplane averages, as they are in all the above cases, it seems to be fair to ascribe the SBBH as an implicit assumption.

While the SBBH is an appealing intuitive explanation for how GAs work, it has never been explicitly proven, and it does not follow from the Schema Theorem. The Schema Theorem describes the expected growth of a hyperplane for a single generation based on its *observed* average fitness, that is, based on the average fitness of current samples of the hyperplane in the population. Over a period of generations, the observed average fitness of a hyperplane does not necessarily reflect the static average fitness of the hyperplane. The SBBH arises when we ignore the crucial distinction between observed average fitness and static average fitness. In particular, we get the SBBH if we drop the time index t from the term $f(H,t) / \overline{f}(t)$ in the Schema Theorem. The SBBH, then, is merely an approximation to the dynamics described by the Schema Theorem.

There are many ways in which such an approximation might be expected to diverge from the dynamics described by the full Schema Theorem. We will examine just two ways in which the SBBH fails to account for the true dynamics of GAs:

1. Failure to account for collateral convergence.

2. Failure to account for fitness variance within schemas.

The effect of collateral convergence is that, once the population begins to converge, even a little, it is no longer possible to estimate the static average fitness of schemas using the information present in the current population. The effect of fitness variance within schemas is that, in populations of realistic size, the observed fitness of a schema may be arbitrarily far from the static average fitness, even in the initial population. In the next two sections, the reasons listed above are used to show why the Static Building Block Hypothesis, and the accompanying notion of Deception, cannot be used to predict how difficult a function may or may not be for a GA to optimize.

3   COLLATERAL CONVERGENCE

The primary reason that the SBBH is a poor model of GAs is that, except possibly for the very first generation, the population contains only a biased sample of representatives from each schema. This is a normal feature of all GAs, and is true regardless of population size, but it can yield results that are exactly the opposite of what one might expect from the SBBH. A simple thought experiment will bring the point home. Consider the following two first-order schema competitions:

		$f(H)$
H_A:	0#...###...	9
H_B:	1#...###...	1
H_C:	##...#0#...	6
H_D:	##...#1#...	4

Assume that the initial population is selected using a uniform distribution, and that the population is sufficiently large that, for each schema, the observed fitness at time $t = 0$ approximates the schema's static average fitness. Then, the Schema Theorem predicts that the expected allocation of trials in the second generation is as follows:

		Pop % in H at time $t = 1$
H_A:	0#...###...	90
H_B:	1#...###...	10
H_C:	##...#0#...	60
H_D:	##...#1#...	40

Note that the Schema Theorem predicts that the population will begin to converge more rapidly with respect to the first competition than with respect to the second competition. Now, can we predict how these two first-order competitions will proceed at time $t = 1$? The answer is no, since we no longer have sufficient information to estimate what the *observed fitnesses* of the competing schema might be. The *collateral convergence* of the population[2] toward H_A will bias the samples taken from H_C and H_D at time $t = 1$. For example, we would expect that 90% the observed representatives of H_C would have a 0 in the first position. The static average fitness does not indicate what the observed fitness of this highly biased sample might be. Taking this collateral convergence into account, we cannot predict, on the basis of the static average fitness of the low-order hyperplanes, whether H_C will continue to grow more rapidly than H_D at time $t = 1$. We have even less reason to predict whether the GA will ultimately converge to H_C or H_D, and yet this is what the SBBH claims to do.

To see how collateral convergence can affect the study of Deception, consider the following problem:

$$\max \ f(x_1, x_2), \text{ where } 0 \le x_i \le 1, \text{ and}$$

$$f(x_1, x_2) = \begin{cases} x_1^2 + 10x_2^2 & \text{if } x_2 < 0.995 \\ 2(1-x_1)^2 + 10x_2^2 & \text{if } x_2 \ge 0.995 \end{cases}$$

Assume that x_1 and x_2 are represented in binary notation on a chromosome of length 20, using 10 bits for each argument. The optimal solution is $(x_1, x_2) = (0, 1)$.

Using the terminology in (Whitley, 1991), this problem has a "fully deceptive subproblem of order-10", meaning that in every schema competition defined over the first 10 bit positions, the "expected winner" of each such competition is the schema containing the "deceptive attractor" 1111111111##########, corresponding to the hyperplane $(x_1, x_2) = (1, \#)$. Despite this high level of Deception[3] a standard GA[4] has

---

[2] *Collateral convergence* refers to the phenomenon that the population converges at different rates to two intersecting hyperplanes, e.g., H_A and H_C, above.

[3] This problem can be made arbitrarily more Deceptive by replicating the gene x_1 in the above example.

[4] GENESIS with population size 200, other parameters set to default values.

no trouble finding the global optimum within a few thousand trials in 10 out of 10 runs. The success of the GA can be predicted by the Schema Theorem, which says that the second 10 bits will rapidly converge toward the value of $x_2 = 1$. Once this collateral convergence occurs, the *observed* fitness of schemas representing values of x_1 near 0 will change from low to high, and the GA will converge toward the global optimum.

Figure 1 illustrates some of the dynamics of the GA on this problem, averaged over 10 independent runs. The graph show the progress of the competition between the schemas:

$$H_0: \ 0\#...\#$$
$$H_1: \ 1\#...\#$$

by indicating the per cent of the population in H_0. The dotted line indicates the convergence of the second 10 bits (x_2). Since $f(H_1) > f(H_0)$, the SBBH predicts convergence to H_1. For $t < 15$, the *observed* fitnesses of the competing schemas agree with the SBBH, that is, $f(H_1, t) > f(H_0, t)$, and the allocation of trials to H_0 declines accordingly. By about generation 20, the value of x_2 has largely converged, and the GA now observes that $f(H_1, 20) < f(H_0, 20)$, thanks to the effects of the second term in the fitness function. In accordance with the Schema Theorem, and in contradiction to the SBBH, the allocation of trials shifts to H_0 and away from H_1.

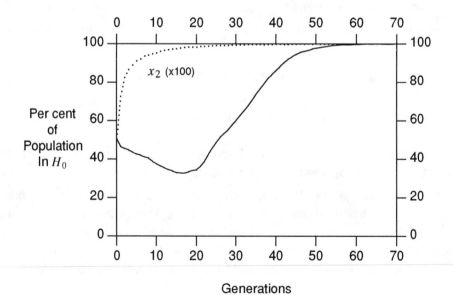

Figure 1: The Schema Theorem in Action.

As this example shows, the ultimate allocation of trials across a given hyperplane partition can be difficult to predict, precisely because the GA makes its allocation decisions based upon the current estimate of payoff associated with hyperplanes. These current estimates are highly influenced by collateral convergence. In general, there is no simple relationship between the observed payoff $f(H,t)$ at time t and the static average payoff $f(H)$. This example shows that some highly Deceptive problems are in fact easy for GAs to optimize. The underlying reason is that the SBBH simply does not model the dynamic behavior of GAs, as described by the Schema Theorem.

It may be thought that having a large population might allow us to ignore the effects of collateral convergence, and use the SBBH as a first approximation to the Schema Theorem. However, the effects of collateral convergence do not really depend on having a small population size. In fact, the thought experiment at the beginning of this section can be applied to an infinite population model. To give a practical example, we repeated the experiment shown in Figure 1, using a population of size 10,000 (instead of 200). The results are shown in Figure 2.

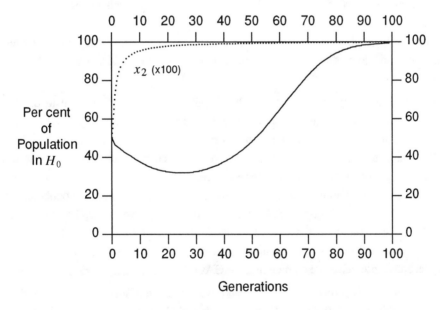

Figure 2: The Schema Theorem in Action with Population Size 10,000

As expected, the convergence is a little slower, since the exponential allocation of trials implies that the time to converge to high values of x_2 increases with the logarithm of the population size. Nevertheless, the overall performance profile is very similar to the small population case. If our analysis of collateral convergence is correct, the results predicted by the SBBH should not be expected for any size population.

4 LARGE VARIANCE WITHIN SCHEMAS

A second shortcoming of the Static Building Block Hypothesis is that it fails to account for the variance of fitness within schemas. With a limited population size and large variance within the schemas, the sampling in the initial, random population will produce errors in the estimate of each schema's static average fitness. This schema variance can lead to results that contradict the SBBH.

To illustrate this, we can define a class of problems that are "easy" in the sense implied by the SBBH -- they have no Deception -- but are, in fact, nearly impossible for GAs to optimize. Consider an L-bit space representing the interval [0, 1] in binary encoding. Let f be defined:

$$f(x) = \begin{cases} 2^{(L+1)} & \text{if } x = 0 \\ x^2 & \text{otherwise.} \end{cases}$$

For any schema H such that the optimum is in H (that is, all the defined positions of H have value 0), $f(H) > 2$, since the sum of the fitness of the points in H is at least $2^{(L+1)}$ and there are at most 2^L points in the hyperplane. For any schema H such that the optimum is not H, $f(S) \leq 1$. So in any schema partition, the schema containing the optimum has the highest static average fitness. That is, there is no Deception at any level in the function. Such functions are often called "GA-easy" (Liepins and Vose, 1991; Wilson, 1991).

Suppose we run a standard GA on f with a population of size polynomial in L. If the optimum is not in the initial population, it will probably never be found. (Of course, it might be created by a lucky crossover or a very lucky multiple mutation.) Why is this function hard for GAs? Because the schemas associated with the optimum have extremely high variance, so the *observed* average fitness for the hyperplanes never reflects their *static* average fitnesses, not even in the initial random population. Of course, this is a "needle-in-a-haystack" function, so we don't expect the GA to solve it on a regular basis. But it does satisfy the commonly used definition of "GA-easy" that follows from the SBBH, so this example provides a counterexample to the claim that only Deceptive problems are challenging for GAs (Das and Whitley, 1991).

There has been some recent work that addresses the (static) fitness variance within schemas (Goldberg and Rudnick, 1988; Rudnick and Goldberg, 1991). This is a step in the right direction, but it is unlikely that an analysis of the static fitness variance will be much more helpful than an analysis of the static fitness averages, for the same reasons as in the previous Section. As the search proceeds, the *observed* variance associated with hyperplanes in the population is unlikely to have any correlation with the *static* variance.

5   AUGMENTED GAs FOR DECEPTIVE PROBLEMS

There is a growing literature on how to make GAs more effective on Deceptive problems. For example, Liepins and Vose (1991) specify representation transformations that render Fully Deceptive problems "fully easy". Goldberg at. al (1991) define "messy GAs" in order to deal with problems of bounded Deception. In this section, we suggest that, assuming that the SBBH applies to GAs, slight changes to the basic GA would be sufficient to solve Deceptive problems.

Much of the work on Deception involves functions for which the bit-wise complement of the global optimum is the Deceptive attractor (Liepins and Vose, 1991; Whitley, 1991). In fact, arguments have been made that all Fully Deceptive problems have this feature (Whitley, 1992). For the purpose of this section, let us accept this argument and suppose that a GA would actually perform according to the SBBH on these fully Deceptive problems. That is, suppose the GA really does converge to the complement of the global optimum. Then one algorithm that finds the global optimum in all such problems is simply:

1. Run a GA until it converges to some string, x
2. Output either x or the complement of x, whichever is better.

At a cost of a single extra evaluation, this algorithm strictly extends the class of problems that can be optimized by GAs to include all Fully Deceptive problems.

However, perhaps the more usual cases are "partially deceptive" problems, that is, problems that have a Deceptive component and a non-Deceptive component. Such problems might be handled by the augmented GA shown in Figure 3. The lines marked with (*) represent the changes to a standard generational GA. In the augmented version, we maintain three separate populations of size N, called P, Q and R. During each generation, we update P according to the original GA, set the members of Q to the bitwise complement of the corresponding elements of P, and create R by crossing over randomly selected parents from P and Q.

Consider any problem for which the original GA (i.e., the one without the (*) lines) finds an acceptable solution in time t, using a population of size N. Then the augmented algorithm finds a solution that is at least as good as the original GA, using at most time $3t$ (assuming that the evaluation time dominates the other operations in the algorithms). Like the simpler variant, this augmented algorithm solves any fully Deceptive problem that has the property that the global optimum is the binary complement of the Deceptive attractor, since as soon as the population P produces a copy of the Deceptive attractor, the population Q produces a copy of the global optimum.

```
            procedure Augmented GA
            begin
                    t = 0;
                    initialize P(t);
                    evaluate structures in P(t);
                    while termination condition not satisfied do
                    begin
                            t = t + 1;
                            select P(t) from P(t-1);
                            recombine structures in P(t);
(*)                         Q(t) = complement(P(t));
(*)                         form R(t) by recombining parents from P(t) and Q(t);
                            evaluate structures in P(t);
(*)                         evaluate structures in Q(t);
(*)                         evaluate structures in R(t);
                            output best structure in P(t) ∪ Q(t) ∪ R(t);
                    end
            end.
```

Figure 3: An Augmented Genetic Algorithm.

By allowing recombination between randomly selected members of populations P and Q, the SBBH predicts that we could generate optimal solution to partially deceptive problems as well. For any component that is not deceptive, the elements in P should converge to the correct component values, according to the SBBH. The deceptive components should converge to the complement of the optimal component values. Thus the complements of the deceptive components -- the optimal component values -- will be stored in population Q. Performing (multi-point) crossover across populations P and Q should eventually produce the optimal components all along the chromosome in population R. Thus the augmented GA can produce a final answer that is never worse than the one produced by the original GA, and it eliminates concern about Deceptive problems, all at a cost of only tripling the computational time.

One might conclude from this discussion is that Deception, as currently defined, is not a serious problem for GAs, since it can be handled by small extensions to the original algorithm. Unfortunately, this discussion is purely academic, since the proposed "solutions", like the notion of Deception itself, are based on the SBBH, and are therefore unlikely to provide useful results for real GAs.

6   SUMMARY

This paper criticizes the Static Building Block Hypothesis as a description of the dynamics of GAs. The SBBH arises by ignoring an important distinction made in the Schema Theorem, between the *observed* fitness of a hyperplane and the *static* fitness of a hyperplane. We have identified two reasons that cause divergent predictions between the SBBH and the Schema Theorem:

1. Failure to account for collateral convergence.

2. Failure to account for fitness variance within schemas.

According to the SBBH, Deceptive problems ought to be difficult for GAs to solve, and "GA-easy" problems ought to be easy for GAs to solve. Taking the above differences between the Schema Theorem and the SBBH into account, it is easy to demonstrate that:

1. Some highly Deceptive problems are easy for GAs to optimize.

2. Some "GA-easy" problems with no Deception are nearly impossible for
   GAs to optimize.

That is, Deception is neither necessary nor sufficient for causing difficulties for GAs. Put another way, the class of Deceptive functions is neither a subset nor a superset of the class of functions that are hard for GAs to optimize, as shown in Figure 4. At the very least, the term *Deceptive* seems to be poorly chosen. More importantly, our arguments show that it is in general impossible to predict the dynamic behaviors of GAs on the basis of the static average fitness of hyperplanes.

Even though the presence or absence of Deception provides no logical implication of problem difficulty, there may be some correlation between Deception and difficulty for GAs. Although we have no way at present to measure such a correlation, it is true that *some* Deceptive problems are difficult for GAs, and *some* "GA-Easy" problems are in fact easy for GAs.[5] An interesting problem would be to characterize *which* Deceptive problems actually cause difficulties for GAs. An approach to this question might involve the design of problems that take advantage of the different rates at which collateral schema competitions converge, as predicted by the Schema Theorem. It seems doubtful that a successful approach will be based on the SBBH.

One might also argue that the work to date that is based on the SBBH has been a preliminary exploration of how to analyze simple distributions of fitness, and was always intended to be replaced by dynamic analysis. In this case, articles using static analysis should qualify their conclusions accordingly, rather than make what appear to be strong

---

[5] For example, the Partially Deceptive problem defined in Section 3 can be made difficult for GAs, as well as Deceptive, if we negate the coefficient on x_2, so that x_2 converges to 0 rather than to 1.

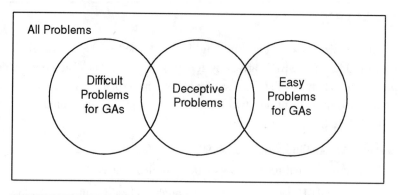

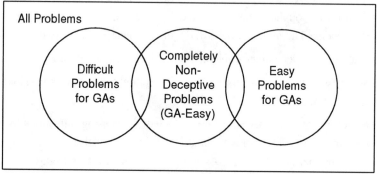

Figure 4: Deception and Problem Difficulty.

statements about GAs and Deception, for example:

> "The use of deceptive functions as test functions is critical to understanding the convergence of any GAs or other similarity-based search technique ... Since these functions are maximally misleading, if an algorithm can solve this class of problem, it can solve anything easier." (Goldberg, Deb, and Korb, 1990)

> "The only challenging problems are deceptive ... Test problems must be used that involve some degree of deception." (Whitley, 1991)

Such conclusions properly apply only to algorithms that satisfy the SBBH. It is not clear that there exist any practical algorithms in this class, but in any case the GA is not one of them.

Some other concerns about the notion of Deception have been raised by others, including Forrest and Mitchell (1993), and De Jong (1992). In particular, measuring GA performance by the ability to find the global optimum seems at odds with the emphasis in (Holland, 1975) on exploiting local information for adaptation. Both De Jong (1975) and Bethke (1981) have proposed other performance metrics that are more appropriate for measuring the effectiveness of adaptive search methods.

It goes without saying that the characterization of hard problems should remain a high priority for the GA research community. However, the characterization must take into account the basic features of the GA, especially its dynamic, biased sampling strategy. Our recommendation is that the efforts currently being expended on the static analysis of functions should be diverted to the dynamic analysis of GAs. It is gratifying to see some recent efforts in this direction (Bridges and Goldberg, 1991; Liepins and Vose, 1991; Nix and Vose, 1992). Nevertheless, it is important that articles on GA theory avoid the implicit assumption of the SBBH. The SBBH is a seductive, but inaccurate, explanation for the power of the GA that can easily mislead (deceive?) newcomers to the field, as well as potential users of the technology.

Acknowledgements

Thanks to Ken De Jong, Bill Spears, and the reviewers for helpful suggestions in improving the presentation of this paper.

REFERENCES

Battle, D. L. and M. D. Vose (1991). Isomorphisms of genetic algorithms. In *Foundations of Genetic Algorithms,* G. J. E. Rawlins (Ed.), San Mateo, CA: Morgan Kaufmann.

Bethke, A. D. (1981). Genetic algorithms as function optimizers. Doctoral dissertation, University of Michigan.

Bridges, C. L. and D. E. Goldberg (1991). The nonuniform Walsh-schema transform. In *Foundations of Genetic Algorithms,* G. J. E. Rawlins (Ed.), San Mateo, CA: Morgan Kaufmann.

Das, R. and D. L. Whitley (1991). The only challenging problems are deceptive. *Proceedings of the Fourth International Conference of Genetic Algorithms* (pp. 116-173). San Mateo, CA: Morgan Kaufmann.

Davidor, Y. (1990). Epistasis variance: suitability of a representation to genetic algorithms. *Complex Systems 4(4),* 369-384.

De Jong, K. A. (1975). *Analysis of the behavior of a class of genetic adaptive systems.* Doctoral dissertation, Department of Computer and Communications Sciences, University of Michigan, Ann Arbor.

De Jong, K. A. (1992). Genetic algorithms are NOT function optimizers. In *Foundations of Genetic Algorithms 2,* D. Whitley (Ed.), San Mateo: Morgan

Kaufmann.

Deb, K. and D. E. Goldberg (1992). Analysing deception in trap functions. In *Foundations of Genetic Algorithms 2*, D. Whitley (Ed.), San Mateo: Morgan Kaufmann.

Dijkstra, E. W. (1968). Go to statement considered harmful. *CACM, 11(3)*, 147-148.

Goldberg, D. E. (1987). Simple genetic algorithms and the minimal deceptive problem. In L. Davis (Ed.), *Genetic algorithms and simulated annealing* (pp. 74-88). London: Pitman.

Goldberg, D. E. (1989a). *Genetic algorithms in search, optimization, and machine learning.* Reading: Addison-Wesley.

Goldberg, D. E. (1989b). Genetic algorithms and Walsh functions: Part I, a gentle introduction. *Complex Systems, 3*, 129-152.

Goldberg, D. E. (1989c). Genetic algorithms and Walsh functions: Part II, deception and its analysis. *Complex Systems, 3*, 153-171.

Goldberg, D. E. (1991). Real-coded genetic algorithms, virtual alphabets, and blocking. *Complex Systems, 5*, 139-167.

Goldberg, D. E. (1992). Construction of high-order deceptive functions using low-order Walsh coefficients. *Annals of Mathematics and Artificial Intelligence, 5*, 35-48.

Goldberg, D. E., K. Deb and B. Korb (1990). Messy genetic algorithms revisited: Studies in mixed size and scale. *Complex Systems, 4*, 415-444.

Goldberg, D. E., K. Deb and B. Korb (1991). Don't worry, be messy. *Proceedings of the Fourth International Conference of Genetic Algorithms* (pp. 24-30). San Mateo, CA: Morgan Kaufmann.

Goldberg, D. E., K. Deb and J. Clark (1992). Accounting for noise in the sizing of populations. In *Foundations of Genetic Algorithms 2*, D. Whitley (Ed.), San Mateo: Morgan Kaufmann.

Goldberg, D. E., B. Korb and K. Deb (1989). Messy genetic algorithms: Motivation, analysis and first results. *Complex Systems, 3*, 493-530.

Goldberg, D. E. and M. Rudnick (1988). Genetic algorithms and the variance of fitness. *Complex Systems, 2*, 265-278.

Grefenstette, J. J. (1990). Genetic algorithms and their applications. In **The Encyclopedia of Computer Science and Technology, Vol. 21,** A. Kent and J. G. Williams (Eds.), New York: Marcel Dekker.

Grefenstette, J. J. (1991). Conditions for implicit parallelism. In *Foundations of Genetic Algorithms*, G. J. E. Rawlins (Ed.), San Mateo, CA: Morgan Kaufmann.

Grefenstette, J. J. and J. E. Baker (1989). How genetic algorithms work: An critical look at implicit parallelism. *Proceedings of the Third International Conference of Genetic Algorithms* (pp. 20-27). San Mateo, CA: Morgan Kaufmann.

Holland, J. H. (1975). *Adaptation in natural and artificial systems.* Ann Arbor: University Michigan Press.

Homaifar, A., X. Qi and J. Fost (1991). Analysis and design of a general GA deceptive problem. *Proceedings of the Fourth International Conference of Genetic Algorithms* (pp. 196-203). San Mateo, CA: Morgan Kaufmann.

Liepins, G. E. and M. D. Vose (1990). Representational issues in genetic algorithms. *J. Exp. Theor. Artificial Intelligence,* 4-30.

Liepins, G. E. and M. D. Vose (1991). Deceptiveness and genetic algorithm dynamics. In *Foundations of Genetic Algorithms,* G. J. E. Rawlins (Ed.), San Mateo, CA: Morgan Kaufmann.

Mason, A. J. (1991). Partition coefficients, static deception and deceptive problems for non-binary alphabets. *Proceedings of the Fourth International Conference of Genetic Algorithms* (pp. 210-214). San Mateo, CA: Morgan Kaufmann.

S. Forrest and M. Mitchell (1993). What makes a problem hard for a genetic algorithms? Some anomalous results and their explanation. To appear in *Machine Learning.*

Nix, A. E. and M. D. Vose (1992). Modeling genetic algorithms with Markov chains. *Annals of Mathematics and Artificial Intelligence, 5,* 79-88.

Rudnick, M. and D. E. Goldberg (1991). Signal, noise, and genetic algorithms. IlliGAL Report No. 91004, Univ. Illinois at Urbana-Champaign. Also available as Oregon Graduate Institute Technical Report No CS/91-013.

Spears, W. M. and K. A. De Jong (1991). An analysis of multi-point crossover. In *Foundations of Genetic Algorithms,* G. J. E. Rawlins (Ed.), San Mateo, CA: Morgan Kaufmann.

Whitley, L. D. (1991). Fundamental principles of deception in genetic search. In *Foundations of Genetic Algorithms,* G. J. E. Rawlins (Ed.), San Mateo, CA: Morgan Kaufmann.

Whitley, L. D. (1992). Deception, dominance and implicit parallelism in genetic search. *Annals of Mathematics and Artificial Intelligence, 5,* 49-78.

Wilson, S. W. (1991). GA-easy does not imply steepest-ascent optimizable *Proceedings of the Fourth International Conference of Genetic Algorithms* (pp. 85-89). San Mateo, CA: Morgan Kaufmann.

Analyzing Deception in Trap Functions

Kalyanmoy Deb and David E. Goldberg
Department of General Engineering
University of Illinois at Urbana-Champaign
117 Transportation Building
104 South Mathews Avenue
Urbana, IL 61801

Abstract

A flat-population schema analysis is performed to find conditions for full deception in trap functions. It is found that the necessary and sufficient condition for an ℓ-bit fully deceptive trap function is that all order $\ell - 1$ schemata are misleading, and it is observed that the trap functions commonly used in a number of test suites are not fully deceptive. Further analysis suggests that in a fully deceptive trap function, the locally optimal function value may be as low as 50% of the globally optimal function value. In this context, the limiting ratio of the locally and the globally optimal function value for a number of fully deceptive functions currently in use are calculated. The analysis indicates that trap functions allow more flexibility in designing a deceptive function. It is also found that the proportion of fully deceptive functions in the family of trap functions is only $O(\ell^{-1} \ln \ell)$, and that more than half of the trap functions are fully easy functions.

1 Introduction

Genetic algorithms (GAs) work by combining low-order, short, and above-average schemata together to form higher order building blocks. It has been shown elsewhere (Goldberg, 1987, 1989a, 1989b, 1990; Liepins & Vose, 1990; Whitley, 1991) that there exists a class of functions—GA-deceptive functions—where low-order building blocks are misleading on average and do not usually combine to form higher order building blocks. In the extreme, there are fully deceptive functions where all low-order schemata containing a suboptimal solution are better than other competing schemata. In solving the problem of finding the global optimum of these functions, simple GAs may diverge from the global solution and converge to the suboptimal solution. GA-deceptive problems strike at the very heart of the theoretical foundations of genetic algorithms, and mechanisms to solve these problems may reveal important insights into the workings of a genetic algorithm.

In this paper, conditions for deception are determined for trap functions (Ackley, 1987). Trap functions are a class of piecewise-linear functions of *unitation* (Goldberg, 1990). Functions of unitation depend only on the total number of 1s in a string and not on the positions of the 1s. For an example, in a three-bit function of unitation, strings 100, 010, and 001 have the same function value, because all these strings have equal number of 1s. These functions are interesting to study because they have only $\ell + 1$ different function values in a total of 2^ℓ strings. The reduction in the number of degrees of freedom makes them easier to understand and manipulate. Once these functions are analyzed for deception, the results can be used to find deception conditions for other functions. In the remainder of this paper, the schema-fitness expressions are derived and a critical condition for a fully deceptive trap function is found. Thereafter, a review of a number of currently used fully deceptive functions is presented. Finally, the proportion of fully deceptive functions in the family of all trap functions is estimated.

2   Trap functions

A trap function is a piecewise-linear function of unitation that divides the total search space into two basins in the Hamming space, one leading to a global optimum and the other leading to a local optimum. In general, the global optimum may be any string, while the local optimum[1] is the one's complement of the global optimum. The unitation, u, may be defined as the Hamming distance of the string from the local optimum. Without loss of generality, the string with all ones and the string with all zeros are assumed to be the globally and the locally optimal strings respectively, and the unitation, u, is defined as the number of ones in a string. A typical trap function is given in equation 1 and is plotted in figure 1.

$$f(u) = \begin{cases} \frac{a}{z}(z - u), & \text{if } u \leq z; \\ \frac{b}{\ell - z}(u - z), & \text{otherwise,} \end{cases} \tag{1}$$

where a and b are constants, and z is the slope-change location. The string with zero unitation has a value equal to a. As unitation increases, the function value decreases linearly and has a value zero when the unitation is equal to a slope-change location z. Thereafter, the function value increases linearly and has a value equal to $b > a$ for the string with the maximum unitation. Trap functions were originally defined by Ackley (1987). In Ackley's trap functions, the parameter z was assigned a fixed value, $z = \lfloor 3\ell/4 \rfloor$. In this paper, this restriction on z is withdrawn and a bound for a fully deceptive trap function is found. In the following section, a deception analysis is performed by calculating the schema average fitness.

---

[1]Whitley (1991) has shown that in a fully deceptive problem, the locally optimal string may not be the complement of the globally optimal string, but must not be more than one-bit away in the Hamming space from the complement of the globally optimal string.

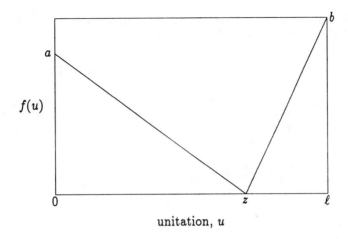

Figure 1: A trap function with a local optimum at $u = 0$ and a global optimum at $u = \ell$.

3   Deception analysis

In this section, schema-fitness expressions are derived for an ℓ-bit trap function. For simplicity, all schemata are grouped into three different categories. Thereafter, conditions for schema partition deception are applied to find the critical condition for a fully deceptive trap function.

3.1   Schema-fitness expressions

The *fitness* of a schema is defined as the average of the function values of all strings contained in the schema. In an order-λ schema, a parameter $u \leq \lambda$ is defined as the number of ones in the fixed positions. Thus, an order-λ schema with unitation u in a problem of size ℓ contains all strings with unitation varying from u to $\ell - \lambda + u$. For convenience, two slope parameters $s_0 = a/z$ and $s_1 = b/(\ell - z)$ are defined. Depending on the parameters u and λ, schemata may be grouped into three different categories.

First, if the number of zeros in the fixed positions in a schema is greater than or equal to $\ell - z$, all strings contained in the schema lie in the interval $0 \leq u < z$. Thus, the fitness of a schema with $\lambda - u \geq \ell - z$ is obtained by calculating the average fitness of strings with unitation varying from u to $\ell - \lambda + u$:

$$f(u, \lambda, \ell, z) = \frac{s_0}{2^{\ell - \lambda}} \sum_{i=0}^{\ell - \lambda} \binom{\ell - \lambda}{i}(z - u - i). \tag{2}$$

Second, if the number of ones in the fixed positions is greater than or equal to z, all strings contained in the schema lie in the interval $z \leq u \leq \ell$. Thus, the fitness for any schema with $u \geq z$ is obtained as follows:

$$f(u, \lambda, \ell, z) = \frac{s_1}{2^{\ell-\lambda}} \sum_{i=0}^{\ell-\lambda} \binom{\ell-\lambda}{i}(u - z + i). \tag{3}$$

Third, for any other schema, strings with as many as $z - 1$ ones lie in the interval $0 \leq u < z$, and strings with at least z ones lie in the interval $z \leq u \leq \ell$. The fitness of a schema is obtained as follows:

$$f(u, \lambda, \ell, z) = \frac{1}{2^{\ell-\lambda}} \left[s_0 \sum_{i=0}^{z-u} \binom{\ell-\lambda}{i}(z - u - i) + s_1 \sum_{i=z-u+1}^{\ell-\lambda} \binom{\ell-\lambda}{i}(u - z + i) \right]. \tag{4}$$

Having calculated the fitness of all schemata, appropriate conditions for deception may now be applied.

3.2   Applying deception conditions

There exists a number of different variations (albeit minor) to the definition of deception in GA literature. In this paper, a schema partition is defined to be deceptive if the schema containing the locally optimal string is no worse than any other schema in the partition. For the trap function given in equation 1, a schema partition of order λ is deceptive if the schema of unitation zero (which contains the locally optimal string) is no worse than any other schema in the partition. A function is defined to be fully deceptive if every schema partition is deceptive. Thus, in order for the above trap function to be fully deceptive, all schemata of unitation zero must have a fitness higher or equal to that of any other schema of the same order.

To compare schemata, a quantity $d(u_1, u_2)$ is defined as the difference in the fitness of two schemata with unitation u_1 and u_2 and of the same order. Thus, for every schema order, λ, the difference in the schema fitness between a schema of unitation, $u = 0$ and a schema of unitation, $1 \leq u \leq \lambda$, may be calculated. By setting the smallest difference in fitness to be greater than or equal to zero, a relation between a, b, and z for a fully deceptive trap function may be obtained. Depending on the parameters, u, λ, and z, all deception conditions may be categorized into three groups.

First, for $\lambda \geq z$, the fitness of the schema with $u = 0$ is calculated from equation 2; the fitness of any other schema with $u \geq z$ is calculated from equation 3. The difference in fitness is written as follows:

$$d(0, u) = \frac{1}{2^{\ell-\lambda}} \left[s_0 \sum_{i=0}^{\ell-\lambda} \binom{\ell-\lambda}{i}(z - i) - s_1 \sum_{i=0}^{\ell-\lambda} \binom{\ell-\lambda}{i}(u - z + i) \right]. \tag{5}$$

Simplifying the above equation by using the identities,

$$\sum_{i=0}^{\ell-\lambda} \binom{\ell-\lambda}{i} = 2^{\ell-\lambda}$$

and

$$\sum_{i=0}^{\ell-\lambda} i\binom{\ell-\lambda}{i} = \frac{\ell-\lambda}{2} 2^{\ell-\lambda},$$

and by setting $d(0, u) \geq 0$, the following inequality is obtained:

$$\frac{s_0}{s_1} \geq \frac{2u - \lambda - (2z - \ell)}{\lambda + (2z - \ell)}. \tag{6}$$

It is interesting to note that the left side of the expression is the ratio of the slopes of two lines in figure 1 and the right side expression increases with increasing u. Thus, the inequality is true for any $u \geq z$, if it is true for $u = \lambda$. Thus, for $z \leq u \leq \lambda$, the following condition is most severe:

$$\frac{s_0}{s_1} \geq \frac{\lambda - (2z - \ell)}{\lambda + (2z - \ell)}. \tag{7}$$

The above inequality is the deception condition for schemata of order λ. In order to find the overall deception condition for $\lambda \geq z$, it is observed that the right side expression increases with λ for fixed z. Arguing as above, the inequality is true for all λ, if it is true for $\lambda = \ell - 1$. Substituting $\lambda = l - 1$ in the above inequality, yields

$$\frac{s_0}{s_1} \geq \frac{2\ell - 2z - 1}{2z - 1}. \tag{8}$$

Thus, if the above condition is true, all schemata with $\lambda \geq z$ and $u \geq z$ are favorable to the local optimum.

Second, for $\lambda > \ell - z$, the fitness of any schema with $u \leq \lambda - (\ell - z)$ is calculated using equation 2. The difference in the fitness of schemata with $u = 0$ and any u in this domain is calculated below:

$$d(0, u) = \frac{s_0}{2^{\ell-\lambda}} \left[\sum_{i=0}^{\ell-\lambda} \binom{\ell-\lambda}{i}(z - i) - \sum_{i=0}^{\ell-\lambda} \binom{\ell-\lambda}{i}(z - u - i) \right]$$

$$= s_0 u. \tag{9}$$

This is always positive. Thus, for $\lambda > \ell - z$ and $u \leq \lambda - (\ell - z)$, the schema with $u = 0$ has a higher fitness than any other schema of the same order.

Third, for any other schema, the fitness is calculated from equation 4. But, before calculating the difference in fitness as above, the sensitivity of equation 4 with respect to u is calculated. Writing $f(u+1, \lambda, \ell, z)$ and subtracting it from $f(u, \lambda, \ell, z)$, and simplifying yields

$$d(u, u + 1) = \frac{1}{2^{\ell-\lambda}} \left[s_0 \sum_{i=0}^{z-u-1} \binom{\ell-\lambda}{i} - s_1 \sum_{i=z-u}^{\ell-\lambda} \binom{\ell-\lambda}{i} \right]. \tag{10}$$

Using the identity $\binom{\ell-\lambda}{i} = \binom{\ell-\lambda}{\ell-\lambda-i}$, the latter summation may be written from $i = 0$ to $i = \ell - \lambda - z + u$. The term $d(u, u + 1)$ in the above equation is positive only when the following inequality is true:

$$\frac{s_0}{s_1} \geq \frac{\sum_{i=0}^{\ell-\lambda-z+u} \binom{\ell-\lambda}{i}}{\sum_{i=0}^{z-u-1} \binom{\ell-\lambda}{i}}. \tag{11}$$

As u is increased and other parameters are kept fixed, the number of terms in the numerator of the right side expression increases and the number of terms in the denominator decreases. Thus, the right side expression increases with increasing u. A parameter u^* is defined as the maximum u satisfying the above inequality. Then, for $0 \leq u \leq u^*$, the function $f(u, \lambda, \ell, z)$ is monotonically decreasing with u. Thus, as u increases, the difference between the fitness of schemata with unitation u and $u + 1$ decreases. Therefore, if this difference is positive at $u = 0$, the fitness of a schema with unitation u is greater than the fitness of a schema with unitation $u + 1$ for any $u \leq u^*$. For $u > u^*$, the above inequality is not true and the fitness of a schema increases with unitation, u. Thus, two limiting cases are sufficient for consideration—the schema with the minimum unitation and the schema with the maximum unitation.

For the minimum unitation, or $u = 1$, the difference, $d(0, 1)$, is the minimum for $\lambda = 1$. Setting the deception condition $d(0, 1) \geq 0$ and simplifying yields

$$\frac{s_0}{s_1} \geq \frac{\sum_{i=0}^{\ell-z-1} \binom{\ell-1}{i}}{\sum_{i=0}^{z-1} \binom{\ell-1}{i}}. \tag{12}$$

For the maximum unitation, or $u = \lambda$, the difference $d(0, \lambda)$ is the minimum for the maximum value of λ in the interval. Thus, setting $f(0, z, \ell, z) \geq f(z, z, \ell, z)$ and using equations 2 and 3 yields

$$\frac{s_0}{s_1} \geq \frac{\ell - z}{3z - \ell}. \tag{13}$$

3.3   Critical condition for deception

The above analysis reveals that there are three different conditions (inequalities 8, 12, and 13) that are to be satisfied for a fully deceptive trap function. Because the left side expressions in all three inequalities are identical, the critical condition occurs for the largest value of the right side expressions. In light of the above analysis, the following theorem may be written.

Theorem 1 *A necessary and sufficient condition for an ℓ-bit trap function to be fully deceptive is that all schemata of order $\ell - 1$ favor the local optimum.*

Proof: It can be easily shown that for $z < \ell - 1$, condition 13 is less severe than condition 8. However, at $z = \ell - 1$, both expressions are identical. Conditions 8 and 12 are compared by calculating the average of a series of ascending numbers. The average of a series of k ascending numbers, c_i, is calculated as $\frac{1}{k} \sum_{i=1}^{k} c_i$. It is a straightforward matter to show that the average of the series is increased if more terms in the series are added. Since $\binom{\ell'}{i+1} > \binom{\ell'}{i}$ for $i < \lfloor \ell'/2 \rfloor$, this principle can be applied to a series of successive binomial coefficients. If there are $k < \lfloor \ell'/2 \rfloor$ terms in the initial series and $(\ell' - 2k - 1)$ successive terms from $i = k + 1$ to $i = \ell' - k - 1$ are added to the series, the resultant average increases:

$$\frac{1}{\ell' - k - 1} \sum_{i=1}^{\ell'-k-1} c_i > \frac{1}{k} \sum_{i=1}^{k} c_i.$$

Setting the first term in the series equal to the sum of coefficients, $\binom{\ell'}{0}$ and $\binom{\ell'}{1}$, or $c_1 = \binom{\ell'}{0} + \binom{\ell'}{1}$, and any other term, c_i, equal to $\binom{\ell'}{i}$, after some rearrangement, the following inequality can be written:

$$\frac{\sum_{i=0}^{k} \binom{\ell'}{i}}{\sum_{i=0}^{\ell'-k-1} \binom{\ell'}{i}} < \frac{k}{\ell' - k - 1}. \tag{14}$$

In order to match the left side of the above expression with the right side expression of inequality 12, $k = \ell - z - 1$ and $\ell' = \ell - 1$ are substituted. Since the above inequality is true for $k < \lfloor \ell'/2 \rfloor$, the following inequality is true only for $z > \lfloor \ell/2 \rfloor$:

$$\frac{\sum_{i=0}^{\ell-z-1} \binom{\ell-1}{i}}{\sum_{i=0}^{z-1} \binom{\ell-1}{i}} < \frac{\ell - z - 1}{z - 1}. \tag{15}$$

Since the expression on either side of the inequality sign is smaller than one for $z > \lfloor \ell/2 \rfloor$, adding 0.5 on both denominator and numerator of the right side expression proves that condition 8 is more severe than condition 12. Although intuitive, it will be shown a little later that no fully deceptive function exists for $z < \lfloor \ell/2 \rfloor$.

Therefore, inequality 8 is the most severe condition for full deception and this occurs for $\lambda = \ell - 1$. Thus, if the condition for deception is satisfied for $\lambda = \ell - 1$, deception conditions for other schemata are also satisfied. This proves the above theorem. $\square$

Thus, the critical condition for a fully deceptive trap function can be obtained by substituting the expressions for two slope values s_0 and s_1 in inequality 8. For convenience, a parameter, r, is defined as the ratio of the locally to the globally optimal function values, or $r = a/b$. Writing the critical condition (inequality 8) in terms of r yields the condition for a fully deceptive trap function:

$$r \geq \frac{2 - 1/(\ell - z)}{2 - 1/z}. \tag{16}$$

4 Critical z for deception

The condition 16 may be rewritten to obtain a critical value of slope-change location, z, for a fully deceptive trap function. But before doing that, a previously ignored issue is considered. In proving theorem 1, it was assumed that no fully deceptive trap function exist for $z < \lfloor \ell/2 \rfloor$. In the remainder of this section, an analysis is performed to investigate this aspect, and then a range of slope-change location, z, is found for a fully deceptive trap function.

To investigate if there are any fully deceptive functions for $z < \lfloor \ell/2 \rfloor$, inequality 14 is further considered. Substituting $k = z - 1$ and $\ell' = \ell - 1$ in inequality 14 and writing the above inequality in terms of the reciprocal of each term, the following inequality is obtained:

$$\frac{\sum_{i=0}^{\ell-z-1} \binom{\ell-1}{i}}{\sum_{i=0}^{z-1} \binom{\ell-1}{i}} > \frac{\ell - z - 1}{z - 1}. \tag{17}$$

Since the expression on the either side of the inequality is greater than one for $z < \lfloor \ell/2 \rfloor$, adding 0.5 on both numerator and the denominator of the right side expression reveals that condition 12 is more severe than condition 8. Taking $k + 1$ terms in the series including $\binom{\ell'}{0}$ and following the calculation as above, it can be shown that the left side expression of the above inequality is greater than $(\ell - z)/z$. Thus, it follows from inequality 12 that $a > b$, which contradicts the basic assumption that the string with unitation of ℓ is not the globally optimal string. Thus, no fully deceptive trap function exists for $z < \lfloor \ell/2 \rfloor$.

As seen from the above analysis, fully deceptive trap function do not exist for values of z smaller than $\lfloor \ell/2 \rfloor$; they do not exist also for all values z greater than $\lfloor \ell/2 \rfloor$. The critical value of z for a fully deceptive trap function may be obtained from inequality 16. Rewriting inequality 16 in terms of z, and solving the quadratic expression for z yields

$$z \geq \frac{\ell}{2} + \frac{1+r}{4(1-r)} \left(\sqrt{1 + 4\ell(\ell - 1)\left(\frac{1-r}{1+r}\right)^2} - 1 \right). \tag{18}$$

For large ℓ, the above expression may be approximated as follows:

$$z_{min} \approx \ell - \frac{1+r}{4(1-r)}. \tag{19}$$

This equation suggests that in a fully deceptive trap function, a large value of z is required for small values of the fitness ratio r. Thus, for a desired r, a fully deceptive function may be designed, using a value of z that satisfies the above inequality.

The deception of Ackley's (1987) trap functions may now be investigated using the above analysis. Ackley considered a fixed value of $r = 0.8$, and a value of $z = \lfloor 3\ell/4 \rfloor$. Substituting these values of r and z in inequality 16, it is observed that Ackley's trap function is fully deceptive only for $\ell \leq 7$. Ackley, however, used four trap functions of size 8, 12, 16, and 20. Thus, none of Ackley's trap functions were fully deceptive. This may be one reason why his hillclimbing method solved those problems fairly well.

5 Limiting values of r

The condition for a fully deceptive trap function is given in inequality 16. For a trap function with given values of a, b, and z, that inequality may be used to test full deception in the function. The inequality 18 may be used to test full deception in a trap function with given values of a and b. In this section, limiting values of the locally to the globally optimal function values are found for all possible values of slope-change location, z. The range of values of this ratio (r) for a class of functions gives an idea about the flexibility in designing a fully deceptive function. In this context, the limiting values of r are found for a number of commonly-used fully deceptive functions.

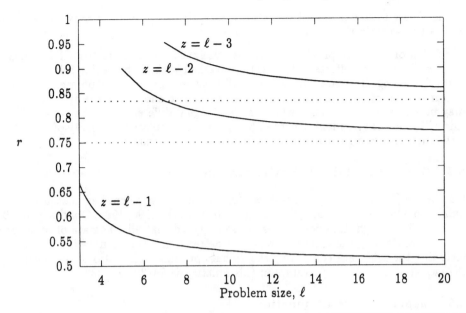

Figure 2: The limiting value of the fitness ratio r is plotted versus problem size at various values of slope-change location z.

5.1 Trap function

The minimum value of r that admits fully deceptive trap functions may be calculated using inequality 16. It is evident from the inequality that the minimum r occurs for the maximum value of z. Thus, substituting $z = \ell - 1$ in the inequality, the limiting value of r in a problem of size ℓ is obtained:

$$r_{min} = \frac{\ell - 1}{2\ell - 3}. \tag{20}$$

It is interesting to note that for large values of ℓ, r_{min} approaches 0.5. Thus, a fully deceptive trap function may be designed so that r approaches 0.5. To illustrate how the critical value of the parameter r depends on the problem length and the parameter z, figure 2 is plotted for various values of ℓ, r, and z. The plot shows that the minimum value of r for full deception occurs at $z = \ell - 1$, and this value approaches 0.5 in the limit. For other values of z, the parameter r approaches a value $1 - 1/(2(\ell - z))$ in the limit.

Recognizing that the globally optimal function value is greater than the locally optimal function value, an upper bound of the fitness ratio r that admits a fully deceptive trap function may also be calculated. Since $b > a$, the maximum value of the fitness ratio r is one. Thus, a fully deceptive trap function may be designed with a fitness ratio r varying from the lower bound given in equation 20 to one.

A number of deceptive problems have been designed for use as test functions in GA studies. These early efforts were largely concerned with (1) achieving full deception and (2) keeping the signal difference small between the two best points. Here, the limiting r values of these functions are compared to show that the trap functions may be adjusted over a much wider range of signal difference. In the next section, this will yield the attendant benefit of being able to calculate the density of deceptive test functions among the set of all possible trap functions.

5.2 Order-three fully deceptive function

Goldberg's (1989a, 1989b) order-three, fully deceptive problem has been used in a number of studies (Deb, 1991; Goldberg, Deb, & Korb, 1990, 1991; Goldberg, Korb, & Deb, 1989). Eight function values are so designed that all schemata of order one and two containing the local optimum are better than other competing schemata, including the schemata containing the global optimum. The ratio of the local and the global optimal function values in this function is 28/30 or 0.933.

5.3 Liepins and Vose's function

Liepins and Vose (1990) introduced a fully deceptive function of any order. Their function has been used in a number of studies (Deb, 1991; Goldberg, Deb, & Korb, 1990, 1991; Liepins & Vose, 1990). This function is a function of unitation, $u(x)$, and is defined for $\ell \geq 3$ as follows:

$$f(x) = \begin{cases} 1, & \text{if } u(x) = \ell; \\ 1 - \frac{1}{2\ell}, & \text{if } u(x) = 0; \\ 1 - \frac{1+u(x)}{\ell}, & \text{otherwise}; \end{cases} \tag{21}$$

The global optimum has a function value equal to 1 and the local optimum has a function value equal to $1 - \frac{1}{2\ell}$. The ratio between the locally and the globally optimal function values is given by

$$r = 1 - \frac{1}{2\ell}. \tag{22}$$

As the problem size increases, the ratio r increases, and for large values of ℓ, the required ratio is close to one. Deb (1991) proved by calculating schema fitness values that this function is fully deceptive for $\ell \geq 3$. Further analysis showed that the second condition can be eliminated for fully deceptive functions of order greater than or equal to 4, and the modified Liepins and Vose's function for $\ell \geq 4$ is as follows:

$$f(x) = \begin{cases} 1, & \text{if } u(x) = \ell; \\ 1 - \frac{1+u(x)}{\ell}, & \text{otherwise}; \end{cases} \tag{23}$$

In this function, the fitness ratio r is given by

$$r = 1 - \frac{1}{\ell}. \tag{24}$$

Though this ratio is smaller than the required ratio in the original function, the ratio approaches one in the limit.

5.4   Whitley's function

Whitley (1991) devised a fully deceptive function that assigns the function value according to a ranking of strings in an ascending order of the Hamming distance from a chosen global optimum. The string at the top of the list is the global optimum and the string at the bottom of the list is the local optimum. The second string in the list is assigned a value B, and successive strings in the list are assigned as

$$f(x) = f(x - 1) + C, \tag{25}$$

where the current string is assigned a value C more than the previous string in the list. The global optimum is assigned a value C more than the locally optimal function value. The parameters B and C are positive quantities. Though there is no bound in the difference between the global and the locally optimal function values, their ratio is bounded. Recognizing that $f(N) = B+(2^\ell-2)C$, and $f(1) = f(N)+C$, the ratio of the locally and the globally optimal function values may be written as follows:

$$r = 1 - \frac{C}{B + C(2^\ell - 1)}. \tag{26}$$

Since, both C and B are positive, the right side of the above equation has a minimum value for $B = 0$, which gives rise to the minimum value of the ratio as follows:

$$r_{min} = 1 - \frac{1}{(2^\ell - 1)}. \tag{27}$$

In this function, the limiting ratio r is greater than that in the previous functions.

5.5   Goldberg's construction from Walsh coefficients

Goldberg (1990) devised a class of fully deceptive functions of unitation by using only low-order Walsh coefficients. All Walsh coefficients of a particular order are assumed to be equal. If w_0, w_1, w_2, and w_3 are order-zero, order-one, order-two, and order-three Walsh coefficients respectively, a fully deceptive function of unitation is written as follows:

$$\begin{aligned} f(u, \ell) &= w_0 + w_1(\ell - 2u) + w_2 \left[\binom{\ell-u}{2} - u(\ell - u) + \binom{u}{2} \right] \\ &\quad + w_3 \left[\binom{\ell-u}{3} - u\binom{\ell-u}{2} + (\ell - u)\binom{u}{2} - \binom{u}{3} \right]. \end{aligned} \tag{28}$$

Setting all deceptive conditions, relations among w_1, w_2, and w_3 are calculated as follows:

$$w_2 \geq \frac{2\ell - 1}{(\ell - 1)(\ell - 2)} w_1; \tag{29}$$

$$w_3 = -\frac{6}{(\ell - 2)^2} w_1. \tag{30}$$

Under these conditions, the minimum value of the fitness ratio r may be simplified as follows:

$$r \approx 1 - \frac{2}{\frac{w_0}{w_1} + \ell}. \tag{31}$$

A value for w_0 is chosen to achieve non-negativity of the function values. Substituting the expressions for w_2 and w_3 from inequality 29 (with the equality condition) and 30 in equation 28 respectively and calculating the unitation u for the minimum function value, and then substituting that minimum function value for zero, an expression for w_0 in terms of w_1 is obtained:

$$\frac{w_0}{w_1} = \frac{29}{54} - \frac{11\ell}{27} + \frac{142 - 271\ell + 96\ell^2}{108(\ell - 2)^2(\ell - 1)} + \left(\frac{4}{27} + \frac{34 - 71\ell + 40\ell^2}{108(\ell - 2)^2(\ell - 1)}\right)\sqrt{16\ell^2 - 24\ell + 21}.$$

(32)

For large values of ℓ, the above equation is approximately linear:

$$\frac{w_0}{w_1} \approx \frac{170}{108} + \frac{5}{27}\ell.$$

(33)

Even though this function demands a smaller fitness ratio r than previous functions, the ratio approaches one in the limit.

The above analyses have shown that in all currently used deceptive functions, the fitness ratio r needs to be high and approaches one for large values of ℓ. From the standpoint of making the problem more difficult from a signal-to-noise perspective (Goldberg, Deb, & Clark, 1991; Goldberg & Rudnick, 1991) this is, of course, desirable. On the other hand, from the standpoint of achieving greater flexibility in test-function design, and more importantly from the standpoint of getting some handle on the density of deceptive functions, the range of parameters over which a function remains deceptive must be understood. In the next section, the r-z boundaries between average-sense easy and hard functions are more carefully delineated.

6   The density of trap-function deception

In this section, the previous analysis of deception is extended to delineate r-z boundaries between functions that are not deceptive, partially deceptive, and fully deceptive in the usual average sense. The analysis is extended to find the proportion of fully deceptive functions in all trap functions.

6.1   Delineation of r-z boundaries

The results of previous analysis can be used to find conditions for trap functions that are fully easy in an average sense. A function is defined to be fully easy if all schemata containing the global attractor is better than any other competing schemata. In previous analysis, the string with $u = 0$ was assumed to be the deceptive attractor, and conditions were found so that all schemata containing the string with $u = 0$ are no worse than other competing schemata. Even though this requires that the ratio r be smaller than or equal to one, in deriving the critical conditions (inequalities 8, 12, and 13) for deception, no restriction was made on the value of r. If the problem is now changed and the string with $u = 0$ is assumed to be the globally best string, the same analysis holds and the same three conditions need to be satisfied for a fully easy function. In proving theorem 1, the critical condition for $z > \lfloor \ell/2 \rfloor$ was found to be inequality 8 and the critical ratio of r was found to be a value smaller than one. In a function with the string with $u = 0$ being the globally optimal string, the ratio r is always greater than one. Thus, in a trap function with the string of zero unitation being the globally optimal string and

$z > \lfloor \ell/2 \rfloor$, a function with any r is fully easy. On the other hand, for $z < \lfloor \ell/2 \rfloor$, inequality 12 was the critical condition. Reflecting function values about $u = \lfloor \ell/2 \rfloor$, such that the string with $u = \ell$ is now the global optimum, it can be concluded that a fully easy function exists for any value of r in the interval $z < \lfloor \ell/2 \rfloor$ and for a value of r satisfying the inequality 12 in the interval $z > \lfloor \ell/2 \rfloor$.

This analysis may be extended to calculate r-z boundaries on trap functions that are partially deceptive to a given order. A function is defined to be *partially* deceptive to an order k if all schema partitions of order less than k are deceptive. Assuming $k = \lambda + 1$, using equations 2, 3, and 4, and recognizing that the critical condition occurs for $u = \lambda$, the r-z boundaries may be calculated for deceptive trap functions to any given order k. For $z < \lfloor \ell/2 \rfloor$, trap functions are not deceptive, and the limiting value of r is one for any given order λ. For $z > \lfloor \ell/2 \rfloor$, the r-z boundaries are calculated for two nonoverlapping cases: $\lambda < z$ and $\lambda \geq z$. For $\lambda < z$, the r-z boundary is given by the equation

$$r = \frac{z \left[\sum_{i=z-\lambda+1}^{\ell-\lambda} \binom{\ell-\lambda}{i} (\lambda - z + i) - \sum_{i=z+1}^{\ell-\lambda} \binom{\ell-\lambda}{i} (i - z) \right]}{(\ell - z) \left[\sum_{i=0}^{z} \binom{\ell-\lambda}{i} (z - i) - \sum_{i=0}^{z-\lambda} \binom{\ell-\lambda}{i} (z - \lambda - i) \right]}, \tag{34}$$

and for $\lambda \geq z$, the r-z boundary is given by the equation

$$r = \frac{z[\lambda - (2z - \ell)]}{(\ell - z)[\lambda + (2z - \ell)]}. \tag{35}$$

These equations are plotted in figure 3 to show the r-z boundaries between easy, partially deceptive, and fully deceptive 50-bit trap functions. The rectangular area for z varying from 1 to $(\ell - 1)$ and for r varying from 0 to 1 covers all possible combinations of r and z. The fitness values are scaled so that the maximum fitness is always one. In the figure, functions falling on a curve with a label k indicate that they are at most order-k deceptive. It is clear that an order-one deceptive trap function has no deceptive schema partition. Thus, the r-z boundary for $\lambda = 1$ obtained from equation 34 is the upper bound for fully easy functions. It is interesting to note that this boundary is identical to that described by inequality 12. Therefore, functions falling in the left side of the order-two curve represent fully easy functions. A minimally deceptive trap function is an order-two deceptive trap function, where all order-one schema partitions are deceptive. Functions falling in the region between the curves for order-two and order-50 are of varying degrees of deception as indicated. The proportion of the area above a curve marked k to the complete rectangular area indicates the proportion of trap functions that are deceptive to an order at least equal to k. Thus, all functions with r and z values falling above the curve marked 40 are deceptive to an order 40 or more. Extending this analysis to the limit, trap functions with r and z values falling above the order-50 curve contain all misleading schemata. Hence, they are fully deceptive. It is interesting that for low-order deceptive functions, a small change of the problem parameters may change the order of deception substantially, but for higher order deceptive problems, a large change in parameters is required to change the order of deception.

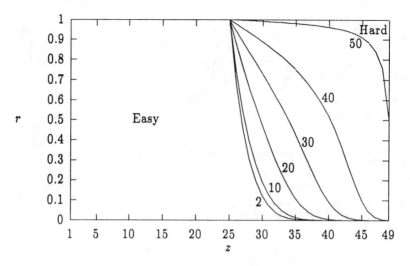

Figure 3: The graph delineates the r-z boundaries between easy, partially deceptive (of different orders), and fully deceptive (hard) 50-bit trap functions.

6.2 Proportion of fully deceptive trap functions

It is clear from figure 3 that the proportion of the functions that are fully deceptive is small. For trap functions with $r \in [0, 1]$ and $z \in \{1, 2, \ldots, \ell-1\}$ chosen uniformly at random, an estimate of the proportion of the fully deceptive functions may be found by calculating the area above the top curve. It was shown before that a fully deceptive trap function does not exist for $z < \lfloor \ell/2 \rfloor$. The condition for the fully deceptive trap functions for $z > \lfloor \ell/2 \rfloor$ is given in inequality 16 and is derived from condition 8. The area above this curve is calculated by integrating r values above the curve for $z = \ell/2$ to $z = \ell - 1$. For large values of ℓ, this area is $(\ln \ell/2)/2$. Since the total area is $\ell - 1$, the proportion of the fully deceptive trap functions is approximately $(\ln \ell/2)/(2\ell)$. Thus, for large values of ℓ, the proportion of fully deceptive functions in all trap functions is $O(\ell^{-1} \ln \ell)$.

It is also shown in this section that for $z < \lfloor \ell/2 \rfloor$, functions with any r are fully easy functions, and for $z > \lfloor \ell/2 \rfloor$, there exist functions with r satisfying the inequality 12 that are fully easy. Thus, more than half the trap functions are fully easy. It can also be shown that as the problem size increases, the area under the bottom curve reduces. Thus, for large ℓ, only half of the trap functions are fully easy and the remaining half are at least partially deceptive. Note that the terms deceptive (or hard) and easy are used to indicate whether or not schemata are misleading in the usual average sense. The actual degree of difficulty increases as the order of deception increases, but the success of GAs in solving problems depends also on a number of other factors such as the underlying linkage of the important building blocks, genetic operators, crossover rate, selection pressure, population size, sampling representation, and others. Nonetheless, average-sense deception remains an important research tool to better understand GA behavior.

7   Conclusions

Piecewise-linear functions of unitation called trap functions have been investigated in this paper. A schema analysis has been performed to find conditions for a fully deceptive trap function. It has been found that the necessary condition is that all order $\ell - 1$ schemata be misleading. Using this condition, it has been shown that Ackley's trap functions are not fully deceptive functions.

Further analyses have shown that a fully deceptive trap function can be designed so that the locally optimal function value is as low as 50% of the globally optimal function value. The ratio between the locally and globally optimal function values has been calculated for a number of currently used fully deceptive functions. It has been found that the limiting value of this ratio in those functions is close to one. Such a ratio is desirable if one is interested in making the function as difficult as possible, but little work exists to investigate how far one can go in the other direction. More importantly, the density of deception of trap functions has been fully investigated. Assuming uniform selection of r and z and examining behavior at large problem size, the proportion of fully deceptive functions decreases in proportion to the ratio of the logarithm of the problem size to the problem size; the proportion of fully easy functions occupies half the r-z plane; and the partially deceptive functions occupy the remainder. More investigation is needed to better understand the density of deception in more generally constructed problems, but this paper has taken some useful first steps in that direction.

Acknowledgments

The authors acknowledge the support provided by the US Army under Contract DASG60-90-C-0153 and by the National Science Foundation under Grant ECS-9022007.

References

Ackley, D. H. (1987). *A connectionist machine for genetic hillclimbing.* Boston, MA: Kluwer Academic Publishers.

Deb, K. (1991). Binary and Floating-point Function Optimization using Messy Genetic Algorithms. (Doctoral dissertation, University of Alabama and IlliGAL Report No. 91004). *Dissertation Abstracts International, 52*(5), 2658B.

Goldberg, D. E. (1987). Simple genetic algorithms and the minimal deceptive problem. In L. Davis (Ed.), *Genetic algorithms and simulated annealing* (pp. 74–88). London: Pitman.

Goldberg, D. E. (1989a). Genetic algorithms and Walsh functions: Part I, a gentle introduction. *Complex Systems, 3,* 129–152.

Goldberg, D. E. (1989b). Genetic algorithms and Walsh functions: Part II, deception and its analysis. *Complex Systems, 3,* 153–171.

Goldberg, D. E. (1990). *Construction of high-order deceptive functions using low-order Walsh coefficients.*, (IlliGAL Report No. 90002). Urbana: University of Illinois, Illinois Genetic Algorithms Laboratory.

Goldberg, D. E., Deb, K., & Clark, J. H. (1991). *Genetic algorithms, noise, and the sizing of populations* (IlliGAL Report No. 91010). Urbana: University of Illinois at Urbana-Champaign, Illinois Genetic Algorithms Laboratory.

Goldberg, D. E., Deb, K., & Korb, B. (1990). Messy genetic algorithms revisited: Studies in mixed size and scale. *Complex Systems, 4*, 415–444.

Goldberg, D. E., Deb, K., & Korb, B. (1991). Don't worry, Be messy. *Proceedings of the Fourth International Conference in Genetic Algorithms and their Applications*, 24–30.

Goldberg, D. E., Korb, B., & Deb, K. (1989). Messy genetic algorithms: Motivation, analysis, and first results. *Complex Systems, 3*, 493–530.

Goldberg, D. E., & Rudnick, M. (1991). Genetic algorithms and the variance of fitness. *Complex Systems, 5*, 265–278.

Liepins, G. E., & Vose, M. D. (1990). Representational issues in genetic optimization. *Journal of Experimental and Theoretical Artificial Intelligence, 2*(2), 4–30.

Whitley, D. (1991). Fundamental principles of deception in genetic search. *Foundations of Genetic Algorithms*, 221–241.

Relative Building-Block Fitness and the Building-Block Hypothesis

Stephanie Forrest
Dept. of Computer Science
University of New Mexico
Albuquerque, NM 87131

Melanie Mitchell*
AI Laboratory
University of Michigan
Ann Arbor, MI 48109

Abstract

The *building-block* hypothesis states that the GA works well when short, low-order, highly-fit schemas recombine to form even more highly fit higher-order schemas. The ability to produce fitter and fitter partial solutions by combining building blocks is believed to be a primary source of the GA's search power, but the GA research community currently lacks precise and quantitative descriptions of how schema processing actually takes place during the typical evolution of a GA search. Another open problem is to characterize in detail the types of fitness landscapes for which crossover will be an effective operator. In this paper we first describe a class of fitness landscapes (the "Royal Road" functions) that we have designed to investigate these questions. We then present some unexpected experimental results concerning the GA's performance on simple instances of these landscapes, in which we vary the strength of reinforcement from "stepping stones"—fit intermediate-order schemas obtained by recombining fit low-order schemas. Finally, we compare the performance of the GA on these functions with that of three commonly used hill-climbing schemes, and find that one of them, "random-mutation hill-climbing", significantly outperforms the GA on these functions.

*Current address: Santa Fe Institute, 1660 Old Pecos Trail, Suite A, Santa Fe, NM 87501

1 INTRODUCTION

Research on the foundations of genetic algorithms aspires to answer two general questions: How do GAs work, and what are they good for? A successful theory of GAs would describe the laws governing the behavior of schemas in GAs and characterize the types of fitness landscapes on which the GA is likely to perform well, especially as compared with other search methods such as hill-climbing. This, of course, requires a statement of what it means for a GA to "perform well". That is, we need a better understanding of what it is the GA is good at doing (e.g., finding a global optimum versus quickly finding a fairly good solution).

Our strategy for answering these questions consists of the following general approach. We begin by identifying *features* of fitness landscapes that are particularly relevant to the GA's performance. A number of such features have been discussed in the GA literature, including local hills, "deserts", deception, hierarchically structured building blocks, noise, and high fitness variance within schemas. We then design simplified landscapes containing different configurations of such features, varying, for example, the distribution, frequency, and size of different features in the landscape. We then study in detail the effects of these features on the GA's behavior. A longer-term goal of this research is to develop statistical methods of classifying any given landscape in terms of our spectrum of hand-designed landscapes, thus being able to predict some aspects of the GA's performance on the given landscape.

It should be noted that by stating this problem in terms of the GA's performance on fitness landscapes, we are sidestepping the question of how a particular problem can best be represented to the GA. The success of the GA on a particular function is certainly related to how the function is "encoded" (Goldberg, 1989b; Liepins & Vose, 1990) (e.g., using Gray codes for numerical parameters can greatly enhance the performance of the GA on some problems), but since we are interested in biases that pertain directly to the GA, we will simply consider the landscape that the GA "sees."

In this paper we describe some initial results from this long-term research program. We began by focusing on the *building-block hypothesis* (Holland, 1975; Goldberg, 1989b), which states that the GA works well when short, low-order, highly-fit schemas ("building blocks") recombine to form even more highly fit higher-order schemas. In Goldberg's words, "...we construct better and better strings from the best partial solutions of past samplings"(Goldberg, 1989b, p. 41). The ability to produce fitter and fitter partial solutions by combining building blocks is believed to be the primary source of the GA's search power. However, in spite of the presumed central role of building blocks and recombination, the GA research community lacks precise and quantitative descriptions of how schemas interact and combine during the typical evolution of a GA search. Thus, we are interested in isolating landscape features implied by the building-block hypothesis, and studying in detail the GA's behavior—the way in which schemas are processed and building blocks are combined—on simple landscapes containing those features.

Other GA researchers have studied these same questions using different techniques. The most prominent approach has been to study the effects of *GA deception* on the GA's performance (e.g., Goldberg, 1987, 1989a; Liepins & Vose, 1990; Whitley,

1991). However, deception is only one among many features of a problem that affect GA performance (e.g., see Liepins & Vose, 1990, and Forrest & Mitchell, 1991). Rather than studying hard problems on which the GA fails, our initial approach has been to examine the GA's behavior on landscapes for which it is likely to perform well. By understanding what features of those landscapes lead to good performance, we hope to better characterize the class of such landscapes.

One major component of this endeavor is to define the simplest class of landscapes on which the GA performs "as expected", thus confirming the broad claims of the building-block hypothesis. However, the task of designing such landscapes has turned out to be substantially more difficult and complex than we originally anticipated. Our initial choices of simple landscapes have revealed some surprising and unanticipated phenomena. The story of how small variations of a basic landscape can make GA search much less effective reveals a great deal about the complexity of GAs and points out the need for a deeper theory of how low-order building blocks are discovered and combined into higher-order solutions.

In the following sections we introduce the *Royal Road* functions, a class of nondeceptive functions in which the building blocks are explicitly defined. We then show how simple variants of these functions can have quite different effects on the performance of the GA, and discuss the reasons for these differences.

2   STEPPING STONES IN THE CROSSOVER LANDSCAPE

The building-block hypothesis suggests two landscape features that are particularly relevant for the GA: (1) the presence of short, low-order, highly fit schemas; and (2) the presence of intermediate "stepping stones"—intermediate-order higher-fitness schemas that result from combinations of the lower-order schemas, and that in turn can combine to create even higher-fitness schemas. Two basic questions about stepping stones are: How much higher in fitness do the intermediate stepping stones have to be for the GA to work well? And how must these stepping stones be configured? To investigate these questions, we first define the Royal Road functions, which contain these features explicitly.

To construct a Royal Road function, we select an optimum string and break it up into a number of small building blocks, as illustrated in Figure 1. We then assign values to each low-order schema and each possible intermediate combination of low-order schemas, and use those values to compute the fitness of a bit string x in terms of the schemas of which it is an instance.

The function $R1$, illustrated in Figure 1, is computed very simply: a bit string x gets 8 points added to its fitness for each of the given order-8 schemas of which it is an instance. For example, if x contains exactly two of the order-8 building blocks, $R1(x) = 16$. Likewise, $R1(111\dots1) = 64$. Stated more generally, the value $R1(x)$ is the sum of the coefficients c_s corresponding to each given schema of which x is an instance. Here c_s is equal to $order(s)$. The fitness contribution from an intermediate stepping stone (such as the combination of s_1 and s_3 in Figure 1) is thus a linear combination of the fitness contribution of the lower-level components. $R1$ is similar to the "plateau" problem described by Schaffer and Eshelman (1991).

$$s_1 = \text{11111111}***; \; c_1 = 8$$
$$s_2 = ********\text{11111111}***; \; c_2 = 8$$
$$s_3 = ****************\text{11111111}***; \; c_3 = 8$$
$$s_4 = ************************\text{11111111}***************************************; \; c_4 = 8$$
$$s_5 = ********************************\text{11111111}*******************************; \; c_5 = 8$$
$$s_6 = **\text{11111111}***********************; \; c_6 = 8$$
$$s_7 = **\text{11111111}***************; \; c_7 = 8$$
$$s_8 = **\text{11111111}; \; c_8 = 8$$
$$s_{opt} = \text{1111111111111111111111111111111111111111111111111111111111111111}$$

Figure 1: An optimal string broken up into eight building blocks. The function $R1(x)$ (where x is a bit string) is computed by summing the coefficients c_s corresponding to each of the given schemas of which x is an instance. For example, $R1(1111111100\ldots0) = 8$, and $R1(1111111100\ldots011111111) = 16$. Here $c_s = \text{order}(s)$.

According to the building-block hypothesis, $R1$'s building-block and stepping-stone structure should lay out a "royal road" for the GA to follow to the global optimum. In contrast, an algorithm such as simple steepest-ascent hill-climbing, which systematically tries out single-bit mutations and only moves in an uphill direction, cannot easily find high values in such a function, since a large number of single bit-positions must be optimized simultaneously in order to move from an instance of a lower-order schema (e.g., $11111111**\ldots*$) to an instance of a higher-order intermediate schema (e.g., $11111111********11111111**\ldots*$). While some random search may be involved in finding the lowest-level building blocks (depending on the size of the initial population and the size of the lowest-level blocks), the interesting aspect of $R1$ is studying how lower-level blocks are combined into higher-level ones, and this is the aspect with which we are most concerned. Part of our purpose in designing the Royal Road functions is to construct a class of fitness landscapes that distinguishes the GA from other search methods such as hill-climbing. This actually turned out to be more difficult than we anticipated, as will be discussed in Section 5.

This class of functions provides an ideal laboratory for studying the GA's behavior:

- The landscape can be varied in a number of ways. For example, the "height" of various intermediate stepping stones can be increased or decreased (e.g., the fitness contribution can be a nonlinear combination of the fitness contributions from the components). Also, the size of the lowest-order building blocks can be varied, as can the degree to which they cover the optimum. Finally, different degrees of deception can be introduced by allowing the lower-order schemas to differ in some bits from the higher-order stepping stones, effectively creating low-order schemas that lead the GA away from the good higher-order schemas. The effects of these variations on the GA's behavior can then be studied in detail.

- Since the global optimum, and, in fact, all possible fitness values, are known in advance, it is easy to compare the GA's performance on different variations of Royal Road functions.

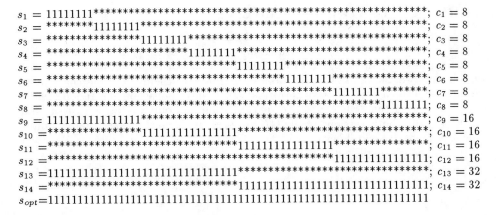

Figure 2: Royal Road Function $R2$. $R2(x)$ is computed in the same way as $R1$: by summing the coefficients c_s corresponding to each of the given schemas of which x is an instance. For example, $R2(1111111100\ldots011111111) = 16$, but $R2(11111111111111111100\ldots0) = 32$. $R2(11111111\ldots1) = 192$.

- All of the desired schemas are known in advance, since they are explicitly built into the function. Therefore, the dynamics of the search process can be studied in detail by tracing the ontogenies of individual schemas.

We are using the Royal Road functions to study a number of questions about the effects of crossover on various landscapes, including the following: For a given landscape, to what extent does crossover help the GA find highly fit schemas? What is the effect of crossover on the waiting times for desirable schemas to be discovered? What are the bottlenecks in the discovery process? How does the configuration of stepping stones and size of steps defined by stepping stones affect the GA's performance? Answering these questions in the context of the idealized Royal Road functions is a first step towards answering them for more general cases.

We first investigated the effect of the step size of the intermediate stepping stones on the GA's performance. To do this, we compared the performance of the GA on $R1$ with its performance on a second function $R2$, where the fitness contributions of certain intermediate stepping stones are much higher. $R2$ is illustrated in Figure 2. $R2$ is calculated in the same way as $R1$: the fitness of a bit string x is the sum of the coefficients corresponding to each schema $(s_1$–$s_{14})$ of which it is an instance.

For example, $R2(1111111100\ldots011111111) = 16$, since the string is an instance of both s_1 and s_8, but $R2(111111111111111100\ldots0) = 32$, since the string is an instance of s_1, s_2, and s_9. Thus, a string's fitness depends not only on the number of 8-bit schemas to which the string belongs, but also on their positions in the string. The optimum string $11111111\ldots1$ has fitness 192, since the string is an instance of each schema in the list.

3 ROYAL ROAD EXPERIMENTS

In an earlier paper (Mitchell, Forrest, & Holland, 1992) we reported some initial results on Royal Road functions. Our main performance measure was the number of generations it took the GA to find the function optimum, although for some experiments we also looked at the discovery time for schemas of different orders. We first confirmed that the GA performs significantly better on $R1$ and $R2$ when crossover is used than when crossover is turned off and only mutation is used, and we showed that both versions of the GA perform significantly better than a simple steepest-ascent hill-climbing algorithm. These results were expected. We then described some unexpected experimental results comparing the GA's performance on $R2$ with its performance on $R1$. Here we extend these experimental results and analyze them in more detail, and compare the GA's performance with that of a more sophisticated hill-climber.

For our initial experiments, we used functions defined over strings of length 64. The GA population size was 128, with the initial population generated at random. In each run the GA was allowed to continue until the optimum string was discovered, and the total number of function evaluations performed was recorded. We used a generational GA with single-point crossover and sigma scaling (Tanese, 1989; Forrest & Mitchell, 1991): an individual i's expected number of offspring is $1 + \frac{F_i - \overline{F}}{2\sigma}$, where F_i is i's fitness, $\overline{F}$ is the mean fitness of the population, and σ is the standard deviation. The maximum expected offspring of any string was 1.5—if the above formula gave a higher value, the value was reset to 1.5. This is a strict cutoff, since it implies that most individuals will reproduce only 0, 1, or 2 times. The effect of this selection scheme is to slow down convergence by restricting the effect that a single individual can have on the population, regardless of how much more fit it is than the rest of the population. Even with this precaution, we observe some interesting premature convergence effects (described in the following section). The crossover probability was 0.7 per pair of parents and the mutation probability was 0.005 per bit.

The probability that a randomly generated string contains one of the bottom-level order-8 schemas is $8 * \frac{1}{2^8} = \frac{1}{32}$. Since the initial population has 128 randomly generated individuals, there were on average $\frac{128}{32} = 4$ total instances of bottom-level schemas in the initial population. That is, there is a 0.5 probability that there will be an instance of any particular block; thus, since there are 8 different lowest-level blocks, there will on average be 4 total instances of lowest-level blocks in the population.

3.1 EXPERIMENTS ON $R1$ AND $R2$

We expected the GA to perform better—that is, find the optimum more quickly—on $R2$ than on $R1$. In $R2$ there is a very clear path via crossover from pairs of the eight initial order-8 schemas (s_1–s_8) to the four order-16 schemas (s_9–s_{12}), and from there to the two order-32 schemas (s_{13} and s_{14}), and finally to the optimum (s_{opt}). We believed that the presence of this stronger path would speed up the GA's discovery of the optimum, but our experiments showed the opposite: the GA performed significantly better on $R1$ than on $R2$. Statistics summarizing the results of 500 runs on each function are given in Table 1. This table gives the mean and

ORIGINAL EXPERIMENT		
	Function Evaluations to Optimum	
500 runs	$R1$	$R2$
Mean	62099 (std err: 1390)	73563 (std err: 1794)
Median	56576	66304

Table 1: Summary of results of running the GA on $R1$ and $R2$. The table gives the mean and median function evaluations taken to find the optimum over 500 runs on each function. The numbers in parentheses are the standard errors.

median number of function evaluations taken to find the optimum over 50 runs each on $R1$ and $R2$.

If we hope to understand the GA's performance in general, we need to understand in detail what are the potential bottlenecks for discovering desirable schemas. This has been studied extensively in the deception literature, but $R2$ is a non-deceptive function that nonetheless contains some features that keep the GA from discovering desirable schemas as quickly as in $R1$.

What slows down the GA in the case of $R2$? To investigate this, we took a typical run of the GA on $R2$ and graphically traced the evolution of each schema in the tree. Figure 3 gives this trace for three sets of schemas: s_1, s_2, and s_9; s_3, s_4, and s_{10}; and s_5, s_6, and s_{11} (see Figure 2). In each plot, the density (% of population) of each schema is plotted against time (generations). The density is sampled every 10 generations.

These plots show a striking phenomenon. In the top plot in Figure 3, s_1 and s_2 appear early and instances of them quickly combine to form s_9. Once each schema is discovered, its density in the population rises quite quickly to over 90% of the population by generation 60 or so. Around generation 220 there is a distinct dip in the densities of these three schemas.

The middle plot shows a very different evolution for s_3, s_4, and s_{10}. The schemas s_3 and s_4 are both present in the initial (randomly generated) population (though s_3's presence at generation 0 is not visible on the plot), but while s_4 rises quickly, s_3 dies out by generation 10, is fleetingly rediscovered (along with s_{10}) at generation 120 (see blip on the x-axis), and does not return until the very end of the run, at which point a mutation brings it (along with s_{10}) back (see blip on the x-axis). This same mutation is responsible for creating s_{opt} at generation 535, when the run ends. The schema s_4, after a quick initial rise, enters a pronounced dip at the same time the milder dip can be seen in the top plot of Figure 3, around generation 220.

What is the cause of these dips, and what prevents s_3 from persisting in the population? A likely answer can be inferred from the bottom plot. Schema s_6 appears around generation 30, rises fairly quickly, taking a sharp upturn around generation 220 and rising to about 95% of the population. Schema s_5 appears briefly around generation 20 (blip on the x-axis) and dies out, but appears again at generation 220. The instance of it in the population is also an instance of s_{11}, and instances of s_{11} rise very quickly. This rise exactly coincides with the minor dip in s_1, s_2, and s_9, and the major dip in s_4. What appears to be happening is the following:

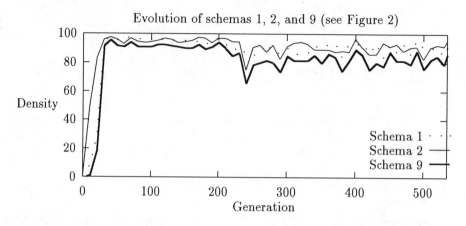

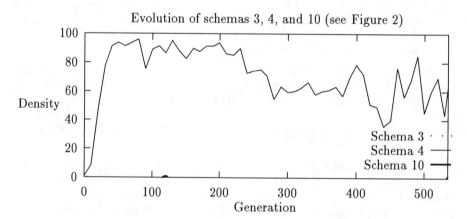

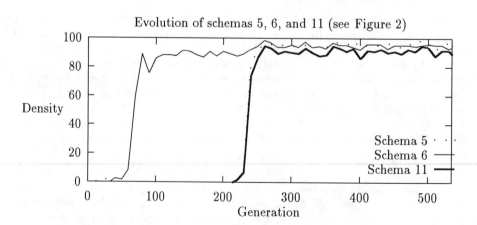

Figure 3: Evolution of three sets of schemas in a typical run of the GA on $R2$. In each plot, the density of each schema (% of population) is plotted against generation. Note that in the middle plot, schemas 3 and 10 are visible only as tiny bumps on the x-axis at generations 120 and 535.

in the first few instances of s_{11}, along with the 16 1's in the fifth and sixth blocks are several 0's in the first through fourth blocks. An instance of s_{11} has fitness $8 + 8 + 16 = 32$, whereas an instance of an order-8 schema such as s_4 has fitness 8. This difference causes s_{11} to rise very quickly compared to s_4, and instances of s_{11} with some 0's in the fourth block tend to push out many of the previously existing instances of s_4 in the population. This phenomenon has been called "hitchhiking", where 0's in other positions in the string hitchhike along with the highly fit s_{11}. The most likely positions for hitchhikers are those close to the highly fit schema's defined positions, since they are less likely to be separated from the schema's defined positions under crossover. Such effects are seen in real population genetics, and have been discussed in the context of GAs by Schraudolph and Belew (1990), and Das and Whitley (1989), among others. Note that this effect is pronounced even with the relatively weak form of selection used in our GA. (We also compared the GA's performance on $R1$ and $R2$ using a linear rank-scaling method (Baker, 1985) instead of the sigma-scaling method described above, and obtained results similar to those given in Table 1.)

The plots given in Figure 3 come from a single run, but this run was typical; the same type of phenomenon was observed on many of the other runs on $R2$ as well. Our hypothesis is that this hitchhiking effect is what causes the relatively slower times (on average) for the GA to find the optimum on $R2$. The power of crossover to combine lower-level building blocks was hampered, since some of the necessary building blocks were either partially or totally suppressed by the quick rise of disjoint building blocks. This suggests that there is more to characterizing a GA landscape than the absolute direction of the search gradients. In these functions, it is the actual differences in relative fitnesses for the different schemas that are relevant.

In $R1$, which lacks the extra fitness given to some intermediate-level schemas, the hitchhiking problem does not occur to such a devastating degree. The fitness of an instance of, say, s_{11} in $R1$ is only 16, so its discovery does not have such a dramatic effect on the discovery and persistence of other order-8 schemas in the function. Contrary to our initial intuitions, it appears that the extra reinforcement from some intermediate-level stepping stones actually harms the GA in these functions.

It might be thought that these results are due in part to sampling error: since the lowest order-building blocks are of length 8, a GA with a population of 128 has no samples of many of the lowest-order building blocks in the initial population (on average, 1/2 of the lowest-order blocks will not be represented), and thus has to wait until the order-8 building blocks are created by random variation. [1] To test the effect of this on our results, we performed two additional experiments: (1) we ran the GA on $R1$ and $R2$ with a population size of 1024, which gives 4 expected instances of each order-8 schema in the initial population; and (2) we ran the GA with population size 128 on modified versions of $R1$ and $R2$ in which the lowest-order building blocks were of length 4 rather than length 8. In this latter case, there are on average 8 instances of each order-4 building block in the initial population. The results from these two experiments are given in Tables 2 and 3. They show that these modifications, although improving the GA's absolute performance (especially

---

[1] Note: this fact caused our experiments (time to find the optimum) to have a higher-than-normal variance, which is why we performed at least 200 runs for most of our experiments.

POPULATION SIZE 1024		
	Function Evaluations to Optimum	
200 runs	$R1$	$R2$
Mean	37453 (std err: 868)	43213 (std err: 1275)
Median	34816	36864

Table 2: Summary of results of 200 runs of the GA with population size 1024 on $R1$ and $R2$.

LOWEST-ORDER SCHEMAS LENGTH 4		
	Function Evaluations to Optimum	
200 runs	$R1$	$R2$
Mean	6568 (std err: 198)	11202 (std err: 394)
Median	5760	9600

Table 3: Summary of results of 200 runs of the GA on modified versions of $R1$ and $R2$, in which the lowest-order building blocks are of length 4.

in the case of order-4 building blocks, which makes the function much easier to optimize), do not change the qualitative difference between the time to optimum for $R1$ and $R2$. This indicates that the difference is not primarily due to sampling error.

These results point to a pervasive and important issue in the performance of GAs: the problem of premature convergence. The fact that we observe a form of premature convergence even in this very simple setting suggests that it can be a factor in any GA search in which the population is simultaneously searching for two or more non-overlapping high fitness schemas (e.g., s_4 and s_{11}), which is often the case. The fact that the population loses useful schemas once one of the disjoint good schemas is found suggests one reason that the rate of effective implicit parallelism of the GA (Holland, 1975; Goldberg, 1989b) may need to be reconsidered. (For another discussion of implicit parallelism in GAs, see Grefenstette & Baker 1989.)

3.2 DO INTRONS SUPPRESS HITCHHIKERS?

In order to understand the hitchhiking behavior more precisely, we performed an experiment that we believed would eliminate it to some degree. Our hypothesis was that hitchhiking occurred in the loci that were spatially adjacent to the high-fitness schemas (e.g., s_{11} above). In order to reduce this effect, we constructed a new function, $R2_{introns}$, by introducing blocks of 8 "introns"—8 additional *'s—in between each of the 8-bit blocks of 1's. Thus in $R2_{introns}$, strings were of length 128 instead of 64. For example, in $R2_{introns}$, s_1 is 11111111**********...*, s_2 is ***************11111111**...*, and their combination, s_9, is 11111111********11111111**...*. The optimum is the string containing each block of 8 1's, where the blocks are each separated by eight loci that can contain either 0's or 1's. The idea here was that a potentially damaging hitchhiker would be at least 8-bits away from the schema on which it was hitchhiking, and would thus be likely to be lost under crossover. (Levenick, 1991, found that insert-

VARIANTS OF R2		
	Function Evaluations to Optimum	
200 runs	$R2_{introns}$	$R2_{flat}$
Mean	75599 (std err: 2697)	62692 (std err: 2391)
Median	70400	56448

Table 4: Summary of results of 200 runs of the GA on two variants of $R2$.

ing introns into individuals improved the performance of the GA in one particular set of environments.)

As shown in column 1 of Table 4, running the GA on $R2_{introns}$ yielded results not significantly different from those for $R2$. This was contrary to our expectations, and the reasons for this result are not clear, but one hypothesis is that once an instance of a higher-order schema (e.g., s_{11}) is discovered, convergence is so fast that hitchhikers are possible even in loci that are relatively distant from the schema's defined positions.

3.3 VARYING THE COEFFICIENTS IN $R2$

It is clear that some intermediate-level reinforcement is necessary for the GA to work. Consider $R1'$, a variant of $R1$, where $R1'(x) = 8$ if x is an instance of at least one of the 8-bit schemas, and $R1'(x) = 64$ if x is an instance of *all* the 8-bit schemas. Here the GA would have no reason to prefer a string with block of 16 1's over a string with a block of 8 1's, and thus there would be no pressure to increase the number of 1's. Intermediate schemas in $R1$ provide *linear* reinforcement, since the fitness of an instance of an intermediate-order schema is always the sum of the fitnesses of instances of the component order-8 schemas. Some schemas in $R2$ provide strong *nonlinear* reinforcement, since the fitness of an instance of, say, s_9 is much higher than the sum of the fitnesses of instances of the component order-8 schemas s_1 and s_2. Our results indicate that the nonlinear reinforcement given by some schemas is too high—it hurts rather than helps the GA's performance.

Does nonlinear reinforcement ever help the GA rather than hinder it? To study this we constructed a new function, $R2_{flat}$, with a much weaker nonlinear reinforcement scheme: for this function, c_1–c_{14} are each set to the flat value 1. Here the reinforcement is still nonlinear (an instance of s_9 will have fitness $1 + 1 + 1$, which is greater than the sum of the two components), but the amount of reinforcement is reduced considerably.

The results of running the GA on $R2_{flat}$ is given in the second column of Table 4. The average time to optimum for this function is approximately the same as for $R1$. Thus the smaller fitness advantage in $R2_{flat}$ does not seem to hurt performance, although it does not result in *improved* performance over that on $R1$.

These phenomena may be related to results by Feldman and his colleagues on the effects of super- and sub-multiplicative fitness functions on the evolutionary viability of crossover (Liberman & Feldman, 1986; Bergman & Feldman, 1990). However, there are several problems with applying Feldman's theorems directly. One problem is that Feldman studies the evolutionary viability of crossover rather than the degree to which crossover helps discover high-fitness individuals. Our work concentrates on the latter. We are currently investigating how these two concerns are related. (This was also studied by Schaffer and Eshelman, 1991.)

4 DISCUSSION

The results described in the previous two sections show that the GA's ability to process building blocks effectively depends not only on their presence, but also on their relative fitness. If some intermediate stepping stones are too much fitter than the primitive components, then premature convergence slows down the discovery of some necessary schemas. Simple introns and a very mild selection scheme do not seem to alleviate the premature convergence and hitchhiking problems.

Our results point out the importance of making the building-block hypothesis a more precise and useful description of building-block processing. While the disruptive effects that we observed (hitchhiking, premature convergence, etc.) are already known in the GA literature, there is as yet no theorem associating them with the building-block structure of a given problem.

In our experiments we have observed that the role of crossover varies considerably throughout the course of the GA search. In particular, three stages of the search can be identified: (1) the time it takes for the GA to discover the lowest-order schemas, (2) the time it takes for crossover to combine lower-order schemas into a higher-order schema, and (3) the time it takes for the higher-order schema to take over the population. In multi-level functions, such as the Royal Road functions, these phases of the search overlap considerably, and it is essential to understand the role of crossover and the details of schema processing at each stage (this issue has also been investigated by Davis, 1989, and by Schaffer & Eshelman, 1991, among others). In previous work, we have discussed the complexities of measuring the relative times for these different phases (Mitchell, Forrest, & Holland, 1992).

5 EXPERIMENTS WITH HILL-CLIMBING

As was mentioned earlier, part of our purpose in designing the Royal Road functions is to construct the simplest class of fitness landscapes on which the GA will not only perform well, but on which it will outperform other search methods such as hill-climbing. In addition to our experiments comparing the GA's performance on $R1$ and $R2$, we compared the GA's performance with that of three commonly used iterated hill-climbing schemes: steepest-ascent hill-climbing, next-ascent hill-climbing (Mühlenbein, 1991), and a scheme we call "random-mutation hill-climbing", that was suggested by Richard Palmer (personal communication). Our implementation of these various hill-climbing schemes is as follows:

- **Steepest-ascent hill-climbing (SAHC):**
    1. Choose a string at random. Call this string *current-hilltop*.
    2. Systematically mutate each bit in the string from left to right, recording the fitnesses of the resulting strings.
    3. If any of the resulting strings give a fitness increase, then set *current-hilltop* to the resulting string giving the highest fitness increase.

HILL-CLIMBING ON R2		
Function Evaluations to Optimum		
200 runs	SAHC	NAHC	RMHC
Mean	> 256,000 (std err: 0)	> 256,000 (std err: 0)	6551 (std err: 212)
Median	> 256,000	> 256,000	5925

Table 5: Summary of results of 200 runs of various hill-climbing algorithms on $R2$.

4. If there is no fitness increase, then save *current-hilltop* and go to step 1. Otherwise, go to step 2 with the new *current-hilltop*.

5. When a set number of function evaluations has been performed, return the highest hilltop that was found.

- **Next-ascent hill-climbing (NAHC):**

  1. Choose a string at random. Call this string *current-hilltop*.

  2. Mutate single bits in the string from left to right, recording the fitnesses of the resulting strings. If any increase in fitness is found, then set *current-hilltop* to that increased-fitness string, without evaluating any more single-bit mutations of the original string. Go to step 2 with the new *current-hilltop*, but continue mutating the new string starting after the bit position at which the previous fitness increase was found.

  3. If no increases in fitness were found, save *current-hilltop* and go to step 1.

  4. When a set number of function evaluations has been performed, return the highest hilltop that was found.

  Notice that this method is similar to Davis's "bit-climbing" scheme (Davis, 1991). In his scheme, the bits are mutated in a random order, and *current-hilltop* is reset to any string having *equal* or better fitness than the previous best evaluation.

- **Random-mutation hill-climbing (RMHC):**

  1. Choose a string at random. Call this string *best-evaluated*.

  2. Choose a locus at random to mutate. If the mutation leads to an equal or higher fitness, then set *best-evaluated* to the resulting string.

  3. Go to step 2.

  4. When a set number of function evaluations has been performed, return the current value of *best-evaluated*.

Table 5 gives results from running these three hill-climbing schemes on $R2$. In each run the hill-climbing algorithm was allowed to continue either until the optimum string was discovered, or until 256,000 function evaluations had taken place, and the total number of function evaluations performed was recorded. As can be seen, steepest-ascent and next-ascent hill-climbing never found the optimum during the allotted time, but random-mutation hill-climbing found the optimum on average more than ten times faster than the GA with population size 128, and more than six times faster than the GA with population size 1024. Note that random-mutation hill-climbing as we have described it differs from the bit-climbing method used by

Davis (1991) in that it does not systematically mutate bits, and it never gives up and starts from a new random string, but rather continues to wander around on plateaus indefinitely. Larry Eshelman (personal communication) has pointed out that the random-mutation hill-climber is ideal for the Royal Road functions—in fact much better than Davis's bit-climber—but will have trouble with any function with local minima. (Eshelman found that Davis's bit-climber does very poorly on $R1$, never finding the optimum in 50 runs of 50,000 function evaluations each.)

In addition to basing our comparison on the number of function evaluations to the optimum, we also compared the average on-line performance (De Jong, 1975) of the GA (population sizes 128 and 1024) with random-mutation hill-climbing, both running on $R2$. Figure 4 plots the results of that comparison. For a given run of the GA or of RMHC, the on-line performance at a given number of function evaluations is defined as the average value of *all* function evaluations made up to that point. We recorded the on-line performance values at intervals of 128 function evaluations, and repeated this procedure for 100 runs. We then averaged the on-line performance values, at each interval of 128 function evaluations, over all the runs. Thus each point on a plot in Figure 4 represents an average of on-line performance values over a number of runs. (The number of values averaged at each point varies: since the GA and RMHC stop when the optimum is found, different runs performed different numbers of function evaluations. We give only averages for which there were a significant number of values to average.)

It can be seen from the plots that RMHC significantly outperforms both versions of the GA under this measure as well.[2]

These results are a striking demonstration that, when comparing the GA with hill-climbing on a particular problem or test-suite, it matters *which* type of hill-climbing algorithm is used. Davis (1991) has also made this point.

The Royal Road functions were originally designed to serve two quite different purposes: (1) as an idealized setting in which to study building-block processing and the role of crossover, and (2) as an example of a simple function that distinguishes GAs from hill-climbing. While we have discovered that certain forms of hill-climbing outperform the GA on these functions (thus, they are inappropriate in exactly this form for the second purpose), they do fulfill the first purpose.

6   CONCLUSIONS AND FUTURE DIRECTIONS

The research described in this paper is an initial step in understanding more precisely how schemas are processed under crossover. By studying the GA's behavior on simple landscapes in which the desirable building blocks are explicitly defined, we have discovered some unanticipated phenomena related to the GA's ability to process schemas efficiently, even in nondeceptive functions. The Royal Road functions capture, in an idealized and clear way, some landscape features that are particularly relevant for the GA, and we believe that a thorough understanding of the GA's behavior on these simple landscapes will be very useful in developing more detailed and useful theorems about GA behavior.

---

[2]It is interesting to note that under this measure, the 128 population GA outperforms the 1024 population GA in the initial stages of the run.

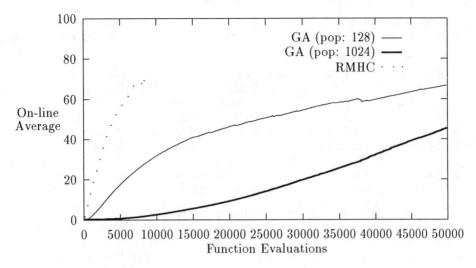

Figure 4: Plots of the average on-line performance of the GA (population sizes 128 and 1024) and of random-mutation hill-climbing (RMHC), over 100 runs. The plot for RMHC stops at around 6000 function evaluations because RMHC had almost always found the function optimum by that time.

The research reported here is work in progress, and there are several directions for future investigation. Here we sketch some of our short and longer range research plans.

In the short term, we plan to study more carefully the bottlenecks in the discovery of desirable schemas, and to quantify more precisely the relationship between the fitness values of the various building blocks and the degree to which these bottlenecks will occur. Hitchhiking is evidently one bottleneck, and we need to understand better in what way it is occurring and under what circumstances. Once we have described the phenomena in more detail, we can begin developing a mathematical model of the schema competitions we observe (illustrated in Figure 3) and how they are affected by different building-block fitness schemes. This model may be related to models proposed by Vose and Liepins (1991).

Our hitchhiking results need to be further analyzed and explained, and we plan a more detailed analysis of the different effects of various nonlinear reinforcement schemes. In particular, more details are needed in the comparison of GA performance on $R1$ with performance on $R2_{flat}$, and on other coefficient schemes.

We believe that there are versions of "royal-road" landscapes that will fulfill our goal of finding simple functions that distinguish GAs from hill-climbing. For example, we plan to try the following variants: adding noise, including all combinations of lower-order schemes in the explicit list of schemas, and allowing schemas to overlap.

The Royal Road functions explore only one type of landscape feature that is of relevance to GAs: the presence and relative fitnesses of intermediate-order building blocks. Our longer-term plans include extending the class of fitness landscapes under investigation to include other types of relevant features; some such features were described by Mitchell, Forrest, and Holland (1992). We are also interested in developing statistical measures that could determine the presence or absence of the features of interest. These might be related to work on determining the correlation structure of fitness landscapes (see Kauffman, 1989; Lipsitch, 1991; and Manderick, de Weger, and Spiessens, 1991). If such measures could be developed, they could be used to help predict the likelihood of successful GA performance on a given landscape.

Acknowledgments

The research reported here was supported by the Center for Nonlinear Studies at Los Alamos National Laboratory, Associated Western Universities, and NSF grant IRI-9157644 (support to S. Forrest); the Michigan Society of Fellows and the University of Michigan EECS Department (support to M. Mitchell); and the Santa Fe Institute (support to both authors). John Holland has collaborated with us in much of this work, and is a constant source of invaluable assistance and advice. We thank Richard Palmer for suggesting the "random-mutation hill-climbing" algorithm, and Larry Eshelman for comments and reports of experiments on various hill-climbing schemes. We also thank Robert Axelrod, Aviv Bergman, Arthur Burks, Michael Cohen, Marcus Feldman, Rick Riolo, and Carl Simon for many helpful discussions on these issues.

References

J. E. Baker (1985). Adaptive selection methods for genetic algorithms. In J. J. Grefenstette (Ed.), *Proceedings of the First International Conference on Genetic Algorithms and Their Applications*. Hillsdale, NJ: Lawrence Erlbaum Associates.

A. Bergman and M. W. Feldman (1990). More on selection for and against recombination. *Theoretical Population Biology, 38(1)*, 68–92.

R. Das and D. Whitley (1991). The only challenging problems are deceptive: Global search by solving order-1 hyperplanes. In R. K. Belew and L. B Booker (Eds.), *Proceedings of the Fourth International Conference on Genetic Algorithms*. San Mateo, CA: Morgan Kaufmann.

L. D. Davis (1989). Adapting operator probabilities in genetic algorithms. In J. D. Schaffer (Ed.), *Proceedings of the Third International Conference on Genetic Algorithms*. San Mateo, CA: Morgan Kaufmann.

L. D. Davis (1991). Bit-climbing, representation bias, and test suite design. In R. K. Belew and L. B Booker (Eds.), *Proceedings of the Fourth International Conference on Genetic Algorithms*. San Mateo, CA: Morgan Kaufmann.

K. A. De Jong (1975). *An Analysis of the Behavior of a Class of Genetic Adaptive Systems*. Unpublished doctoral dissertation, University of Michigan, Ann Arbor, MI.

S. Forrest and M. Mitchell (1991). The performance of genetic algorithms on Walsh polynomials: Some anomalous results and their explanation. In R. K. Belew and L. B. Booker (Eds.), *Proceedings of the Fourth International Conference on Genetic Algorithms*. San Mateo, CA: Morgan Kaufmann.

D. E. Goldberg (1987). Simple genetic algorithms and the minimal deceptive problem. In L. D. Davis (Ed.), *Genetic Algorithms and Simulated Annealing* (Research Notes in Artificial Intelligence). Los Altos, CA: Morgan Kaufmann.

D. E. Goldberg (1989a). Genetic algorithms and Walsh functions: Part II, Deception and its analysis. *Complex Systems*, 3:153–171.

D. E. Goldberg (1989b). *Genetic Algorithms in Search, Optimization, and Machine Learning*. Reading, MA: Addison Wesley.

J. J. Grefenstette and J. E. Baker (1989). How genetic algorithms work: A critical look at implicit parallelism. In J. D. Schaffer (Ed.), *Proceedings of the Third International Conference on Genetic Algorithms*. San Mateo, CA: Morgan Kaufmann.

J. H. Holland (1975). *Adaptation in Natural and Artificial Systems*. Ann Arbor, MI: The University of Michigan Press.

S. A. Kauffman (1989). Adaptation on rugged fitness landscapes. In D. Stein (Ed.), *Lectures in the Sciences of Complexity*, 527–618. Reading, MA: Addison-Wesley.

J. R. Levenick (1991). Inserting introns improves genetic algorithm success rate: Taking a cue from biology. In R. K. Belew and L. B. Booker (Eds.), *Proceedings of the Fourth International Conference on Genetic Algorithms*. 123–127. San Mateo, CA: Morgan Kaufmann.

U. Liberman and M. W. Feldman (1986). A general reduction principle for genetic modifiers of recombination. *Theoretical Population Biology, 30(3)*, 341–371.

G. E. Liepins and M. D. Vose (1990). Representational issues in genetic optimization. *Journal of Experimental and Theoretical Artificial Intelligence, 2*, 101-115.

M. Lipsitch (1991). Adaptation on rugged landscapes generated by local interactions of neighboring genes. In R. K. Belew and L. B. Booker (Eds.), *Proceedings of the Fourth International Conference on Genetic Algorithms*. San Mateo, CA: Morgan Kaufmann.

B. Manderick, M. de Weger, and P. Spiessens (1991). The genetic algorithm and the structure of the fitness landscape. In R. K. Belew and L. B. Booker (Eds.), *Proceedings of the Fourth International Conference on Genetic Algorithms*. San Mateo, CA: Morgan Kaufmann.

M. Mitchell, S. Forrest, and J. H. Holland (1992). The royal road for genetic algorithms: Fitness landscapes and GA performance. In *Proceedings of the First European Conference on Artificial Life*. Cambridge, MA: MIT Press/Bradford Books.

H. Mühlenbein (1991). Evolution in time and space—The parallel genetic algorithm. In G. J. E. Rawlins (Ed.), *Foundations of Genetic Algorithms*, 316–337. San Mateo, CA: Morgan Kaufmann.

J. D. Schaffer and L. J. Eshelman (1991). On crossover as an evolutionarily viable strategy. In R. K. Belew and L. B. Booker (Eds.), *Proceedings of the Fourth International Conference on Genetic Algorithms*, 61–68. San Mateo, CA: Morgan Kaufmann.

N. N. Schraudolph and R. K. Belew (1990). Dynamic parameter encoding for genetic algorithms. CSE Technical Report CS 90-175. Computer Science and Engineering Department, University of California, San Diego.

R. Tanese (1989). *Distributed Genetic Algorithms for Function Optimization*. Unpublished doctoral dissertation, University of Michigan, Ann Arbor, MI.

M. Vose and G. Liepins. (1991). Punctuated equilibria in genetic search. *Complex Systems 5*, 31–44.

L. D. Whitley (1991). Fundamental principles of deception in genetic search. In G. Rawlins (Ed.), *Foundations of Genetic Algorithms*. San Mateo, CA: Morgan Kaufmann.

PART 4

CONVERGENCE AND GENETIC DIVERSITY

Accounting for Noise in the Sizing of Populations*

David E. Goldberg, Kalyanmoy Deb, and James H. Clark
Department of General Engineering
University of Illinois at Urbana-Champaign
117 Transportation Building
104 South Mathews Avenue
Urbana, IL 61801

Abstract

This paper considers the effect of noise on the quality of convergence of genetic algorithms (GAs). A population-sizing equation is derived to ensure that minimum signal-to-noise ratios are favorable to the discrimination of the best building blocks required to solve a problem of bounded deception. In five test problems of varying degrees of nonlinearity, nonuniform scaling, and nondeterminism, the sizing relation proves to be a conservative predictor of average correct convergence. These results suggest how the sizing equation may be viewed as a coarse delineation of a boundary between two distinct types of GA behavior. Besides discussing a number of extensions of this work, the paper discusses how these results may one day lead to rigorous proofs of convergence for recombinative GAs operating on problems of bounded deception.

1 Introduction

This paper considers a single question that has puzzled both novice and experienced genetic-algorithm (GA) users alike: how can populations be sized to promote the selection of correct (global) building blocks? The answer comes from statistical

---

*Portions of this paper are excerpted from a paper by the authors entitled "Genetic Algorithms, Noise, and the Sizing of Populations" (Goldberg, Deb, & Clark, 1991).

decision theory and requires us to examine building-block signal differences in relation to population noise. In the remainder, a simple population-sizing equation is derived and is used to calculate population sizes for a sequence of test functions displaying varying degrees of nonlinearity, nonuniform fitness scaling, and nondeterminism. The simple sizing equation is shown to be a conservative yet rational means of estimating population size. Extensions of these calculations are also suggested, with the possibility that these methods may be used to develop fully rigorous convergence proofs for recombinative GAs in problems of bounded deception.

2   Population Sizing in the Presence of Noise

Holland's (1970) identification of schemata (building blocks) as the unit of selection and specification of a bound on their expected growth (Holland, 1975) has provided a clear picture of the conditions necessary for successful discovery. With a better understanding of the workings of GAs, we are at some point led to consider the accuracy of the decision making of selection as partial solutions are forced to compete with one another. When viewed in this way, the problem of choosing between competing building blocks becomes a fairly well posed problem in statistical decision theory. A historical review of past efforts connected to building-block decision making and population sizing is presented in the original study (Goldberg, Deb, & Clark, 1991). Here, we are concerned primarily with deriving and using the population-sizing equation.

We start by considering two competing building blocks; call them H_1 (with mean fitness f_{H_1} and fitness variance $\sigma^2_{H_1}$) and H_2 (with mean fitness f_{H_2} and fitness variance $\sigma^2_{H_2}$). With enough samples, the mean fitness approaches a normal distribution as guaranteed by the central limit theorem. Pictorially, the situation we face with a single sample of each of two normally distributed schemata is displayed in figure 1. Clearly schema H_1 is the better of the two, and assuming that the problem is not deceptive or that we are considering a sufficiently high-order schema in which deception is no longer an issue, we hope to choose strings that represent H_1 more often than those that represent H_2. With a single sample in the pictured event, we can calculate the probability that the worse schema is better than a particular fitness value f' by finding the area of the shaded region. The overall probability that the sample fitness of the second-best schema is higher than the sample fitness of the best schema may be calculated by accumulating the above probability for all possible values of f'. This computation is called the *convolution* of the two distributions; conveniently, the convolution of the two normal distributions is itself normal: the mean of the convolution is calculated as the difference in the means of the two individual distributions and the variance of the convolution is simply the sum of the individual variances. Thus, defining the *signal difference* $d = f_{H_1} - f_{H_2}$ and calculating the mean variance of the two building blocks as $\sigma^2_M = (\sigma^2_{H_1} + \sigma^2_{H_2})/2$, the probability of making an error on a single trial of each schema may be calculated by finding the probability α such that $z^2(\alpha) = d^2/(2\sigma^2_M)$, where $z(\alpha)$ is the ordinate of a unit, one-sided, normal deviate. Henceforth, we will drop the α and simply recognize z as the tail deviate value at a specified error probability.

If one sample of each building block were all we were permitted, it would be difficult to discriminate between all but the most widely disparate building blocks.

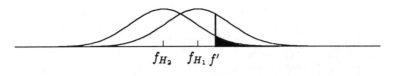

Schema fitness

Figure 1: Overlapping distributions of competing schemata permit the possibility of making errors in decisions, especially when only one sample of each schema is taken.

In population-based approaches such as genetic algorithms, we are able to sample simultaneously multiple representatives of building blocks. Thus, as we take more samples, the standard deviation of the mean difference becomes tighter and tighter, meaning that we can become more confident in our ability to choose better building blocks as the population size increases. To put this into practice for particular competitors in a partition of given cardinality, we recognize that the variance of the mean goes as the variance of a single trial divided by the number of samples. Because the likely number of samples in a uniformly random population of size n is simply the population size divided by the number of competing schemata κ in the partition to which the two schemata belong. The corresponding relationship to obtain discrimination with an error rate α may be written as

$$z^2 = \frac{d^2}{2\sigma_M^2/n'},\tag{1}$$

where $n' = n/\kappa$. Calling z^2 the coefficient c (also a function of α) and rearranging, we obtain a fairly general population-sizing relation as follows:

$$n = 2c\kappa\frac{\sigma_M^2}{d^2}.\tag{2}$$

Thus, for a given pairwise competition between schemata, the population size varies inversely with the square of the signal difference that must be detected and proportionally to the product of the number of competitors in the competition partition, the total building-block error, and a constant that increases exponentially with decreasing permissible error (Goldberg, Deb, & Clark, 1991). Thus, to use this equation conservatively, we must size the population for those schemata that may be deceptive and have the highest value of $\kappa\sigma_M^2/d^2$.

Many readers will recognize the previous derivation as containing elements of Holland's (1973) 2^k-armed bandit argument and De Jong's (1975) decision-theory calculation, with the important exception that here we do not try to optimize the trial allocation. Instead, we concentrate on getting the decision making, and thus the convergence, correct, and later ask how many function evaluations are required. We believe this apparently small difference in emphasis is crucially important in answering the questions of interest to the genetic algorithms community.

2.1 Other sources of noise

The equation derived above is fairly general; however, we have assumed that all the noise faced by the schemata comes from the variance of fitness within the population. Although this is largely true in many problems, GAs may face noise from a variety of sources, including inherently noisy problems, noisy selection algorithms, and the variance of other genetic operators. The sizing equation remains valid even in cases where these sources are significant with respect to the collateral noise if we adjust the variance by including a multiplier for each of the additional sources of stochasticity. For the ith source of noise (call it n_i) with magnitude $\sigma_{n_i}^2$, we can define the relative noise coefficient

$$\rho_{n_i}^2 = \frac{\sigma_{n_i}^2}{\sigma_M^2}. \tag{3}$$

Thereafter, the total additional relative noise coefficient may be calculated $\rho_T^2 = \sum_i \rho_{n_i}^2$, and the modified population-sizing relation may be obtained:

$$n = 2c(1 + \rho_T^2)\kappa\gamma^2, \tag{4}$$

where $\gamma^2 = \sigma_M^2/d^2$, the mean squared inverse overall signal-to-noise ratio.

2.2 Specializing the sizing equation

The general relationship derived above is widely applicable—perhaps too widely applicable if one of our aims is to see how the error-limiting population size varies with the difficulty or length of the problem. To understand these factors better, we specialize the equation somewhat. Consider strings of length ℓ over alphabets of cardinality χ, and assume that the function is of bounded deception in that building blocks of some order $k \ll \ell$ containing the global optimum are superior to their competitors. Focusing on the highest order partitions is conservative, and each one contains $\kappa = \chi^k$ competitors. It is convenient (but not necessary) to view the function as the sum of m independent nonoverlapping subfunctions, f_i, each of the same order, k, of the most deceptive partition, thus giving $m = \ell/k$ (for simplicity we are assuming that ℓ is divisible by k). The root-mean-squared (RMS) subfunction variance may be calculated as follows:

$$\sigma_{rms}^2 = \frac{1}{m}\sum_{i=1}^{m}\sigma_{f_i}^2. \tag{5}$$

Then we estimate the variance of the average order-k schema by multiplying the RMS value by $m - 1$:

$$\sigma_M^2 = (m - 1)\sigma_{rms}^2. \tag{6}$$

Using $m - 1$ recognizes that the fixed positions of a schema do not contribute to variance, although the conservative nature of the bound would not be upset by using m. Substituting this value together with the cardinality of the partition into the sizing equation yields

$$n = 2c\beta^2(1 + \rho_T^2)m'\chi^k, \tag{7}$$

where $m' = m - 1$ and $\beta^2 = \sigma_{rms}^2/d^2$, the squared RMS subfunction inverse signal-to-noise ratio. The bounds on subfunction variance for a number of fitness cases are derived in the original study (Goldberg, Deb, & Clark, 1991). It suffices here to note that the variance of a function lies anywhere from a minimum value $(f_{max} - f_{min})^2/(2\chi^k)$ to a maximum value $(f_{max} - f_{min})^2/4$.

Assuming fixed c, β, and ρ_T, we note that the sizing equation 7 is $O(m\chi^k)$. If the problems we wish to solve are of bounded and fixed deception (fixed k for given alphabet cardinality regardless of string length), we note that population sizes are $O(m)$, and recalling that $m = \ell/k$, we concluded that $n = O(\ell)$. Elsewhere (Goldberg & Deb, 1991), it has been shown that the typical scaled or ranked selection schemes used in GAs converge in $O(\log n)$ generations, and unscaled proportionate schemes converge in $O(n \log n)$ time. For the faster of the schemes, this suggests that GAs can converge in $O(\ell \log \ell)$ function evaluations even when populations are sized to control error. Moreover, even if we use the slower of the schemes, and imagine that the m building blocks converge one after another in a serial fashion and require α to decrease as m^{-1}, GA convergence should be no worse than an $O(\ell^2 \log^3 \ell)$ affair. We will examine the rapid and accurate convergence that results from appropriate population sizing in the next section.

Further detail and a gentler derivation are available elsewhere (Goldberg, Deb, & Clark, 1991). In the subsequent sections, we refer to that study as the 'complete study'. In the next section, we apply a simple GA to a sequence of test functions designed to test the efficacy of the population sizing.

3   Testing the Population-sizing Equation

In this section, we test the hypothesis that the population-sizing equation derived in the previous section is a conservative aid to reducing errors in building-block selection. We do this by first drawing a somewhat tighter connection between average generational decision error and building-block convergence. We then discuss the design of a suite of problems that test the population-sizing relation across a range of problems that are linear or nonlinear, deterministic or inherently noisy, or uniformly or nonuniformly scaled.

3.1   Connection between generational error and ultimate convergence

Earlier, we took a generational viewpoint of decision making and calculated a population size to control the error of decision for a pair of competing building blocks. We must proceed from this generational perspective to the viewpoint at the end of a run. Calling S the event that we succeed in converging to the right competing building block at the end of a run, M the event in which we make a mistake in choosing the correct building block during the initial generation, and C the event that we choose correctly during the initial generation, we can calculate the success probability as follows:

$$P(S) = P(S|M)P(M) + P(S|C)P(C). \tag{8}$$

The interaction between ultimate success and initially correct or incorrect decision making is fairly complex, but we can reason simply as follows. If we choose correctly initially, the probability that we converge correctly is nearly one. On the other hand, the greatest chance for making a mistake comes after an initial error, because we have stepped in the wrong direction. Although it is possible (and sometimes even fairly probable) to recover from such initial mistakes, we conservatively ignore such recovery, and get a straightforward bound on ultimate success probability. Setting $P(S|M) = 0$ and $P(S|C) = 1$, and recognizing that $P(C)$ is at least as large as $1 - \alpha$, we obtain

$$P(S) = 1 - \alpha. \tag{9}$$

We define the confidence factor $\zeta = 1 - \alpha$ and plot various convergence measures (usually proportion of building blocks correct) against ζ. Since the chances of getting better than $P(S) = \zeta$ convergence is substantial, the measure of whether the population sizing is conservative will simply be that empirical data fall somewhere above the 45 degree line. In what follows, we call the $P(S) = \zeta$ line the expected lower bound (or expected LB), but we recognize here that it is fairly coarse.

3.2 Test suites and simulation results

To test the population-sizing equation, we consider a simple GA run using various population sizes on a test suite of five real-valued functions over bit strings with various levels of stochasticity, nonlinearity, and fitness scaling[1]. Here, we present representative graphical results from three of the five problems and discuss important highlights of other results.

We choose our simple GA carefully to bound the results expected in a range of GAs used in practice (Goldberg, 1989). To examine whether the type of selection significantly affects the quality of convergence, we try a number of schemes to start, including many of those in wide use. In subsequent tests, we restrict our experiments to tournament selection as a good compromise between quick answers and quality convergence. In all runs, simple, one-point crossover has been adopted and no mutation ($p_m = 0$) was used to ensure that initial diversity provided the only means of solving a problem. All runs are terminated when the population converges completely, and to obtain a statistically valid result, all simulations are performed ten times, each starting with different random-number-generator seeding.

3.3 Test function F1: A uniform, linear problem

The initial function chosen to test the population-sizing relation is the uniform linear problem:

$$f_1(\mathbf{x}) = \sum_{i=1}^{\ell} x_i \tag{10}$$

where $x_i \in \{0, 1\}$. This is, of course, the so-called one-max function, and its solution is the string with all ones. Since the problem is linear, the critical building block is of order one; the signal we wish to detect has magnitude $1 - 0 = 1$, and the variance of the order-1 variance is simply $(1 - 0)^2/4 = 0.25$, using the variance estimates of the previous section. Thus $\beta^2 = 0.25/1 = 0.25$, and the overall sizing relation becomes $n = c(\ell - 1)$.

---

[1]The term fitness scaling refers to the relative contribution of building blocks in the fitness function.

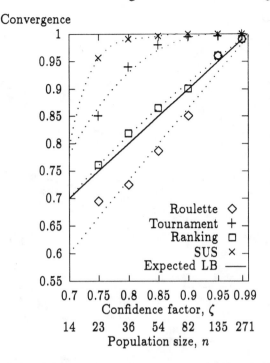

Figure 2: Simulation results for F1 with $\ell = 50$ are presented on a graph of convergence as measured by the average number of correct alleles versus confidence and population size. On all but unranked roulette-wheel selection, the graph shows that the sizing equation is conservative even when no additional sources of stochasticity are considered.

In the complete study, we used three string-length values, $\ell = 20, 50, 200$. In this paper, complete simulation results with $\ell = 50$ are presented. To give the GA a good workout, we have tested F1 with a variety of selection operators: roulette-wheel selection (roulette); roulette-wheel selection with ranking (ranking); stochastic universal selection (SUS); and binary tournament selection without replacement (tournament). Roulette-wheel selection is the usual Monte-Carlo scheme with replacement, where the selection probability $p_i = f_i / \sum_j f_j$. The ranked selection scheme uses linear ranking (where the population is linearly ranked according to the fitness of strings so that the best string is assigned two copies and the worst string is assigned zero copy) and Monte-Carlo selection, and the SUS scheme uses this well-known low-noise procedure. Tournament selection is performed without replacement in an effort to keep the selection noise as low as possible.

Figure 2 shows convergence versus confidence factor (and population size) for $\ell = 50$. The increasing conservatism of the sizing relation with increased n is not unexpected. The lower bound, relating confidence and ultimate convergence ignores all possibility of correcting an initial error. As n increases, drift time for poorly discriminated building blocks increases (Goldberg & Segrest, 1987), thereby increasing the probability that a correction can be obtained.

Function evaluations

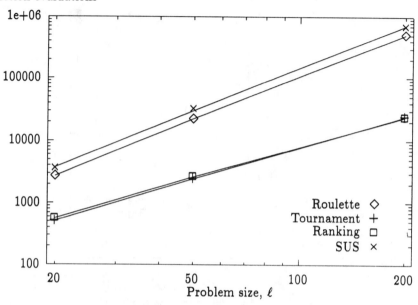

Figure 3: The total number of function evaluations for each selection scheme is graphed versus ℓ value on log-log axes at $\zeta = 0.9$ for function F1. The total number of function evaluations varies approximately as $\ell^{1.7}$ in the pushy (ranking and tournament) selection schemes and $\ell^{2.3}$ in the purely proportionate (SUS and roulette) schemes.

Among the most striking features of these results is that the roulette-wheel traces fall below the 45-degree line. Since roulette-wheel selection is most certainly noisy, the predictive performance of the sizing equation can be improved through a ρ_f^2 adjustment that accounts for the noise of selection (Goldberg, Deb, & Clark, 1991).

The second most striking feature of the F1 results is the high performance of the two quiet selection schemes, SUS and tournament. This is not unexpected, but the reason for the superiority of SUS in most of the cases is unclear without further investigation. Figure 3 shows the total number of function evaluations versus confidence factor for all schemes and string lengths $\ell = 20, 50, 200$. Clearly, the superiority of SUS is bought at high computational cost. It is well known (Goldberg & Deb, 1991) that purely proportionate schemes tend to slow as average fitness rises, but this has a beneficial effect on the quality of convergence, because less pressure is applied to force bad decisions. On the other hand, this increases substantially the total number of function evaluations, and in the remainder of the study we will concentrate on tournament selection as a good compromise between quality and speed of convergence. Looking at these results more closely, the number of function evaluations grows proportionally to $\ell^{1.7}$ for the two pushy schemes (ranking and tournament), and the number of function evaluations grows roughly as $l^{2.3}$ for the two purely proportionate schemes (SUS and roulette). Recall (Goldberg &

Deb, 1991) that ranked and tournament schemes tend to converge in something like $O(\log n)$ generations and that purely proportionate schemes tend to converge in $O(n \log n)$ time; overall, we should expect a total number of function evaluations of $O(\ell \log \ell)$ to $O(\ell \log^2 \ell)$ for the pushy schemes, which is consistent with the observed $\ell^{1.7}$, and we should expect convergence of $O(\ell^2 \log \ell)$ to $O(\ell^2 \log^3 \ell)$ for the two purely proportionate schemes, which is consistent with the observed $\ell^{2.3}$. The consistency of these results gives us some hope that these suggestions about convergence and its time complexity can be taken to theoremhood, a matter to be discussed somewhat later.

3.4   Test function F2: A nonuniform, linear problem

In the complete study, function F2 is a 50-bit linear function with nonuniform scaling of bits. Five bits are assigned a value $\delta < 1$ each to test if the sizing equation can pick up the small signal amidst a large collateral noise. Using tournament selection with all other GA parameters and operators as discussed earlier, blocks of simulations are run for $\delta = 0.4, 0.6,$ and 0.8. In all runs at each value of ζ, the sizing equation is found to be a conservative predictor of population size.

3.5   Test function F3: A uniform, linear function with added noise

For the third test function, we consider another linear function, except this time we add zero-mean Gaussian noise:

$$f_3(\mathbf{x}) = \sum_{i=1}^{50} x_i + g(\sigma_n^2), \tag{11}$$

where $x_i \in \{0, 1\}$ and $g(\sigma_n^2)$ is a generator of zero-mean Gaussian noise of specified variance σ_n^2.

The sizing relation is the same as in F1, except that a factor ρ_T^2 must be used to account for the noise. Four different levels of noise $\sigma_n^2 = 12.25, 24.5, 49.0,$ and 98.0 were added, and these correspond to ρ_T^2 values of 1, 2, 4, and 8. Convergence (over all bits) versus confidence factor ζ is shown in figure 4, for blocks of ten simulations on each σ_n^2-ρ_T^2 case. The sizing relation is conservative in all four cases; as before, increasing conservatism is observed with increasing n.

Table 1: Copies of this subfunction are used in functions F4 and F5. Here u is the *unitation* or the number of ones in the subfunction's substring.

$u(x)$	$f_4'(x)$
0	3
1	2
2	1
3	0
4	4

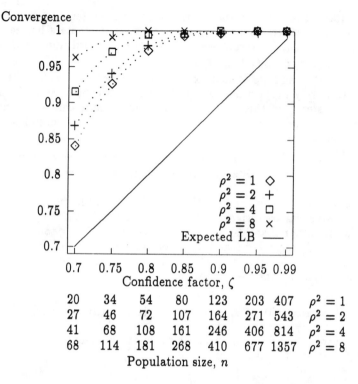

Figure 4: F3 convergence as measured by the average number of ones versus confidence value shows that the population-sizing equation adequately handles noisy problems when adjustment is made for the additional stochasticity.

3.6 Test function F4: A uniformly scaled, nonlinear function

In order to study variance-based population sizing in nonlinear problems, a 40-bit, order-four deceptive problem has been designed.

$$f_4(\mathbf{x}) = \sum_{i=1}^{10} f_4'(x_{I_i}), \qquad (12)$$

where the subfunction f_4' is shown in table 1, and the sequence of index sets is the ten sets containing four consecutive integers each: $I_1 = \{1, 2, 3, 4\}$, and $I_{i+1} = I_i + 4$. The function is a function of unitation, $u(x)$ (a function of the number of ones in the substring argument), and elsewhere (Deb & Goldberg, 1991) this function has been shown to be fully deceptive by calculating schema average fitness values. The variance of the subfunction may be calculated directly and is found to be 1.215. Recognizing that there are ten subfunctions ($m = 10$), each binary subfunction is of order four ($\chi = 2$, $k = 4$), and the fitness difference between the best and the second best substring is one ($d = 1$), the population-sizing equation reduces to $n = 2c(1.215)(10 - 1)2^4/(1^2) = 350c$.

To eliminate building-block disruption as a concern, each subfunction is coded

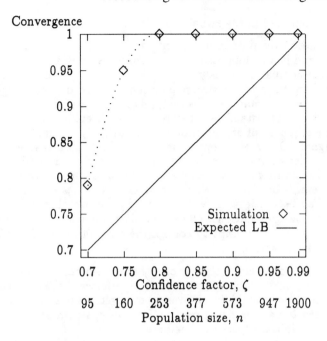

Figure 5: F4 convergence as measured by the average number of correct building blocks versus the confidence factor shows that the sizing equation conservatively bounds the actual convergence in a fairly difficult, albeit uniformly scaled, deceptive problem.

tightly, and tournament selection is used with all other GA operators and parameters set as in previous runs. Figure 5 shows convergence measured by the average number of correct building blocks versus the confidence factor. Once again, the sizing equation conservatively bounds final convergence.

3.7 Function F5: A nonuniformly scaled, nonlinear problem

In the complete study, function F5 is identical to F4, except that the fifth subfunction is scaled down by a factor four. Starting with $\zeta = 0.7$, in all runs at each value of ζ, the GA converges to the correct (all-ones) string. In each case, the sizing equation is found to be a conservative population-sizing tool as well.

4 Extensions

The simple population-sizing equation presented in this paper has proven to be a usefully conservative estimate of the population size required to make a controllably small number of building-block errors at the end of a run. A number of applications and extensions to this work are suggested in the complete study. A few of the more important are outlined in the remainder of this section.

The sizing relation requires some (albeit minimal) knowledge about the problem being solved, and it may be possible to get online estimates of the necessary values through online population measurements. Specifically, the sizing relation requires information about the problem size, population variance, minimum signal, and order of deception. Variance may be measured directly and used straightaway. Minimum desired signal can be established beforehand, or keeping track of the change of fitness after a sequence of one-position mutations can give an adequate estimate of minimum signal. Order of deception is more difficult to measure. Again, a prior limit on the order of maximum deception to be uncovered can be established, or it may be possible to get some estimate of deception by doing recursive updates of schema averages or Walsh coefficients as more samples are taken. The schema averages or Walsh coefficients may then be used to see whether there is any evidence of deception in past populations. Once these data are available, the population size may be adjusted in an attempt to control the error of decision, yet keep no more copies than is necessary.

The sizing equation deserves testing on other-than-binary codings, although the assumptions used in its derivation are so straightforward that the success demonstrated in this paper should carry over to other structures without modification. We have started to use the sizing relation in problems with permutation operators and codings; our initial experience has been positive.

The relation between specified error and ultimate convergence adopted herein is conservative, but it should be possible to develop a more fundamental relation between the two. One thing that aids convergence is that variance in the first generation is something of a worst case. As positions converge, less fitness variance is felt by the remaining competitors, and the environment of decision is much less noisy. Also, as population sizes are increased, convergence is aided, because drift times increase linearly with size (Goldberg & Segrest, 1987); those building blocks in the noise soup—those with relatively unfavorable signal-to-noise ratios—have a longer time to drift around before randomly converging to one value or another. It should be possible to construct asymptotic models that more closely relate these effects without resorting to full Markov equations.

This paper has simply scratched the surface in its investigation of sources of noise other than collateral or building block noise. Beyond the additive Gaussian noise considered here lie other noisy objective functions, and these should be examined to see if the simple variance adjustment is sufficient. The prior expectation is that the adjustment should work because the central limit theorem works; however the question deserves less flippancy and closer inquiry. Also, the noise generated by various selection schemes should be investigated, as should the noise generated by other genetic operators. A crossover operator that disrupts a short schema more than expected can be deleterious to convergence and can also cause errors of decision. Similarly, a mutation operator that hits a low-order schema more often than average can be a problem. These effects should be studied more carefully, and ultimately they can be incorporated into a variance-adjusted schema theorem.

Finally, by getting the decision making in GAs right, we feel we have opened the door to straightforward, yet rigorous, convergence proofs of recombinative GAs.

Elsewhere (Goldberg & Rudnick, 1991), it was pointed out that the schema theorem could be made a rigorous lower bound on schema growth if only the various terms were adjusted conservatively for variance effects. We stand by that claim here and suggest that the result can be pushed even further to obtain proofs of polynomial convergence within an epsilon of probability one in problems of bounded deception. The actual proofs will resemble those of computational learning theory, and although a number of technical details appear fairly tricky, getting the decision making correct in a probabilistic sense is a critical piece of this important puzzle.

5   Conclusions

This paper has developed and tested a population-sizing equation to permit accurate statistical decision making among competing building blocks in population-oriented search schemes such as genetic algorithms. In a suite of test functions from linear to nonlinear, from deterministic to stochastic, and from uniformly scaled to poorly scaled, the population-sizing relation conservatively has bounded the actual accuracy of GA convergence when necessary sources of stochasticity are properly considered and the worst-case signal-to-noise ratio is used in sizing. The paper has also examined the total number of function evaluations required to solve problems accurately. Depending whether purely proportionate selection or more pushy schemes such as ranking and tournament selection have been used, convergence is no worse than a quadratic or cubic function of the number of building blocks in the problem. These results are consistent with previous theoretical predictions of GA time complexity and open the door to formal proofs of polynomial GA convergence in problems of bounded deception, using the basic approach of this paper together with methods similar to those established in computational learning theory.

Acknowledgments

The authors acknowledge the support provided by the US Army under Contract DASG60-90-C-0153 and by the National Science Foundation under Grant ECS-9022007.

References

De Jong, K. A. (1975). An analysis of the behavior of a class of genetic adaptive systems. (Doctoral dissertation, University of Michigan). *Dissertation Abstracts International, 36(*10), 5140B. (University Microfilms No. 76-9381)

Deb, K., & Goldberg, D. E. (1991). *Analyzing deception in trap functions* (IlliGAL Report No. 91009). Urbana: University of Illinois at Urbana-Champaign, Illinois Genetic Algorithms Laboratory.

Goldberg, D. E. (1989). *Genetic algorithms in search, optimization, and machine learning.* Reading, MA: Addison-Wesley.

Goldberg, D. E., & Deb, K. (1991). A comparative analysis of selection schemes used in genetic algorithms. *Foundations of Genetic Algorithms*, 69–93.

Goldberg, D. E., Deb, K., & Clark, J. H. (1991). *Genetic algorithms, noise, and the sizing of populations* (IlliGAL Report No. 91010). Urbana: University of Illinois at Urbana-Champaign, Illinois Genetic Algorithms Laboratory.

Goldberg, D. E., & Rudnick, M. (1991). Genetic algorithms and the variance of fitness. *Complex Systems, 5*, 265–278.

Goldberg, D. E., & Segrest, P. (1987). Finite Markov chain analysis of genetic algorithms. *Proceedings of the Second International Conference on Genetic Algorithms*, 1–8.

Holland, J. H. (1970). Hierarchical descriptions of universal spaces and adaptive systems. In A. W. Burks (Ed.), *Essays on cellular automata* (pp. 320–353). Urbana: University of Illinois Press.

Holland, J. H. (1973). Genetic algorithms and the optimal allocations of trials. *SIAM Journal of Computing, 2*(2), 88–105.

Holland, J. H. (1975). *Adaptation in natural and artificial systems.* Ann Arbor, MI: University of Michigan Press.

Syntactic Analysis of Convergence in Genetic Algorithms

Sushil J. Louis
Department of Computer Science
Indiana University
Bloomington, IN 47405
louis@cs.indiana.edu

Gregory J. E. Rawlins
Department of Computer Science
Indiana University
Bloomington, IN 47405
rawlins@cs.indiana.edu

Abstract

We use the average hamming distance of a population as a syntactic metric to obtain probabilistic bounds on the time convergence of genetic algorithms. Analysis of a *flat* function provides worst case time complexity for static functions and gives a theoretical basis to the problem of premature convergence. We suggest simple changes that mitigate this problem and help fight deception. Further, employing linearly computable syntactic information, we can provide upper limits on the time beyond which progress is unlikely on an arbitrary function. Preliminary results support our analysis.

1 INTRODUCTION

A Genetic Algorithm (GA) is a randomized parallel search method modeled on natural selection (Holland, 1975). GAs are being applied to a variety of problems and are becoming an important tool in machine learning and function optimization (Goldberg, 1989). Their beauty lies in their ability to model the robustness, flexibility and graceful degradation of biological systems.

Natural selection uses diversity in a population to produce adaptation. Ignoring the effects of mutation for the present, if there is no diversity there is nothing for natural selection to work on. Since GAs mirror natural selection, we apply the same principles and use a measure of diversity for estimating time to stagnation or convergence. A GA converges when most of the population is identical, or in other words, the diversity is minimal. Using the average hamming distance (hamming average) sampled between all members of a population as a measure of population diversity we derive an upper bound for the time to minimal diversity, when the genetic algorithm may be expected to make no further progress (hamming convergence).

Previous work in GA convergence by Ankenbrandt, and Goldberg and Deb focuses on the time to convergence to a particular allele using fitness ratios to obtain bounds on time complexity (Ankenbrandt 1991; Goldberg and Deb 1991). Since GAs use syntactic information to guide their search, it seems natural to use syntactic metrics in their analysis. Such analysis, using hamming averages, can predict the time beyond which qualitative improvements in the solution are unlikely, however it cannot predict the quality of the converged solution. Our analysis gives a theoretical basis for the popular notion among GA practitioners that selection is overly exploitative. We suggest some remedies based on this analysis.

The next section defines our model of a genetic algorithm and identifies the effect of genetic operators on our diversity metric, which is the hamming average. Subsequently we derive an upper bound on the expected time to convergence. This suggests syntactic remedies to the problem of premature convergence and, as a side effect, how to mitigate deception in GAs. Since crossover does not affect the hamming average we extrapolate the change in the hamming average sampled during the first few generations to predict the hamming average in later generations. Results presented in section seven on a test suite of functions indicate that surprisingly accurate predictions are possible. The last section covers conclusions and directions for further research.

2   GENETIC ALGORITHMS AND HAMMING DISTANCE

A genetic algorithm works with a population and encodes a problem's parameters in a binary string. The initial population is formed by a randomly generated set of strings. Our model of a genetic algorithm assumes proportional selection, n-point crossover and the usual mutation operator. With the GA operators defined we can analyze their effects on the average hamming distance of a population.

The average hamming distance of a population is the average distance between all pairs of strings in a population of size N. As each member of the population is involved in $N - 1$ pairs, the sample size from which we calculate the average is:

$$\frac{N(N - 1)}{2}$$

Let the length of the member strings in the population be l. The hamming average of the initial population is well approximated by the normal distribution with mean h_0 where

$$h_0 = \frac{l}{2}$$

and standard deviation s_0 given by

$$s_0 = \frac{\sqrt{l}}{2}$$

Ignoring the effect of mutation, the hamming average of a converged population is zero.[1] Given that the initial hamming average is $l/2$ and the final hamming average is zero, the effects of selection and crossover determine the behavior of a genetic algorithm in the intervening time.

3 CROSSOVER AND AVERAGE HAMMING DISTANCE

Assuming that offspring replace their parents during a crossover, all crossover operators can be partitioned into two groups based on whether or not they change the hamming average. If one parent contributes the same alleles to *both* offspring (as in masked crossover (Louis and Rawlins 1991)) the hamming distance between the children is less than the hamming distance between their parents. This leads to a loss of genetic material, reducing population hamming average and resulting in faster hamming convergence. We do not consider such operators in this paper. The vast majority of traditional operators, like one-point, two-point, ... l-point, uniform and punctuated crossover (De Jong and Spears 1991; Schaffer and Morishima 1987, 1988; Syswerda 1989), do not affect the hamming average of a population. Before proving this we introduce some notation. Let the two parents A and B, and their offspring C and D, be denoted by:

$$A = a_1, a_2, \ldots, a_l$$
$$B = b_1, b_2, \ldots, b_l$$
$$\text{and}$$
$$C = c_1, c_2, \ldots, c_l$$
$$D = d_1, d_2, \ldots, d_l$$

Traditional crossover, realized by a binary mask $M = (m_1, m_2, \ldots, m_l)$ of length l, is defined by

$$A \times B \xrightarrow{M} C$$

$$\text{where } c_i \leftarrow \left\{ \begin{array}{ll} a_i & \text{if } m_i = 1 \\ b_i & \text{if } m_i = 0 \end{array} \right\} \forall\, i \mid (1 \leq i \leq l)$$

Similarly

$$B \times A \xrightarrow{M} D$$

$$\text{where } d_i \leftarrow \left\{ \begin{array}{ll} b_i & \text{if } m_i = 1 \\ a_i & \text{if } m_i = 0 \end{array} \right\} \forall\, i \mid (1 \leq i \leq l)$$

M determines the implemented traditional crossover operator. With this definition, we prove the following lemma.

Lemma 1 *Traditional crossover operators do not change the average hamming distance of a given population.*

Proof: We prove that the hamming average in generation $t + 1$ is the same as the hamming average at generation t under the action of crossover alone. Assuming

---

[1]Mutation increases the hamming average of a converged population by an amount $\epsilon > 0$ depending on the probability of mutation.

a binary alphabet $\{a, b\}$, we can express the population hamming average at generation t as the sum of hamming averages of l loci, where l is the length of the chromosome. Letting $h_{i,t}$ stand for the hamming average of the ith locus we have:

$$h_t = \sum_{i=1}^{l} h_{i,t} \qquad (1)$$

The hamming average in the next generation is

$$h_{t+1} = \sum_{i=1}^{l} h_{i,t+1}$$

In the absence of selection and mutation, crossover only changes the order in which we sum the contributions at each locus. That is:

$$h_{i,t} = h_{i,t+1}$$

Therefore

$$h_t = h_{t+1}$$

Q.E.D

Having eliminated crossover from consideration since it causes no change in the hamming average, we look to selection as the force responsible for hamming convergence.

4 SELECTION AND AVERAGE HAMMING DISTANCE

Selection is the domain-dependent part of a genetic algorithm. But, independent of the domain, we would like to prove that selection with probability greater than $1/2$ reduces the hamming average in successive generations, and then obtain an upper bound on the time to convergence. Obtaining an upper bound is equivalent to assuming the worst possible search space and estimating the time required for finding a point in this space. For a static function, a space on which an algorithm can do no better than random search satisfies this criterion. The *flat* function defined by

$$f(x_i) = \text{constant}$$

contains no useful information for an algorithm searching for a particular point in the space. Thus no algorithm can do better than random search on this function. However, a simple GA without mutation loses diversity and eventually converges. An expression for the time convergence on such a function gives an upper bound on the time complexity of a genetic algorithm on any static function. The GA's convergence on the flat function is caused by random genetic drift where small random variations in allele distribution cause a GA to drift and eventually converge. To derive an upper bound we start with the time for a particular allele to become fixed due to random genetic drift.

If a population of size N contains a proportion p_i of a binary allele i, then the probability that k copies of allele i are produced in the following generation is given by the binomial probability distribution

$$\begin{pmatrix} N \\ k \end{pmatrix} p_i^k (1 - p_i)^{N-k}$$

Using this distribution we have to calculate the probability of a particular frequency of occurrence of allele i in subsequent generations. This is a classical problem in population genetics. Although the exact solution is complex, we can approximate the probability that allele frequency takes value p in generation t. Wright's approximation for intermediate allele frequencies and population sizes, as given in Gayle (Gayle 1990), is sufficient for our purposes. Let $f(p, t)$ stand for the probability that allele frequency takes value p in generation t where $0 < p < 1$ then

$$f(p, t) = \frac{6p_0(1 - p_0)}{N} \left(1 - \frac{2}{N}\right)^t$$

This specifies the probability that an allele has *not* converged. Therefore the probability that an allele is fixed (converged) at generation t is

$$\mathcal{P}(t) = 1 - f(p, t)$$

Applying this to a genetic algorithm, assuming a randomly instantiated population at $t = 0$, we have

$$\mathcal{P}(t) = 1 - \frac{6p_0(1 - p_0)}{N} \left(1 - \frac{2}{N}\right)^t$$

If we assume that alleles consort independently, which is true for a flat function, then the expression for the probability that all alleles are fixed at generation t for a chromosome of length l alleles, is given by

$$\mathcal{P}(t, l) = \left[1 - \frac{6p_0(1 - p_0)}{N} \left(1 - \frac{2}{N}\right)^t\right]^l \tag{2}$$

Equation 2 gives us the time to convergence for a genetic algorithm on a flat function and is therefore an upper bound for any static function. For example, on a 50 bit chromosome and a population size of 30, this gives an upper bound of 92% on the probability of convergence in 50 generations. Experimentally, we get between 92% and 73% convergence starting with an initial hamming average of $50/2 = 25$. Previous work by Goldberg and Segrest on genetic drift gives a more exact albeit more computationally taxing expression for time to convergence due to genetic drift (Goldberg and Segrest 1987). They also include the effect of mutation in their analysis, which once again is directly applicable to our problem. Finally, that GAs converge so quickly due to random drift gives a theoretical basis to the often observed problem of premature convergence. We suggest some methods of mitigating this problem in the next section.

4.1 HANDLING PREMATURE CONVERGENCE

Nature uses large population sizes to "solve" the premature convergence problem. This is expensive, furthermore we need not be restricted to nature's methods but can do some genetic engineering of our own. Mutation seems a likely candidate, and in practice, is the usual way of maintaining diversity. However, although high mutation rates may increase diversity, its random nature raises problems. Mutation is as likely to destroy good schemas as bad ones and therefore elitist selection is needed to preserve the best individuals in a population. This works quite well in practice, but is unstructured and cannot insure that all alleles are always present in the population.

Instead of increasing mutation rates, we pick an individual and add its *bit complement* to the population. This ensures that every allele is present and the population spans the entire encoded space of the problem. We can pick the individual to be complemented in a variety of ways depending on the assumptions we make about the search space. Randomly selecting the individual to be complemented makes the least number of assumptions and may be the best strategy in the absence of other information. We could also select the best or worst individual, or use probabilistic methods in choosing whom to complement. Instead of complementing one individual, we can choose to complement a set of individuals, thus spanning the encoded space in many directions. The most general approach is to maintain the complement of every individual in the population, doubling the population size.

The presence of complementary individuals also makes a GA more resistant to deception. Intuitively, since the optimal schema is the complement of the deceptively optimal schema, it will be repeatedly created with high probability as the GA converges to the deceptive optimum (Goldberg et al. 1989). In our experiments we replaced the minimum fitness individual with either the complement of a randomly picked individual in the current population, or the complement of the current best individual. Let l_i represent the number of bits needed to represent variable i in a deceptive problem, then the functions we used can be described as follows

$$\text{Deceptive}(x_i) = \sum_{i=1}^{n} \left\{ \begin{array}{ll} x_i & \text{if } x_i \neq 0 \\ 2^{l_i+2} & \text{if } x_i = 0 \end{array} \right\}$$

We used *Dec1* and *Dec2* which are 10-bit and 20-bit deceptive problems respectively. Letting superscripts denote the number of bits needed for each variable, Dec1 can be described by

$$\text{Dec1: Deceptive}(x_1^2, \ x_2^8)$$

and Dec2 by

$$\text{Dec2: Deceptive}(x_1^2, \ x_2^8, \ x_3^5, \ x_4^5)$$

Figure 1 compares the average fitness after 100 generations of a classical GA (CGA) and a GA with complements (GAC) on a Dec1. The GAC easily outperforms the classical GA, both in average fitness and the number of times the optimum was found. In most experiments the classical GA fails to find the optimum in 11 runs of the GA with a population size of 30 in a 100 generations. Since the converged hamming average for the CGA is very small, we do not expect it to be ever able to find the global optimum. GAC on the other hand easily finds the optimum for Dec1 within a 100 generations and usually finds the optimum for Dec2 within a few

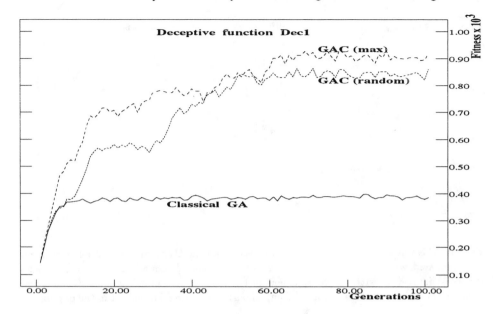

Figure 1: Average fitness over 100 generations of classical GA and GA with complements. GAC (max) replaces the worst individual with the complement of the current best individual. GAC (random) replaces the worst individual with the complement of a random individual in the population

hundred generations. Figure 2 shows the number of times the optimum was found for Dec1 and Dec2 within a total of 100 and 1000 generations respectively.

Although this genetically engineered algorithm can mitigate certain kinds of deception, it is not a panacea and cannot guarantee an optimal solution on all problems.Messy Genetic Algorithms (Goldberg et al. 1989) can make much stronger guarantees but their computational complexity is daunting. Not surprisingly, GAC also does better than a classical GA on Ackley's One Max and no worse on the De Jong test suite (Ackley 1987; De Jong 1975). In problems consisting of mixed deceptive and easy subproblems the GAC still does much better than the classical GA. Finally, GA-hard problems, that are both deceptive and epistatic, may need masked crossover or other length independent recombinant operators to be solved successfully using complements (Louis and Rawlins 1991).

5 PREDICTING TIME TO CONVERGENCE

We now have an upper bound on time to convergence on static functions, but predicting performance on an arbitrary function is more difficult because of the non-linearities introduced by the selection intensity. However, computing the rate of decrease in the hamming average while a GA is working on a particular problem allows us to predict roughly the time to hamming convergence.

We first assume that string similarity implies fitness similarity (the similarity as-

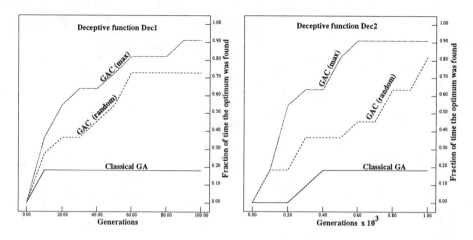

Figure 2: Number of times the optimum was found on **Dec1** and **Dec2** for a classical GA compared with the same statistic for GAs with complements. GAC (max) replaces the worst individual with the complement of the current best individual. GAC (random) replaces the worst individual with the complement of a random individual in the population

sumption). Unimodal functions satisfy this assumption. Even if the function is multimodal, genetic drift, unless countered by niching or other schemes, will ultimately cause behavior that is predictable by the following model. We start with a general equation for the change in hamming average per generation:

$$h_{t+1} = f(h_t)$$

relating h_t, the hamming average in generation t, to the hamming average in the next generation. Two observations guide our choice of the function $f(h_t)$.

1. The schema theorem along with our similarity assumption indicate $f(h_t)$ is linear:

$$h_{t+1} = ah_t + b$$

2. Without mutation the final hamming average is zero, which implies that $b = 0$.

This gives us the equation:

$$h_{t+1} = ah_t$$

We can solve this simple recurrence, and the general solution is given by

$$h_t = \left\{ \begin{array}{ll} l/2 & t = 0 \\ a^t h_0 & t > 0 \end{array} \right\} \tag{3}$$

We can estimate a by keeping track of the hamming average while running a GA. We want to stop when the hamming average gets under what is computationally feasible to explore with enumeration. Convergence to a hamming average of x with a standard deviation of y where $x + y \approx 10$ will suffice since it is not unreasonable to exhaustively search $2^{10} = 1024$ points in the search space. Mutation however

Table 1: Table comparing actual and predicted hamming averages

Function	Observed			Predicted	
	h_0	h_{50}	std. dev.	$h_{50}\ O(N^2)$	$h_{50}\ O(N)$
Flat	25	4.6	2.4	4.1	4.7
F1	15	1.9	1.2	1.1	3.2
F2	12	2.5	1.1	3.2	7.7
F3	25	3.2	1.9	5.5	3.7
F4	120	20.4	7.6	14.8	17.2
F5	17	2.7	1.9	3.1	2.4
One Max	25	3.7	2.3	2.3	2.1
One Max	100	15.7	8.1	8.5	17.5

may cause the hamming convergence value to be very large (> 10). The value of the hamming average at convergence will then depend on the chromosome length and the probability of mutation.

6   RESULTS

We present results comparing our analytical predictions with empirical performance on the following problems:

1. Flat: $f(x_1, x_2) = 10$, with a 50 bit chromosome.

2. DeJong's five functions: $F1 \ldots F5$

3. One Max : $f(X) = |X|$ where $|X|$ stands for the norm of the bit string X, for both a 50 and a 200 bit chromosome.

The GA population size in all experiments was 30, using roulette wheel selection and two-point crossover with no mutation. Estimating the value for a in equation 3 from the first 10 generations of a run, we predict the hamming average at generation 50. The fourth column uses a computed over all $N(N-1)/2$ pairs and the last column computes a using $\lceil \sqrt{N} \rceil$ strings in the population. The results are summarized in table 1. The experimental values are averages over 10 runs.

The predicted results are very close to the actual values even for this rough approximation. All but two predictions are within one standard deviation of observed value. When this is not the case, the predicted values are greater than the observed ones. This is surprising considering the roughness of our approximation. That good results are obtainable with simplified analysis clearly indicates the validity of our approach.

7   CONCLUSIONS

Analyzing a GA is complicated because of its dependence on the application domain and our lack of understanding of the mapping between our view of the application domain and the GA's view of the same domain. The GA's view of the domain is implicit in the chosen encoding. However, useful analysis can be done on the syntactic information in the encoding, giving us a surprising amount of knowledge. An upper bound on the time complexity followed immediately from considering the effects of drift on allele frequency. The high rate of convergence even on the flat function indicates that selection is overly exploitative. This led us to suggest maintaining complements of individuals in the population thus preserving diversity. Preliminary results indicate that not only do complements maintain diversity but they also mitigate the effects of GA-deception. Fitness-based recombination operators which are more susceptible to deception and premature convergence, but which are not bound to short building blocks for progress, can use complements to maintain diversity and become less susceptible to deception.

Results show that good predictions of time complexity are possible, even from a rough model that uses easily computable syntactic information. The implications of the results on other GA parameters such as population size and mutation rate, both affecting the rate of hamming convergence, are being investigated. Bounds on population size, derived from an analysis of the variance in hamming convergence with variance in population size, can be compared with bounds estimated by other methods.

Refining our model includes using more terms in the hamming average equation. Mutation's effect cannot be underestimated and should be incorporated into our model. The rate of mutation determines the amount of diversity and markedly affects hamming convergence in later generations.

Finally, although we have a bound on the time complexity, qualitative predictions may not be possible with purely syntactic information. The schema theorem links schema proportions to fitness and may thus give us a handle on qualitative predictions.

References

Ackley, D. A., *A Connectionist Machine for Genetic Hillclimbing.* Kluwer Academic Publishers, 1987.

Ankenbrandt, C. A., "An Extension to the Theory of Convergence and a Proof of the Time Complexity of Genetic Algorithms," *Foundations of Genetic Algorithms.* Rawlins, Gregory J. E., Editor, Morgan Kauffman, 1991, 53-68.

De Jong, K. A., "An Analysis of the Behavior of a class of Genetic Adaptive Systems," Doctoral Dissertation, Dept. of Computer and Communication Sciences, University of Michigan, Ann Arbor, 1975.

De Jong, K. A., and Spears, W. M., "An Analysis of Multi-Point Crossover," *Foundations of Genetic Algorithms.* Rawlins, Gregory, J. E., Editor, Morgan Kauffman, 1991, 301-315.

Freund, John. E., *Statistics: A First Course.* Prentice-Hall, 1981.

Gayle, J. S., *Theoretical Population Genetics.* Unwin Hyman, 1990, 56-99.

Goldberg, David E., *Genetic Algorithms in Search, Optimization, and Machine Learning.* Addison-Wesley, 1989.

Goldberg, David E., Korb, Bradley, and Deb, Kalyanmoy. "Messy Genetic Algorithms: Motivation, Analysis, and First Results", TCGA Report No. 89002, Tuscaloosa: University of Alabama, The Clearinghouse for Genetic Algorithms, 1989.

Goldberg, D. E., and Deb, Kalyanmoy., "A Comparative Analysis of Selection Schemes Used in Genetic Algorithms," *Foundations of Genetic Algorithms.* Rawlins, Gregory J. E., Editor, Morgan Kauffman, 1991, 69-93.

Goldberg, D. E., and Segrest, Philip., " Finite Markov Chain Analysis of Genetic Algorithms," *Proceedings of the Second International Conference on Genetic Algorithms,* Lawrence Erlbaum Associates, 1987, 1-8.

Holland, John H., *Adaptation In Natural and Artificial Systems.* Ann Arbor: The University of Michigan Press. 1975.

Louis, Sushil J., and Rawlins, Gregory J. E. "Designer Genetic Algorithms: Genetic Algorithms in Structures Design," *Proceedings of the Fourth International Conference on Genetic Algorithms,* Morgan Kauffman, 1991, 53-60.

Schaffer, David. J., and Morishima, Amy, "An Adaptive Crossover Distribution Mechanism for Genetic Algorithms," *Proceedings of the Second International Conference on Genetic Algorithms,* Lawrence Erlbaum Associates, 1987, 36-40.

Schaffer, J. David, and Morishima, Amy, "Adaptive Knowledge Representation: A Content Sensitive Recombination Mechanism for Genetic Algorithms," *International Journal of Intelligent Systems,* John Wiley & Sons Inc., 1988, Vol 3, 229-246.

Syswerda, G., "Uniform Crossover in Genetic Algorithms," *Proceedings of the Third International Conference on Genetic Algorithms,* Morgan Kauffman, 1989, 2-8.

Population Diversity
in an Immune System Model:
Implications for Genetic Search

Robert E. Smith
Dept. of Engineering Mechanics
The University of Alabama
P.O. Box 870278
Tuscaloosa, AL 35487
rob@comec4.mh.ua.edu

Stephanie Forrest
Dept. of Computer Science
University of New Mexico
Albuquerque, NM 87131
forrest@unmvax.cs.unm.edu

Alan S. Perelson
Theoretical Division
Los Alamos National Laboratory
Los Alamos, NM 87545
asp@receptor.lanl.gov

Abstract

In typical applications, *genetic algorithms* (GAs) process populations of potential problem solutions to evolve a single population member that specifies an "optimized" solution. The majority of GA analysis has focused these optimization applications. In other applications (notably *learning classifier systems* and certain connectionist learning systems), a GA searches for a population of *cooperative* structures that jointly perform a computational task. This paper presents an analysis of this type of GA problem. The analysis considers a simplified genetics-based machine learning system: a model of an immune system. In this model, a GA must discover a set of pattern-matching *antibodies* that effectively match a set of *antigen* patterns. Analysis shows how a GA can automatically evolve and sustain a diverse, cooperative population. The cooperation emerges as a natural part of the antigen-antibody matching procedure. This emergent effect is shown to be similar to *fitness sharing*, an explicit technique for multi-modal GA optimization. The results imply that procedures like those in the immune system model could promote diverse, cooperative populations without the explicit, global calculations required by fitness sharing.

1   Introduction

Maintaining diversity of individuals within a population is necessary for the long term success of any evolutionary system. Genetic diversity helps a population adapt quickly to changes in the environment, and it allows the population to continue searching for productive niches, thus avoiding becoming trapped at local optima. In genetic algorithms (GAs), it is difficult to maintain diversity because the algorithm assigns exponentially increasing numbers of trials to the observed best parts of the search space (cf. Schema Th. (Holland, 1975)). As a result, the standard GA has strong convergence properties. For optimization problems, convergence can be an advantage, but in other environments it can be detrimental. Further, even in optimization, strong convergence can be problematic if it prematurely restricts the search space.

In optimization, when the GA fails to find the global optimum, the problem is often attributed to *premature convergence*, which means that the sampling process converges on a local rather than the global optimum. Several methods have been proposed to combat premature convergence in conventional GAs (DeJong, 1975; Goldberg, 1989; Booker, 1982; Deb, 1989a). These include restricting the selection procedure (crowding models), restricting the mating procedure (assortative mating, local mating, etc.), explicitly dividing the population into subpopulations (common in parallel GAs), and modifying the way fitnesses are assigned (fitness sharing).

In many settings, convergence by the GA on a global optimum is not appropriate. For example, in a classifier system (Holland et al., 1986) genetic operators are a natural way to search for a useful set of rules that collectively performs well in a task environment, each rule playing a unique and complementary role. Thus, the system needs to evolve a set of rules that are specialized to various tasks (or niches) rather than producing a homogeneous (converged) population of similar rules. As a second example, consider a computational model of the immune system in which a population of antibodies is evolving to cover a set of antigens (Forrest & Perelson, 1991). If the antibody population is sufficiently large, it clearly makes sense to evolve antibodies that are specialized to recognize different classes of antigens instead of evolving one generalist antibody that weakly matches all antigens. For these more ecological environments, genetic operators are clearly relevant for evolving a good solution, if inappropriate convergence can be avoided.

To date, most GA analysis has focused on problems in which each population member's fitness is independent of other population members, thus excluding coevolutionary systems such as classifier systems. In this paper, we introduce a simple model based on a genetic algorithm simulation of the immune system in which each individual's fitness is functionally dependent on the rest of the population, thus capturing one important aspect of ecological problems. The functional dependence is introduced through the use of a simplified bidding mechanism similar to that of classifier systems. We show mathematically that the use of the simple bidding procedure combined with a traditional GA is sufficient for the population to discover and maintain independent subpopulations. Further, our analysis shows that the model implements a form of *implicit* fitness sharing which we relate to previous (explicit) models of fitness sharing. This implies that procedures like those used in the model could promote diversity without the explicit, global calculations required for fitness sharing. These implications are particularly strong for classifier systems that previously employed fitness sharing (Smith, 1991) or some other explicit niching mechanism (Booker, 1985), since the immune system model has many similarities to a simplified classifier system.

2 GA Simulations of the Immune System

Our immune system protects us from a wide variety of different viruses, bacteria, toxins, and other pathogenic organisms. Although there are many different host defense mechanisms employed, the first step in any of these mechanisms is the recognition of the foreign cell or molecule, which we call *antigen*. Recognition in the immune system occurs by a chemical interaction between antigen and a specific host defense molecule such as *antibody*. Thus, the problem that the immune system faces is the generation of a repertoire of antibodies with sufficient diversity to recognize any antigen.

Forrest, et al. (in preparation) study an abstract version of this problem in which antigens and antibodies are represented by bit strings of a fixed length ℓ. Recognition is assessed via a string matching procedure. The antigens are considered fixed, and a population of N antibodies is evolved to recognize the antigens using a GA. For any set of antigens, the goal is to obtain a covering set of antibodies such that each antigen is recognized by at least one antibody in the population. Maintaining antibody diversity is crucial to obtaining a cover.

The model is based on a universe in which both antigens and receptors on B cells and T cells are represented by binary strings (Farmer, Packard, & Perelson, 1986). This is certainly a simplification from the real biology in which genes are specified by a four-letter alphabet and recognition between receptors and antigens is based on their three-dimensional shapes and physical properties. Further, the model does not distinguish between receptors on B cells and the soluble, secreted form of the receptor, which is antibody. However, the universe of binary strings is rich enough to allow one to study how a relatively small number of recognizers (the antibodies) can evolve to recognize a much larger number of different patterns (the antigens).

The initial model makes the important simplification that a bit string represents both the genes that code for a receptor and the phenotypic expression of the receptor molecule. The model includes only recognition of our idealized antigens by receptors and does not consider how the immune system neutralizes an antigen once it is recognized.

A receptor, or "antibody," is said to *match* an antigen if their bit strings are complementary (maximally different). Since each antibody may have to match against several different antigens simultaneously, we do not require perfect bit-wise matching. There are many possible match rules that are plausible physiologically (Perelson, 1989). The degree of match is quantified by a match score function $M : Antigen \times Antibody \rightarrow \Re$. This function identifies contiguous regions of complementary bitwise matches within the string, computes the lengths (l_i) of the regions, and combines them such that long regions are rewarded more than short ones. Using this basic idea, many different specific functions can be defined that are linear or nonlinear in l_i. We have studied the behavior of the model under several different functions (Forrest et al., in preparation).

Using the bit string representation for antibodies, we construct one population of antigens and one of antibodies. Antibodies are matched against antigens, scored according to a fitness function M, and replicated using a conventional genetic algorithm. Figure 1 illustrates

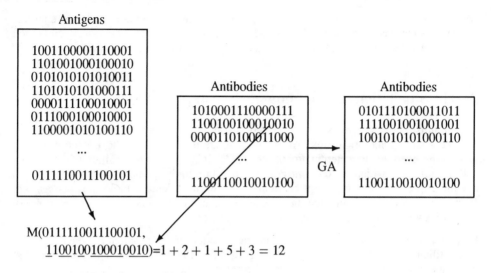

Figure 1: A schematic illustration of the immune model. The match score, M, of an antigen-antibody pair is the number of bits that are complementary.

the basic immune model.

From this basic model, many variations can be created by changing the details of how antibodies are chosen to be matched against antigens. For example, we have used the basic model to study antigen populations that cannot be matched by one antibody type. Suppose the population of antigens is:

$$50\% - 000\ldots000$$
$$50\% - 111\ldots111.$$

In order for an antibody population to match these antigens perfectly, there would need to be some antibodies that are all 1s and others that are all 0s. Thus, a solution to this problem would require the GA to maintain two different solutions simultaneously. This is a simple example of a multiple-peak problem in which there are only two peaks and they are maximally different. Multiple-peak problems are difficult for the GA because of its strong convergence tendencies. Typically, on multiple-peak problems, genetic drift will lead the GA to (randomly) converge on one of the peaks. However, for this simplified version of the immune problem, a solution requires a population of antibodies that contains strings that are all 1s and all 0s. The hybrids formed by crossover are not useful.

With a fixed set of antigens, the antibodies are initialized either to be completely random (to see if the GA can learn the correct antibodies) or initially given the answer by setting the population to include the correct antibodies (000...000 and 111...111 in the example). By giving the answer initially, the stability of the answer can be tested. Fitness scoring is as follows:

1. A single antigen is randomly selected from the antigen population.

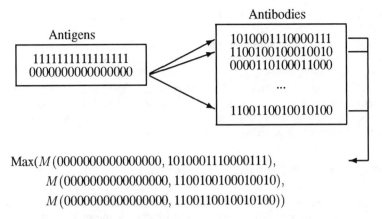

$$\text{Max}(M(0000000000000000, 1010001110000111),$$
$$M(0000000000000000, 1100100100010010),$$
$$M(0000000000000000, 1100110010010100))$$

Figure 2: A schematic illustration of the two-peak problem. Each antigen corresponds to one peak. An antigen and a subset of the antibody population are each selected randomly; the antibody (from the subset) that matches the antigen best (Max) has its fitness incremented.

2. From the population of N antibodies a randomly selected sample of size σ is taken without replacement.

3. For each antibody in the sample, *match* it against the selected antigen, determine the number of bits that match, and assign it a *match score*[1].

4. The antibody in the sample group with the highest match score is determined. Ties are broken at random.

5. The match score of the winning antibody is added to its fitness.

6. This process is repeated for C cycles.

In this scheme, the fitness values of antibodies are interdependent, since an antibody's proportion is only increased if it is the best matching antibody in the sample. Numerical experiments reported in Forrest et al. (in preparation) have shown that this scheme can maintain a diverse population of antibodies that cover a set of antigens, that the antibodies will occur with frequency proportional to the sampling rate for each antigen, and that the system performance seems relatively insensitive to the Hamming distance between these antigens. We shall show that this procedure implicitly embodies fitness sharing (Deb, 1989b; Deb & Goldberg, 1989; Goldberg & Richardson, 1987). The process is iterated so that each antigen has a chance of being selected and each antibody will receive a fair evaluation of its fitness. Figure 2 illustrates the model.

This model corresponds quite closely to certain features of learning classifier systems. In

---

[1]For the purposes of this discussion, details of the matching and scoring procedures are unimportant.

effect, each antibody in our model is a highly simplified classifier rule with only a condition part (defined over the alphabet $\{0, 1\}$ rather than the more traditional $\{1, 0, \#\}$). The message-processing cycle is similar to classifier systems in that it uses a bidding mechanism. In our model, the bidding is deterministic (the closest match always wins), whereas many different bidding schemes are often used with classifier systems. Since we only allow one winner, this is analogous to a classifier system with a message list of size 1. Obviously, there are many aspects of classifier systems that our model does not incorporate (e.g., posting messages, bucket brigade learning, etc.), but even in this simplified form we can study the interaction between the genetic algorithm and bidding.

3   Emergent Fitness Sharing in the Immune System Model

To understand the mechanisms of maintaining population diversity in the immune system model, we first calculate an antibody's expected fitness. Some new notation is required. Let the distance between an antibody i and an antigen j be called d_{ij}. The following discussion assumes that d_{ij} is the number of bits of antibody i that do not match (i.e. are not complementary to) those in antigen j, although other distance metrics can be used in the following developments. Under this distance metric, antibody i and antigen j are said to *perfectly match* if $d_{ij} = 0$. The maximum distance possible between an antibody and an antigen is ℓ, the bit string length. Let $s(d_{ij})$ be the match score assigned to antibody i when it is matched against antigen j. Let $N_j(m)$ be the number of antibodies in the population that are at distance m from antigen j. Also let α_j be the probability of selecting antigen j for matching. Finally, let f_i be the expected fitness of antibody i.

Consider a given antigen j. Assume that the population is of size N, and that it contains $N_j(m)$ antibodies at distance m from antigen j. The probability that w antibodies at exactly distance m from antigen j are in a sample of size σ taken without replacement from this population, $p(w; \sigma, N, N_j(m))$, is given by the hypergeometric distribution (Freund, 1962; Hines & Montgomery, 1980)

$$p(w; \sigma, N, N_j(m)) = \frac{\left(\begin{array}{c} N_j(m) \\ w \end{array} \right) \left(\begin{array}{c} N - N_j(m) \\ \sigma - w \end{array} \right)}{\left(\begin{array}{c} N \\ \sigma \end{array} \right)} , w = 0, 1, \ldots, \sigma . \quad (1)$$

Since the hypergeometric distribution will play an important role in subsequent calculations, it is important to understand how it arises. Think of the $N_j(m)$ antibodies at distance m as "successes" and the remaining $N - N_j(m)$ antibodies as "failures." We choose a sample of size σ without replacement and are interested in the probability of picking w success elements and hence necessarily $\sigma - w$ failure elements. There are $\binom{N}{\sigma}$ possible ways of picking a sample of size σ. The number of ways of picking w successes from a total of $N_j(m)$ elements is $\binom{N_j(m)}{w}$, whereas the number of ways of picking $\sigma - w$ failures from $N - N_j(m)$ elements is $\binom{N-N_j(m)}{\sigma-w}$. Thus, the fraction of times a sample is drawn with w success elements and $\sigma - w$ failure elements is given by Eq. (1).

It will be useful for later discussions to note two special cases. If the sample size $\sigma = 1$, then the probability that the sample contains an antibody at distance m is

$$p(1; 1, N, N_j(m)) = \frac{N_j(m)}{N} ,$$

and the probability that it does not contain such an antibody is

$$p(0; 1, N, N_j(m)) = \frac{N - N_j(m)}{N} ,$$

If the sample size $\sigma = N$, then

$$p(w; N, N, N_j(m)) = \begin{cases} 1 & \text{if } w = N_j(m) \\ 0 & \text{otherwise.} \end{cases}$$

3.1 Expected fitness of an antibody when perfect matching is required

To introduce the method of calculating the expected fitness of an antibody, we first consider the case in which an antibody receives a non-zero score only if it perfectly matches the antigen. Thus, the match score $s(d_{ij}) \neq 0$ if and only if $d_{ij} = 0$. Let the score received for a perfect match be s_p. For antibody i that perfectly matches antigen j to receive a score s_p at time t, the following conditions must be met:

(i) Antigen j must be the antigen selected for matching against. This occurs with probability α_j.

(ii) Antibody i must be in the sample of size σ. One must consider this event given that i is one of w perfectly matching antibodies in the sample.

(iii) Antibody i must be chosen to be the tie breaker from amongst the w. This occurs with probability $1/w$.

Note from Eq. (1) that the probability that w antibodies in a sample of size σ are at distance 0 from antigen j is $p(w; \sigma, N, N_j(0))$. Given that w perfect matches against antigen j appear in the sample, the probability that this sample contains *one particular* antibody i that perfectly matches j is $p(1; w, N_j(0), 1) = w/N_j(0)$. Since events (i), (ii) and (iii) are independent, the probability that antibody i receives a non-zero score is given by the following sum over all possible values of w:

$$\alpha_j \sum_{w=1}^{\sigma} \frac{1}{w} \frac{w}{N_j(0)} p(w; \sigma, N, N_j(0)) = \alpha_j \frac{1}{N_j(0)} (1 - p(0; \sigma, N, N_j(0))).$$

Hence, the expected fitness of antibody i after one cycle is

$$f_i = \frac{s_p \alpha_j}{N_j(0)} (1 - p(0; \sigma, N, N_j(0))) .$$

Note that the expected fitness for C cycles would simply be Cf_i. Since C will be a common factor for all expected fitness values, it will not have a bearing on the expected behavior of fitness-proportionate selection, and, therefore, it is not considered in the subsequent discussion.

The term $s_p \alpha_j$ roughly corresponds to the height of the fitness function at point j in the sequence space, i.e. the ℓ-dimensional hypercube. The expected fitness calculation indicates that this value is divided by the proportion of individuals at that point, $N_j(0)$. This is similar to a limiting case of fitness sharing (Deb, 1989b; Deb & Goldberg, 1989; Goldberg & Richardson, 1987), where the parameter $\sigma_s = 0$, and an individual's fitness is divided by the proportion of identical individuals in the population. The final hypergeometric term in the calculation is due to the sampling scheme. Its role in the analogy to fitness sharing will be clarified in the following discussion.

3.2   Expected fitness of an antibody when partial matching is allowed

We now consider the general case in which an antibody receives a score for a partial match with an antigen at distance $d_{ij} = m$, where m ranges from $m = 0$ (perfect match) to $m = \ell$ (perfect mismatch). As before, in each cycle of the algorithm an antigen is picked at random, with replacement. Assume that each antigen j is selected with probability α_j. Let $\mathcal{S}_i(m)$ be the set of all antigens j at distance m from antibody i ($d_{ij} = m$). For antibody i to receive the match score in a cycle when antigen j at distance m is selected, we require

(i) No antibody at distance less than m from antigen j occurs in the sample. Recall that only the closest antibody in the sample receives the match score.

(ii) If w antibodies in the sample are all at distance m, antibody i must be in the sample and be chosen as the tie breaker. This latter event occurs with probability $1/w$.

Events (i) and (ii) are not independent. Thus to compute the probability that events (i) and (ii) are both true, we use the well known formula (Freund, 1962):

$$P(E_1 \cap E_2) = P(E_2|E_1)P(E_1) , \qquad (2)$$

where E_i denotes event (i).

To compute $P(E_1)$, the probability of event (i), we again use the hypergeometric distribution. Recall that $N_j(k)$ is the number of antibodies at distance k from antigen j. Thus there are a total of

$$V_j(m) = \sum_{k=0}^{m-1} N_j(k) \qquad (3)$$

antibodies at distance less than m from antigen j. The probability that none of these are in the sample of size σ is $p(0; \sigma, N, V_j(m))$. For later reference we define $V(0) = 0$, i.e., there are no antibodies closer than distance 0.

To compute $P(E_2|E_1)$, note that given that none of the $V_j(m)$ antibodies appears in the match sample, the probability that w of the $N_j(m)$ antibodies at distance m from antigen j appear in the match sample is $p(w; \sigma, N - V_j(m), N_j(m))$. Not all antibodies at distance m from antigen j need be copies of antibody i. If w antibodies at distance m are in the sample, then the probability that antibody i is one of the w is $p(1; w, N_j(m), 1) = w/N_j(m)$. The probability antibody i is chosen as the tie breaker is $1/w$. Since any value of w between 1 and σ is possible, we find

$$P(E_2|E_1) = \sum_{w=1}^{\sigma} p(w; \sigma, N - V_j(m), N_j(m))/N_j(m)$$

$$= (1 - p(0; \sigma, N - V_j(m), N_j(m))/N_j(m) . \qquad (4)$$

Finally, combining the terms discussed above yields:

$$f_i = \sum_{m=0}^{\ell} \sum_{j \in \mathcal{S}_i(m)} \frac{s(d_{ij})\alpha_j}{N_j(m)} \left[p(0; \sigma, N, V_j(m))(1 - p(0; \sigma, N - V_j(m), N_j(m))) \right] , \qquad (5)$$

where the first summation, $\sum_{m=0}^{\ell}$, considers all possible distances m from antibody i, and the second summation, $\sum_{j \in \mathcal{S}_i(m)}$, considers all antigens that are at distance m from antibody i. As in the previous simplified example, the terms $s(d_{ij})\alpha_j$ are related to the fitness available to a given antibody i, but in this case, the antibody can share in the finite fitness resources available at many distant antigens. However, like in fitness sharing, each of these resources is divided amongst the individuals that share it, as is indicated by the divisor $N_j(m)$.

Thus far the correspondence to fitness sharing is relatively straight-forward. However, the meaning of the bracketed hypergeometric terms,

$$[p(0; \sigma, N, V_j(m))(1 - p(0; \sigma, N - V_j(m), N_j(m)))] \ , \tag{6}$$

should be explained. The first term represents the probability that no antibody within distance $m - 1$ of antigen j will be selected in the sample. The second term represents the probability that, given the previous condition, at least one copy of an antibody at distance m from antigen j will be in the sample.

3.3 Relation to Explicit Fitness Sharing

To clarify the role of the hypergeometric terms in fitness sharing, we examine two special cases. Consider sample size $\sigma = 1$. In this case, the hypergeometric term (6) becomes

$$\left(\frac{N - V_j(m)}{N}\right)\left(1 - \frac{N - V_j(m) - N_j(m)}{N - V_j(m)}\right) = \frac{N_j(m)}{N} \ .$$

Thus, for $\sigma = 1$

$$f_i = \sum_{m=0}^{\ell} \sum_{j \in \mathcal{S}_i(m)} \frac{s(d_{ij})\alpha_j}{N} \ .$$

In this special case, there is no fitness sharing, and the fitness values are independent. Essentially, the relative, expected fitness values are equivalent to those one would expect under a standard genetic algorithm. Under these conditions one would expect fitness-proportionate selection to converge to a single type of antibody, due to *genetic drift* (Goldberg & Segrest, 1987). Note that this corresponds to fitness sharing (Deb, 1989b; Deb & Goldberg, 1989; Goldberg & Richardson, 1987) with the parameter σ_s set to a value that spans the entire search space.

As a second special case, consider $\sigma = N$. If one assumes that a perfectly matching antibody exists for every available antigen in the population, the expected fitness reduces to

$$f_i = \frac{s_p \alpha_i}{N_i(0)} \ .$$

In this case, each antibody is only divided by its own effective proportion in the population. Like fitness sharing with $\sigma_s = 0$, one would expect fitness-proportionate selection to distribute the population based on relative fitness in one step, without a search for peak antibodies (Smith et al., 1992).

These special cases show that the limiting behavior of σ is similar to the limiting behavior of σ_s. To investigate the effects of other values of σ, consider the term

$$R = [p(0; \sigma, N, V_j(m))(1 - p(0; \sigma, N - V_j(m), 1))] \ , \tag{7}$$

in which $N_j(m)$ is set to one. This simulates the situation where one antibody at distance m from the selected antigen is competing with $V_j(m)$ closer antibodies for fitness resources available from that antigen.

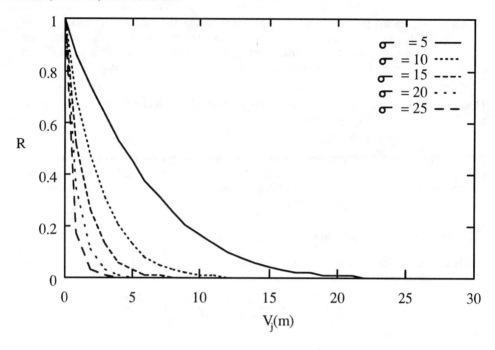

Figure 3: The hypergeometric term R, given by Eq. (12), versus V_j for various sample sizes σ.

Figure 3 shows R plotted versus $V_j(m)$ for $N = 30$ and various values of σ.[2] Note that R is zero for $V_j(m) > (N - \sigma)$, and that R is near zero for a range of values of $V_j(m)$ lower than $N - \sigma$. These curves are similar to sharing functions used in explicit fitness sharing with $\alpha < 1$ (Deb, 1989b; Deb & Goldberg, 1989; Goldberg & Richardson, 1987).

This graph and previous arguments imply that the hypergeometric terms correspond to a sharing function, and that σ plays a role in the immune system algorithm that is similar to that of σ_s in fitness sharing. Its value essentially implies a cutoff beyond which no sharing can occur. However, there is an important distinction to be drawn between fitness sharing and the implicit sharing in the immune system simulations. In fitness sharing, σ_s is a strict cutoff based on d_{ij}, which Deb (Deb, 1989b) recommends setting based on the volume of a hypersphere around a given peak. In the immune system algorithm, σ dictates a cutoff based on $V_j(m)$, which is the *proportion of the population within* a hypersphere of radius $m - 1$ around a given antigen. As the proportion of antibodies close to an antigen increases, the likelihood of more distant antibodies winning the match competition decreases. Thus, effective antibodies block sharing by less effective antibodies. The sample size σ is a control on this effect. Under this scheme, the boundaries of sharing are a function of the proportion of antibodies clustered around given antigens. These emergent sharing boundaries explain much of the resilience shown by the GA in the immune system experiments.

---

[2]Since only relative values of fitness terms are significant, curves in this figure are linearly scaled between zero and one.

4 Conclusions

Maintaining diversity in a population is important in many computational models. For example, in classifier systems it is important to maintain a diverse set of rules that collectively perform complex tasks in varying situations. One mechanism for maintaining diversity in genetic algorithms is fitness sharing (Deb, 1989b; Deb & Goldberg, 1989; Goldberg & Richardson, 1987). Although effective in some situations, fitness sharing has several important limitations:

- It requires a comparison of every population member to every other population member in each generation (N^2 comparisons).

- Setting the critical parameter σ_s requires knowledge about the number peaks in the space (Deb, 1989b).

- Setting σ_s is also dependent on a uniform distribution of peaks in the search space. Although Deb (1989b) shows experiments where sharing succeeds on a with problem mildly non-uniform peak separation, it is likely fitness sharing would overlook peaks that were distributed with less uniformity (K. Deb, personal communication, July 2, 1991). Goldberg (personal communication, May 11, 1992) has suggested that new mechanisms could be added to fitness sharing to allow for maintenance of multiple, non-uniformly distributed peaks. However, these techniques have yet to be fully investigated.

Here we have shown that an algorithm developed by Forrest et al. (in preparation) to study pattern recognition in the immune system has emergent properties that are similar to those of explicit fitness sharing. However, the fitness sharing in this algorithm is implicit, the number of peaks is determined dynamically by the algorithm and there is no specific limitation on the distance between peaks.

In an extended version of this study (Smith et al., 1992), we have observed that in addition to the maintaining diversity, the immune system model exhibits the ability to generalize. Simulations show that σ is a control of generalization in the antibody population. As σ is reduced, selective pressures are altered such that a population of more general antibodies emerges.

The fitness sharing effects in the immune system algorithm are a result of the sample-and-match procedures employed. Whether similar effects can be obtained as an alternative to explicit fitness sharing in multi-modal optimization problems is unclear. However, in other ecological problems the analysis performed here may have a straight forward application. For instance, in learning classifier systems, where ideas like "partial matching," "best match," and "generalization" have natural interpretations, techniques like those used here may prove useful.

In the context of classifier systems, some of the results presented here can be thought of as an analytical confirmation of arguments presented by Wilson (1987). Wilson's experiments and approximate analyses suggest that division of reward between like-acting classifiers can lead to stable subpopulations of different classifier "concepts". Booker (1982) employs a similar technique. Although the mechanisms are different, the emergent effects in these systems are similar to those in the immune system simulations with $\sigma = N$.

The techniques presented here may also prove useful in other systems that form computational networks via genetic learning. A particular instance is in the development of polynomial networks for system modeling (Kargupta & Smith, 1991), where explicit fitness sharing was previously required.

Whether or not the techniques used in the immune system simulations can be explicitly transferred to more prescriptive applications, analysis of the type used in this paper can aid in understanding GA behavior in settings that require diversity. Improved understanding in this area is necessary if GAs are to be actively employed in systems that require cooperating sets of interacting individuals.

5   Acknowledgements

This work was partially done under the auspices of the U.S. Department of Energy. We thank the Center for Nonlinear Studies, Los Alamos National Laboratory and the Santa Fe Institute for support. Forrest also acknowledges the support of the Association of Western Universities and National Science Foundation (grant IRI-9157644). Perelson acknowledges the support of the National Institutes of Health (grant AI28433). Brenda Javornik programmed the simulations referred to in this paper. Javornik and Ron Hightower both provided helpful comments on the manuscript.

References

Booker, L. B. (1982). *Intelligent behavior as an adaptation to the task environment.* PhD thesis, The University of Michigan, Ann Arbor, MI.

Booker, L. B. (1985). Improving the performance of genetic algorithms in classifier systems. In *Proceedings of an International Conference on Genetic Algorithms and their Applications*, 80–92, Pittsburgh, PA. Lawrence Erlbaum.

Deb, K. (1989a). *Genetic algorithms in multimodal function optimization* (Tech. Rep.). Department of Engineering Mechanics, University of Alabama, Tuscaloosa, AL: The Clearinghouse for Genetic Algorithms. (Master's Thesis)

Deb, K. (1989b). *Genetic algorithms in multimodal function optimization* (TCGA Report No. 89002). Tuscaloosa: The University of Alabama, The Clearinghouse for Genetic Algorithms.

Deb, K., & Goldberg, D. E. (1989). An investigation of niche and species formation in genetic function optimization. *Proceedings of the Third International Conference on Genetic Algorithms*, 42–50.

DeJong, K. A. (1975). *An analysis of the behavior of a class of genetic adaptive systems.* PhD thesis, The University of Michigan, Ann Arbor, MI.

Farmer, J. D., Packard, N. H., & Perelson, A. S. (1986). The immune system, adaptation, and machine learning. In D. Farmer, A. Lapedes, N. Packard, & B. Wendroff (Eds.), *Evolution, games and learning.* Amsterdam: North–Holland. (Reprinted from *Physica*, 22D, 187–204)

Forrest, S., Javornik, B., Smith, R., & Perelson, A. S. (in preparation). *Using genetic algorithms to explore pattern recognition in the immune system.*

Forrest, S., & Perelson, A. S. (1991). Genetic algorithms and the immune system. In H. Schwefel, & R. Maenner (Eds.), *Parallel Problem Solving from Nature*, 320–325, Berlin. Springer-Verlag (Lecture Notes in Computer Science).

Freund, J. E. (1962). *Mathematical statistics*. Englewood Cliffs, NJ.: Prentice-Hall.

Goldberg, D., & Richardson, J. (1987). Genetic algorithms with sharing for multimodal function optimization. In *Proceedings of the Second International Conference on Genetic Algorithms*, 148–154, San Mateo, CA. Morgan Kaufmann.

Goldberg, D. E. (1989). *Genetic algorithms in search, optimization, and machine learning*. Reading, MA: Addison Wesley.

Goldberg, D. E., Deb, K., & Horn, J. (1992). *Massive multimodality, deception, and genetic algorithms* (IlliGAL Technical Report No. 92005). Urbana, Illinois: University of Illinois at Urbana-Champaign.

Goldberg, D. E., & Segrest, P. (1987). Finite Markov chain analysis of genetic algorithms. In *Proceedings of the Second International Conference on Genetic Algorithms*, 1–8, Pittsburgh, PA. Lawrence Erlbaum.

Hines, W. W., & Montgomery, D. C. (1980). *Probability and statistics in engineering and management science* (2nd ed.). New York: Wiley.

Holland, J., Holyoak, K., Nisbett, R., & Thagard, P. (1986). *Induction: Processes of inference, learning, and discovery*. Cambridge, MA: MIT Press.

Holland, J. H. (1975). *Adaptation in natural and artificial systems*. Ann Arbor, MI: The University of Michigan Press.

Kargupta, H., & Smith, R. E. (1991). System identification with evolving polynomial networks. In *Proceedings of the Fourth International Conference on Genetic Algorithms*, 370–376, San Mateo, CA. Morgan-Kaufmann.

Perelson, A. S. (1989). Immune network theory. *Immunol. Rev., 110*, 5–36.

Smith, R. E. (1991). *Default hierarchy formation and memory exploitation in learning classifier systems* (TCGA Report No. 91003). Tuscaloosa: University of Alabama. (Ph.D dissertation).

Smith, R. E., Forrest, S., & Perelson, A. S. (1992). *Searching for diverse, cooperative populations with genetic algorithms* (TCGA Report No. 92002). Tuscaloosa: University of Alabama.

Wilson, S. W. (1987). Classifier systems and the animat problem. *Machine Learning, 2*, 199–228.

Remapping Hyperspace During Genetic Search: Canonical Delta Folding

Keith Mathias and Darrell Whitley
Department of Computer Science
Colorado State University
Fort Collins, Colorado 80523
mathiask, whitley@cs.colostate.edu

Abstract

Delta coding, an iterative strategy for genetic search, has been empirically shown to improve genetic search performance for some optimization problems. However, detailed analyses reveal that the remapping of hyperspace produced by delta coding is rudimentary. The underlying mechanisms of delta coding are examined and a more robust remapping method is introduced. *Canonical delta folding* provides a canonical ordering of the strings in the search space and then employs a powerful folding mechanism for remapping hyperspace. The behavior of *canonical delta folding* for remapping deceptive functions is examined using a set of exact equations for modeling the computational behavior of a simple genetic algorithm.

1 Introduction

Delta coding is a strategy that applies genetic algorithms in an iterative fashion (Whitley, Mathias and Fitzhorn, 1991). After the initial run of a genetic algorithm, the best solution parameters are saved as the *interim solution*. The genetic algorithm is then restarted with a new random population. The parameter substrings in subsequent runs of the genetic algorithm are decoded such that they represent a distance or *delta value* ($\pm\delta$) away from the interim solution parameters. The delta values are combined with the interim solution so that the resulting parameters are evaluated using the same fitness function. This treatment forms a new hypercube with the interim solution at its origin. Delta coding not only provides a mechanism for remapping hyperspace, it also allows one to reduce and enlarge the size of the hypercube being searched.

While delta coding remaps hyperspace from iteration to iteration, it does so only in a limited fashion. A more general strategy for remapping hyperspace is introduced in this paper. *Canonical delta folding* retains the iterative search strategy of original delta coding by using random population reinitialization and by constructing a new hypercube with the interim solution at its origin. Unlike delta coding, which uses a numeric distance metric to construct the new hypercube, canonical delta folding first shifts the strings in Hamming space to locate the interim solution at the origin, then uses a "fold mechanism" to remap hyperspace. Together, these processes remap hyperspace at each iteration.

2   Background

The goal of genetic search is to exploit hyperplane information feedback by evaluating the relative fitness of strings. This information is used to allocate reproductive opportunities so as to direct the search toward particular partitions of hyperspace that contain above average solutions. However, it is precisely this hyperplane feedback that can mislead the search so that it converges on a solution which is locally good, but is not the global optimum.

Delta coding remaps the search space by constructing a new hypercube with the interim solution at the origin. Iteratively remapping hyperspace is an interesting approach to genetic search for several reasons. First, each new hypercube creates different sets of hyperplane competitions. Remapping hyperspace can also increase or decrease the number of local minima that exist in Hamming space. Remapping hyperspace is not a new idea. In fact it has been shown that for select cases of fully deceptive problems an exact transformation exists which *remaps* hyperspace and causes all hyperplane competitions to correctly lead toward the global optimum (Liepins and Vose, 1990; Battle and Vose, 1991). This result is fragile however, because the notion of a fully deceptive problem is brittle (Whitley 1991). Currently there is no known general method for making an arbitrary function easier to optimize. The space of all possible transformations is also much larger than the solution space we wish to explore because the transformation requires a matrix for mapping the 2^L vectors of length L from one mapping of hyperspace to another. Delta coding exploits the idea that while it may not be possible to find an optimal mapping of hyperspace, different mappings of the space can be explored with the expectation that some mappings will be more useful than others.

Delta coding, although developed independently, shares similarities with other strategies, especially ARGOT (Shaefer, 1987) and Dynamic Parameter Encoding (Schraudolph and Belew, 1990). These genetic search strategies explicitly restrict search to promising areas in hyperspace. Other strategies such as micro-GAs (Goldberg, 1989b; Krishnakumar, 1989) and CHC (Eshelman, 1991) are also similar to delta coding in that the diversity needed to sustain search is regenerated by restarting the genetic algorithm in one way or another.

2.1   The Delta Coding Algorithm

Our delta coding implementation uses GENITOR (Whitley, 1989) as the basic engine for genetic search. On the first iteration, the genetic algorithm evaluates the problem in a normal fashion except that the population diversity is monitored by measuring the Hamming distance between the best and worst strings in the population. If the Hamming distance is more than one, the search continues; otherwise the search is temporarily stopped. The best solution is saved as the *interim solution*. A completely new population is then created randomly and GENITOR is restarted. Each parameter, when decoded for fitness evaluation, is applied as a *delta value* ($\pm\delta$) to the interim solution saved from the previous iteration. We are, in effect, searching a new hypercube with the interim solution at its origin. This cycle of converging to a new interim solution and then restarting the search is repeated until a maximum number of trials have been exhausted or until an acceptable solution (Fitness < Threshold) has been found. Each cycle is referred to as a *delta iteration*.

NORMAL GENITOR PHASE:
    While (Hamming Distance Between Best and Worst Population Members > 1)
    {
        Apply Recombination
        Evaluate Fitness and Insert Offspring
        IF (Fitness < Threshold) THEN **HALT**
    }

TRANSITION PHASE:
    Save Best Solution in Population as *INTERIM SOLUTION*
    Re-initialize Population
    Encode Parameters Using ($X - 1$) Bits, Use Extra Bit as Sign

DELTA ITERATION PHASE: /* Apply GENITOR in Delta Mode */
    While ((Trials < MAX_TRIALS) && (Fitness < Threshold))
    {
        While (Hamming Distance Between Best and Worst Population Members > 1)
        {
            Apply Recombination
            Decode All Parameter String Values
            Add All Decoded String Values to *INTERIM SOLUTION* Parameters
            Evaluate Fitness and Insert Offspring
        }
        Save Best New Solution as *INTERIM SOLUTION*
        Re-initialize Population
        IF ((Delta Values == 0) && (Bit String_length < Initial Encoding Length))
            THEN Encode Parameters Using 1 Additional Bit
        ELSE
                Encode Parameters Using 1 Less Bit
    }

Figure 1: The *Delta Coding* Algorithm.

Each delta parameter is decoded and applied to its corresponding interim solution parameter by the following method: if the sign bit of the delta value is zero then directly add the delta value to the interim solution parameter; if the sign bit of the delta value is one, then complement all of the other bits in the parameter and then add the result to the interim solution parameter. All addition is performed mod 2^b, where b is the number of bits in the original parameter encoding. Complementing and adding bits to the interim solution is equivalent to subtracting their numeric value from the interim solution.

Upon completion of the first iteration of genetic search the number of bits used to encode the problem is not altered. Only the mapping is changed, using the first bit in each parameter as a sign bit and the remaining bits as the delta value. However, after the first iteration, the size of the hypercube can also be altered by reducing the number of bits used to represent the space. The algorithm uses information about the interim solution to trigger operations that expand and contract the solution space which is being explored. This is done by varying the number of bits used to represent each parameter in one of two ways.

First, if the new best solution is different than the previous interim solution the number of bits used to encode each delta value (parameter) is reduced by one bit. This effectively shrinks the hypercube. Reductions are limited to some minimal number of bits for the delta values. The new interim solution need not have a higher fitness than the last, only different. This might seem in conflict with the elitist nature of the search but is consistent with the notion that the search should not be driven by the single best solution. Second, if a delta run converges to exactly the same solution (i.e. delta values of all zeros), the number of bits used to encode the delta values ($\pm\delta$) is increased by one bit, expanding the hypercube. The expansion mechanism is bounded by the initial parameter encoding length.

The reduced hypercube produces a more focused search. However, use of this mechanism requires an increased number of parameters to govern the reduction/expansion strategies. It also introduces the possibility of excluding the global optimum from the reduced search space. However, parts of the solution space that are excluded from the reduced search space at one delta iteration can be reintroduced later. Even if the search space is not expanded, each new interim solution repositions the hypercube and defines a new version of the search space.

2.2 Empirical Performance: An Example Application

Delta coding has been used to find near optimal rigid body transformations to map a set of points in three dimensions (3-D) onto a target set of points in a 2-D domain. Eight transformation parameters were encoded using 12 to 14 bits each, for a total of 104 bits with transformation parameters of translation along and rotation around three axes, a uniform scale, and perspective. A fitness function designed to measure Euclidean distance between the target and transformed points was used to evaluate the transformation parameter sets (Mathias, 1991). Delta coding was evaluated against GENITOR and GENITOR II, which is a distributed island model genetic algorithm using GENITOR as its basis (Whitley and Starkweather, 1990).

Genetic Algorithm	Number of SubPopulations	Total Population	Total Trials	Average Error	Best Error
GENITOR	1	5,000	500,000	1.45	0.4900
GENITOR II	10	5,000	500,000	1.22	0.3270
GENITOR II	100	5,000	500,000	0.88	0.1950
Delta Coding	1	500	500,000	0.10	0.0037

Table 1. *Delta Coding* Compared with GENITOR and GENITOR II.

A very brief summary of the comparison experiments for the 3-D to 2-D transformation problem is provided in Table 1. All results are averaged over 30 experiments. Delta coding performs better than the other genetic algorithms tested and converges faster using an overall population 10% the size of the others (Whitley et al., 1991). Our previous research has shown that a reduced population size can translate to a significant savings with respect to both time and total computation effort (Starkweather, Whitley and Mathias, 1991). Performance comparisons over time are shown in Figure 2. The "spiking" behavior represents the restarts of the algorithm using delta coding. These restarts are also the points where the search space has been remapped.

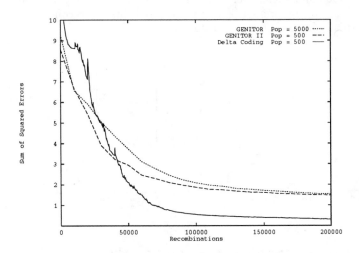

Figure 2: Delta Coding Performance Comparisons.

2.3 Limitations of Remapping with Delta Coding

Analysis of the remapping strategy in delta coding reveals that hyperspace is transformed on each iteration but in an unpredictable fashion since the mapping is based on a numeric shift which is unrelated to Hamming space. This is a direct result of adding or subtracting the delta values to the interim solution parameters. Table 2 shows the delta coding for a 3 bit space where the interim solution has converged to 000. Delta coding interprets the binary string as a signed magnitude number.

numeric parameters	0	1	2	3	4	5	6	7
binary coding	000	001	010	011	100	101	110	111
numeric shifts	0	1	2	3	-3	-2	-1	-0
simple delta coding	000	001	010	011	111	110	101	100

Table 2. Example of the Numeric Shift within Delta Coding.

Note that the numeric parameter 7 is represented by 100 in the delta coding, which is adjacent in Hamming space to 000. In general, *if the interim solution is located at the origin of the hypercube,* then delta coding folds the space around the sign bit, so that the string which would be the complement of the interim solution in a normal binary encoding is placed adjacent to the interim solution in Hamming space when the search switches to delta coding.

Figure 3a illustrates a 4 bit fully deceptive function, referred to here as function *df0*, where the fitness of the strings is represented on the Y axis. All order-1, order-2 and order-3 hyperplanes lead the search toward the string 0000, referred to here as the *deceptive attractor* (Whitley 1991; 1992a). The global optimum is at 1111. The strings in Figure 3a are organized according to Hamming distance with respect to 0000 and then ascending numeric value. This means strings are sorted first according to the number of 1 bits they contain, and second according to what would normally be their numeric value.

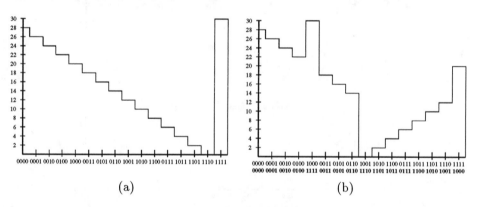

(a)   (b)

Figure 3: Remapping Search Space with Delta Coding; Interim Solution 0000. Figure 3b includes the delta encoding (row 1) and the original encoding (row 2).

Applying the folding principles illustrated in Table 2 to function *df0* (Figure 3a) produces the delta coding version of the function which is shown in Figure 3b. Figure 3b is sorted using the Hamming distance ordering as described above. While a one dimensional ordering of strings does not represent the extent of the remapping, it does provide insight into the way the space has been remapped. There are still two local minima in Hamming space: one located at 1000 and another at 1111. (Note that 0000 is no longer a local minima). However, most lower order hyperplanes now lead toward the global optimum and analytical results based on the work of Vose and Liepins (1990) and Whitley (1992b) indicate that a traditional simple genetic algorithm (Goldberg, 1989a) will now converge to the global optimum.

It is easy to see that the remapping achieved in Table 2 and in Figure 3 might be a useful one. The real question we would like to address is whether one can *consistently* remap hyperspace such that the remapping is helpful.

Delta coding does not always remap the complement of the interim solution to a point adjacent in Hamming space to the interim solution. This only occurs if the interim solution is at the origin of the hypercube. For example, a new function, referred to here as *df1*, can be produced by shifting function *df0*, shown in Figure 3a, so that the *deceptive attractor* corresponding to the local minima is now located at 0101, while all other Hamming distance relationships are maintained. The global optimum in the new function, *df1*, is located at 1010. This "shift" can be achieved by doing a bitwise exclusive-or between 0101 and all strings in the space corresponding to the function in Figure 3a. (Note that the numerical ordering relationships are not maintained). Assuming that a genetic algorithm converges to 0101 on the first iteration while trying to optimize function *df1*, delta coding will remap the string 0101 to 0000 but now remaps its complement, 1010, to 0101 as shown in Figure 4. In this case analytical results indicate that delta coding has remapped hyperspace such that the genetic algorithm still will not converge to the global optimum.

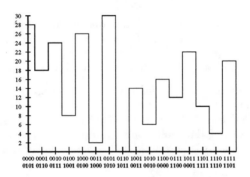

Figure 4: Delta Coding Remapped Space for Function *df1*; Interim Solution 0101. Figure 4 includes the delta encoding (row 1) and the original encoding (row 2).

We would argue that a remapping policy should be invariant to simple shifts in the function space. Since both functions *df0* and *df1* are really the same function the remapping policy should treat them in exactly the same way.

3   Canonical Delta Folding

Canonical delta folding employs the same iterative mechanisms present in the original delta coding strategy but includes a more powerful and predictable method for remapping hyperspace. The key difference between the original delta coding strategy and canonical delta folding is in the method by which parameters are remapped during the delta iterations. Original delta coding remapped parameters based on positive or negative offsets. This created a shift of all of the strings in numeric space.

The remapping performed in *canonical delta folding* is applied in two phases. First, the space is placed in a canonical ordering with respect to the interim solution. This produces an ordering of strings consistent with respect to Hamming distance relationships regardless of the interim solution that is chosen. Second, the search space is folded. The canonical ordering is accomplished by using *exclusive or* (denoted ⊕) to "shift" and thereby reorder all strings in the search space so that the interim solution is mapped to the string composed of all zeros. In a 3 bit problem space, this is equivalent to rotating the cube so that the interim solution is moved to the origin. At this point, *the hypercube has not been remapped* in the sense that exactly the same set of hyperplane partitions are preserved and all of the Hamming distances between specific solutions are still the same. For purposes of genetic search, this is still exactly the same search problem.

In canonical delta folding, it is actually the fold which remaps the strings in the new search space, changing the relationships between strings and hyperplane partitions in the space. To fold, if the first bit (or fold bit) of the string in canonical space is one, then all other bits in the string are complemented. The combined ordering and folding operation is represented as:

$$\delta_S = [C_{f1}(S \oplus I)]$$

where S is the string from the population being remapped, δ_S is its *canonical delta fold* representation, I is the bit string of the interim solution, and C_{f1} is the bit-wise complement function applied conditional upon the value of the "fold bit" being one. The subscript "f1" denotes that the first bit is used as the fold bit.

The remapping produced by *delta folding* is identical to the remapping produced by delta coding if 1) the interim solution is the string of all zeros and 2) the first bit is used as the fold bit. Folding now occurs in Hamming space and not in the numeric space as before. The following example shows the canonical delta fold transformation as applied to the string 1010 where the interim solution is 0101:

$$(S \oplus I) = (1010 \oplus 0101) = 1111 \qquad [C_{f1}(S \oplus I)] = (C_{f1}(1111)) = 1000$$

The complement of the interim solution is now remapped to a point adjacent to the interim solution in the new hypercube. This strategy remaps hyperspace in a consistent fashion regardless of the value of the interim solution. Empirical tests indicate that *canonical delta folding* produces a predictable remapping of hyperspace at each iteration. Thus, we conjecture that the remapping employed in *delta folding* gives a different mapping of the space at each iteration provided that the interim solution is different.

Delta folding provides an invertible point-to-point mapping. The inversion of the mapping (which is typically what we want to do: that is, go from the delta fold hypercube back to the original parameter space) is performed as follows:

$$S = [(C_{f1}(\delta_S)) \oplus I].$$

Table 3 demonstrates the one stage remapping used by the original delta coding as compared to the two stage remapping of canonical reordering and *delta folding* using C_{f1}. The example is a simple 3 bit problem with a local optima (LOC) at 010 and the global optimum (GLO) at 101. It also assumes that the genetic algorithm converges to the interim solution, 010. Note that 010 in the original space is moved to 000 in the remapped hypercube and its complement, 101, is now located at the adjacent position, 100.

	LOC				GLO			
original mapping	000	001	010	011	100	101	110	111
numeric shifts	-1	-0	0	1	2	3	-3	-2
simple delta coding	101	100	000	001	010	011	111	110
canonical reordering: $S \oplus I$	010	011	000	001	110	111	100	101
delta folding: $C_{f1}(S \oplus I)$	010	011	000	001	101	100	111	110

Table 3. Example of Delta Coding and Canonical Delta Folding.

Table 4 illustrates a 3 bit remapping where 000 is the local optima (LOC) and the first bit is used as a fold bit. The global optimum (GLO) is 111. In this situation delta coding and delta folding remap the space in exactly the same way.

	LOC							GLO
original mapping	000	001	010	011	100	101	110	111
numeric shifts	0	1	2	3	-3	-2	-1	-0
simple delta coding	000	001	010	011	111	110	101	100
canonical reordering: $S \oplus I$	000	001	010	011	100	101	110	111
delta folding: $C_{f1}(S \oplus I)$	000	001	010	011	111	110	101	100

Table 4. Delta Coding and Delta Folding; Interim Solution 000.

3.1 Folding Hyperspace

Assuming that the interim solution is zero, or that canonical delta folding is being used with C_{f1}, then the remapping used in Tables 3 and 4 can be expressed using a matrix that maps strings from one coding of the space to another. For a 3 bit problem the following relationship holds between S, a string in the original coding of the space, and δ_S, a string in the delta fold space.

$$(I \oplus S) \begin{bmatrix} 1 & 1 & 1 \\ 0 & 1 & 0 \\ 0 & 0 & 1 \end{bmatrix} = \delta_S$$

All vectors are row vectors. All addition is mod 2, which is equivalent to exclusive-or. In general, the matrix for mapping strings of arbitrary length from the original coding of the space to the delta version of the search space has 1 bits in the row corresponding to the fold bit, 1 bits on the diagonal and 0 bits everywhere else. In this case the fold bit is the first bit. (Our thanks to Michael Vose for suggesting the essential form of this matrix.) The same matrix also maps δ_S back to S.

$$\left\{ \delta_S \begin{bmatrix} 1 & 1 & 1 \\ 0 & 1 & 0 \\ 0 & 0 & 1 \end{bmatrix} \right\} \oplus I = S$$

Liepins and Vose (1990) show that a matrix exists which will transform select fully deceptive functions into a form in which no deception exists; Battle and Vose (1991) explore the foundations of these transformations. Delta coding and delta folding are similar in as much as the matrix shown here removes *most* of the deception when select fully deceptive functions are remapped. Specifically, this remapping strategy will remove most of the deception for any fully deceptive function with 2 local minima because the local minima must be in a complementary or near complementary position in hyperspace (Whitley, 1991). Delta folding can never remove all of the deception in a fully deceptive function because the order-1 hyperplane that corresponds to the fold-bit is not remapped.

Remapping the search space with regard to the previous examples is beneficial because we know the structure of the underlying function. However, such *a priori* knowledge about the function is typically not available. Furthermore, understanding the problem well enough to know the best way to remap it almost certainly means the problem has already been solved. What can be the benefit of blindly remapping the space then? Or do the remappings necessarily have to be blind?

Liepins and Vose have argued that there exists some transformation that will make the genetic search "easy" for any problem. However, for a string of length L there exists $2^{(L^2)}$ different transformations. Therefore, the search space for finding an "easy" transformation is larger than the search space for the problem itself, which is only 2^L. This makes the search for an optimal mapping appear to be almost hopeless. While the remapping of hyperspace has the potential for increasing the number of local minima in the Hamming space and making the problem "harder" for the genetic algorithm to search, it also has the possibility of decreasing the number of local minima. However, the process of transforming the search space need not have the goal of finding an optimal mapping of hyperspace; in general, for remapping to be useful we need only to find some mappings of the space that are "easier" than the original mapping.

There are two key ideas that motivate the use of delta folding. First, it is important that the algorithm explores many mappings of hyperspace, since it is reasonable to assume that some mappings will be found that are easier than others. Second, there may be general heuristic principles which can be used to direct the search toward more effective mappings.

One important heuristic behind delta folding is the idea that the remapping strategy should attempt to resample those points in the search space discriminated against in the previous iteration. This is done by folding those points that are dissimilar to the interim solution back into the search space in positions closer to the interim solution in Hamming space.

Another important notion in the *canonical delta folding* remapping scheme is that any bit in a string or substring can serve as the fold bit. This allows the exploration of other mappings from iteration to iteration by simply choosing a new bit to fold the space. Later in this paper we explore ways to heuristically select fold bits that may be more likely to lead to useful remappings of hyperspace.

3.2 Reducing the Search Space with Canonical Delta Folding

Another important component of delta coding is the ability to focus the search in smaller and possibly more promising regions of the search space. This is done by reducing the number of bits used to encode the delta parameters that are added to the interim solution in the iterative stages of the search. The most apparent weakness in the delta coding reduction scheme is that the reduced space being searched was always a subset located numerically adjacent to the interim solution.

Using canonical delta folding, the first requirement of a hypercube reduction scheme is that the reduced mapping must be one-to-one and onto with respect to the original mapping of the function space. Three different mapping schemes are proposed here. All the reduction schemes presented in this section use the first bit as the fold bit. Each scheme maps the reduced strings onto the rightmost bit positions of the string being constructed and then pads the leftmost bits with either 1 or 0.

This first folding strategy presented is the one already used (e.g., in Tables 3 and 4) to show that delta folding and delta coding can produce the same remapping when the interim solution is zero. This mapping generalizes to other interim solutions for delta folding, assuming that the original problem space is canonically reordered with respect to the interim solution. But it does not generalize to any other mapping for delta coding. The pseudocode for mapping the reduced delta space string, δ_S, back to the original function space is as follows:

IF (bit1 of $\delta_S == 1$)	$\Longrightarrow$	$(C_{f1}(\delta_S))$ and pad to the left with 1's
ELSE	$\Longrightarrow$	Pad δ_S to the left with 0's
XOR Padded-string with current interim solution		

The points sampled in canonic space over multiple reductions of the search space for this strategy are represented in Figure 5. The interim solution is located at 0. The points sampled by the reduced mapping are denoted by the raised portions of the line. Going from top to bottom, each line represents a further reduction of the search space by a single bit. While this strategy has weaknesses it is instructive for understanding the behavior of the delta coding and delta folding algorithms and has been shown to be useful in some search environments.

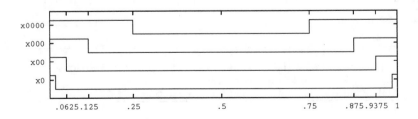

Figure 5: Points Sampled for Remapping Strategy 1.

This first strategy samples both information local to the interim solution and wraps around the space to sample information about the global complement of the interim solution. However, there are now different complements to be considered. Besides the global complement which exists in the original problem space, there is also a local complement within the reduced search space. The following algorithm is designed to sample both the global complement in the original problem space and the local complement in the reduced space. References to "bit1" and "bit2" refer to the first and second bit of δ_S, respectively.

IF (bit1 of δ_S == 0)	$\Longrightarrow$	Pad δ_S to the left with 0's
ELSE IF (bit1 == 1) && (bit2 == 0)	$\Longrightarrow$	$(C_{f1}(\delta_S))$ and Pad to the left with 1's
ELSE IF (bit1 == 1) && (bit2 == 1)	$\Longrightarrow$	Pad δ_S to the left with 0's
XOR Padded string with current interim solution		

In this second remapping strategy the points sampled and folded into the reduced search space are spread over the entire search space. Further inspection reveals that both the global complement (complement of the interim solution), and the local complement (complement of the interim solution within the reduced space) are always included in the sampling of the reduced search space. This strategy still folds the global complement of the interim solution to a position adjacent to the interim solution in Hamming space. Figure 6 shows the sampling pattern of this strategy over multiple reduction iterations.

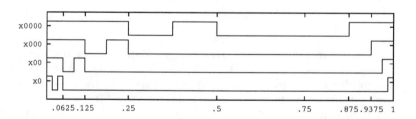

Figure 6: Points Sampled for Remapping Strategy 2.

The third strategy is another simple mapping strategy which has been found to be useful.

IF (bit1 == 1) & & (bit2 == 1)	$\Longrightarrow$	Prefix with 1
IF (bit1 == 1) & & (bit2 == 0)	$\Longrightarrow$	$(C_{f1}(\delta_S))$ and Prefix with 0
Pad prefixed-string to the left with 0's		
XOR Padded string with current interim solution		

The points sampled in canonic space over multiple reductions of the search space for this strategy are shown in Figure 7. The algorithm is quite similar to Strategy 2 except that instead of always incorporating the global complement of the interim solution into the reduced search space, the search space is narrowed more. The points sampled favor the current interim solution but also include the current complement of the reduced form of the interim solution and the complement of the interim solution at the level above (i.e. with one additional bit added to the representation). Therefore, in the first sampling the global and local complements are sampled. At the next level of reduction the local complement at the current level and the local complement of the current interim solution with one additional bit added are sampled. A possible weakness of this strategy is that the complements are not folded so as to be any closer to the interim solution in Hamming space.

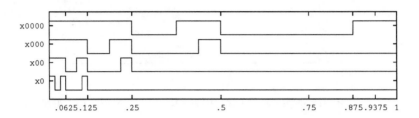

Figure 7: Points Sampled for Remapping Strategy 3.

3.3 Behavior Analysis Using an Executable Model

The executable equations developed by Whitley (1992b) are used to model the behavior of a simple genetic algorithm. These equations are a special case of the models introduced by Vose and Liepins (1991) in as much as the equations do not include mutation. For small problems (e.g., less than 15 bits) these equations allow the representation of each string in the space to be tracked with respect to its current representation in the population of a standard genetic algorithm using fitness proportionate reproduction. This model assumes an infinite population and no mutation.

The problems we have used to test remapping strategies are a series of fully deceptive and partially deceptive functions. These were chosen in part because if there is no deception in a function, the executable equations indicate that the genetic algorithm will always converge to the global optimum. Only a few of the functions used in the testing and analysis are provided here as examples.

Figures 8 and 9 show the tracking of the frequency of strings in the population over time for the fully deceptive function introduced earlier as *df1* in Figure 4. When the equations are executed for the original problem the simple genetic algorithm converges to the local optima, 0101 as expected. The string competitions for the delta coding space are shown in Figure 8. In this version of the space the genetic algorithm again converges to the local optima, now represented by the string 0000. The global optimum, 1010 in the original encoding, is represented as 0101 in the delta coding mapping. Delta coding reduces the number of bits in conflict between the global and local optima but there is a loose linkage between the common bits.

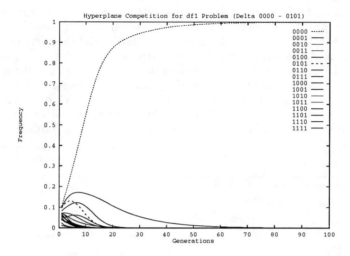

Figure 8: Hyperplane Competitions for Delta Coding of *df1*.

In the *delta folded* mapping, shown in Figure 9, the global optimum is remapped to the string 1000. When the global optimum is shifted to a position adjacent to the interim solution and the number of bits in conflict is reduced to one, it gains enough underlying hyperplane support to survive until the string can directly compete with the interim solution. This allows a simple genetic algorithm to converge to the global optimum.

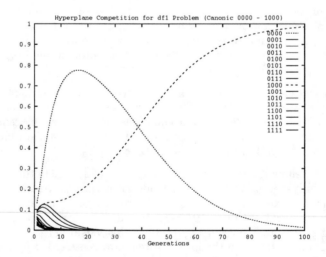

Figure 9: Hyperplane Competitions for Delta Folding of *df1*.

To illustrate the effectiveness of the reduction strategies a 4 bit function, referred to as *df2*, is introduced in Table 5. This function is deceptive in all order-2 and order-3 hyperplanes and all order-1 hyperplanes except one. The function has a local optima at 0011 and the global optimum is located at 1111. Tracking of the hyperplanes in the executable model shows that the genetic algorithm converges to the local optima at 0011 in the original encoding. Table 5 shows the fitness values for each of the strings and also shows the 3 bit representation resulting from the application of reduction strategy 1 as described above.

3 Bit Encoding	4 Bit Encoding	Fitness	3 Bit Encoding	4 Bit Encoding	Fitness
011	0000	25		1000	27
010	0001	24		1001	20
001	0010	25		1010	5
000	0011	28		1011	2
	0100	26	111	1100	4
	0101	23	110	1101	1
	0110	6	101	1110	0
	0111	3	100	1111	30

Table 5. Example Optimization Encoding (*df2*).

Remapping of the function using delta coding and delta folding locates the local optima at 0000 and the global optimum at 1011. The executable model shows that both versions of the remapped space converge to the local optima at 0000. While the tracking graphs are slightly different for the two mappings, they both exhibit similar behavior and result in the same fate. Figure 10 shows the hyperplane competitions for the canonical delta folding remapping. In both cases the hyperplane support needed to enable the genetic algorithm to converge to the global optimum is not provided.

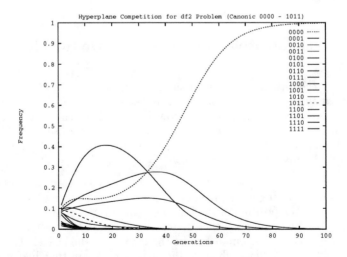

Figure 10: Hyperplane Tracking for Delta Folding on *df2*.

When reduction strategy 1 is applied to the original function the local optima, originally 0011, is mapped to 000 and the global optimum, originally 1111 is remapped to

100. The reduction of conflicting bits between the local and global optima provide the necessary underlying hyperplane support to allow the simple genetic algorithm to converge on the global optimum. The hyperplane competition for the 3 bit version of the search space is shown in Figure 11.

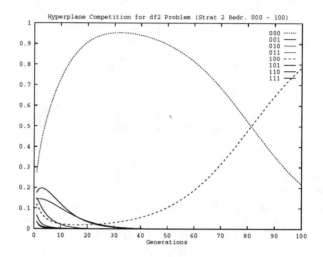

Figure 11: Hyperplane Tracking for Reduction Strategy 1 on *df2*.

4 Additional Adaptive Mechanisms in Delta Folding

The adaptive mechanisms described in the previous sections are useful for remapping the search space and for further focusing the search in a more restricted area corresponding to a reduced hypercube. However, another degree of freedom exists in all of the strategies that have been presented here. The bit used to fold the space (as well as the bits used to determine the prefixes used to pad the reduced strings) do not necessarily have to be the first bit(s).

The C_{f1} operation used for the delta folding operation restricted the folding trigger to the first bit in the string representation. This function can be generalized as C_{fx}, where x represents the position of the fold bit. Changing the fold bit merely means that the space is folded along another dimension. The general pattern of folding in canonical space will always remap half of the strings in the space. The bit which is chosen to fold on will affect which of the strings make up that half of the space which is remapped. For example, folding on the first bit causes the strings in the last half of the canonical ordering to be remapped. However, chosing the second bit will cause the second and last quartiles to be remapped, while chosing the last bit position for folding will remap every other string.

4.1 Behavior Analysis Using C_{fx}

The executable model used to track string representation during genetic search can be used again to study the effectiveness of changing fold bits. The function *df2*, defined in Table 5, was solved using a 3 bit strategy but re-converged to the local optima when mapped at the 4 bit level using the C_{f1} function. Table 6 describes the remapping using C_{f3} to fold the space for the *df2* function.

C_{f3} Encoding	Original Encoding	Fitness	C_{f3} Encoding	Original Encoding	Fitness
1110	0000	25	0110	1000	27
1111	0001	24	0111	1001	20
0001	0010	25	1001	1010	5
0000	0011	28	1000	1011	2
1010	0100	26	0010	1100	4
1011	0101	23	0011	1101	1
0101	0110	6	1101	1110	0
0100	0111	3	1100	1111	30

Table 6. C_{f3} Encoding for Function *df2*.

In the C_{f3} delta folded space the global optimum, originally 1111, is represented by the string 1100. There are no more bits in common between the local and global optima in the new mapping than were in the original representation. There are actually less bits in common than in the original folding using the C_{f1} mapping. However, the schema fitness support has been rearranged such that the global optimum now receives enough lower order schema fitness support to solve the problem. This is shown in Figure 12.

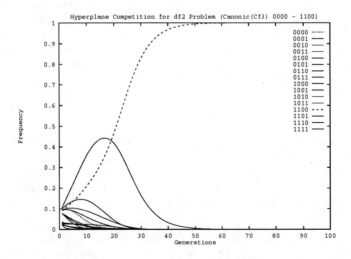

Figure 12: Hyperplane Competitions for C_{f3} Delta Folding on *df2*.

4.1.1 Doing the Right Thing

Showing that there is a way to fold the search space in order to improve the effectiveness of a simple genetic algorithm on specific problems does not necessarily demonstrate that using different folding strategies is generally useful. One might argue that if the problem can be analyzed to the point where a successful strategy can be chosen, then enough is known about the problem that it might solvable by other, more straightforward methods. Also choosing the correct point to apply a particular strategy may not be obvious.

Some clues are available during search which might be generally useful for narrowing the choice of folding strategies. For example, after running the genetic algorithm over multiple iterations and examining the various interim solutions we found that the most advantageous fold bits appear be those which are commonly shared between the various interim solutions. This is partly related to the fact that the order-1 hyperplane that corresponds to the fold bit is the only order-1 hyperplane that is not remapped by the folding process. Furthermore, it has the effect of reshuffling those areas in the search space in which there is less agreement about the overall solution with respect to the various interim solutions. At this point such a strategy can only be considered a heuristic; this work is too preliminary to make any general claims about the effectiveness of this strategy. So far, however, this heuristic has lead to a solution for all of the various deceptive and partially deceptive problems we have examined.

5 Discussion

While canonical delta folding has been designed to include the same advantageous mechanisms found in delta coding, delta folding has several mechanisms aimed at correcting the weaknesses in delta coding. One advantage of canonical delta folding is that multiple strategies for remapping hyperspace are available and if one strategy does not result in progress toward solving an optimization problem, another can be tried. Mechanisms for remapping hyperspace can be continually applied until a satisfactory solution is reached or until the predetermined amount of work the user is willing to expend is exhausted. However, with this new flexibility comes complexity. Where there was once only a single strategy now choices must be made on which strategy to use and when a strategy has outlived its usefulness. The best choices may not be obvious in most situations and strategies must be developed as discussed in section 4 to make good decisions.

Canonical delta folding not only provides more strategies than delta coding for remapping the space but has more flexibility in how they are applied. For instance, the folding strategies need not be applied on a parameter by parameter basis but can be applied over the entire string. Canonical delta folding also does not restrict the folding to a single bit as in delta coding.

Canonical delta folding also provides a mechanism to deal with oscillation between a set of different interim solutions. Oscillation can occur when the remappings converge back to a previous interim solution after re-increasing the number of bits in the encoding. Oscillating between two (or more) interim solutions as the search space is reduced or expanded can effectively stall the search. In delta coding, the single remapping strategy provided no way to avoid this situation. However, with delta folding oscillation can typically be stopped, once detected, by switching the remapping strategy or folding bit.

It is also important to realize that the problems presented here as test functions are very simplistic and are used to make it easy to observe and understand the behaviors of the remapping strategies. In these simple functions the genetic algorithm typically converges to the same local optima. In real problem environments we would expect the genetic algorithm to converge to different points in the search space, especially in multi-modal functions. Each time a new interim solution is converged upon the remapping of the search space is extended providing more search power for the canonical delta folding algorithm.

Acknowledgements

This work was supported in part by NSF grant IRI-9010546.

References

[1] Battle, D. and Vose, M. (1990) "Isomorphisms of Genetic Algorithms." In, *Foundations of Genetic Algorithms.* G. Rawlins, ed. Morgan Kaufmann.

[2] Eshelman, L., (1991) "The CHC Adaptive Search Algorithm: How to have Safe Search When Engaging in Nontraditional Genetic Recombination." *Foundations of Genetic Algorithms.* G. Rawlins, ed. Morgan Kaufman.

[3] Goldberg, D., (1989a) *Genetic Algorithms in Serach, Optimization and Machine Learning.* Addison-Wesley.

[4] Goldberg, D., (1989b) "Sizing Populations for Serial and Parallel Genetic Algorithms." *Proceedings of the Third International Conference on Genetic Algorithms.* Morgan Kaufmann.

[5] Krishnakumar, K., (1989) "Micro-Genetic Algorithms for Stationary and Non-stationary Function Optimization." *SPIE's Intelligent Control and Adaptive Systems Conference*, Paper # 1196-32.

[6] Liepins, G. and Vose, M. (1990) "Representation Issues in Genetic Algorithms." *Journal of Experimental and Theoretical Artificial Intelligence*, 2:101-115.

[7] Mathias, K.(1991) "Delta Coding Strategies for Genetic Algorithms." *Thesis.*, Department of Computer Science, Colorado State University, Fort Collins, Colorado.

[8] Schraudolph, N., Belew, R., (1990) "Dynamic Parameter Encoding for Genetic Algorithms." *CSE Technical Report* #CS 90-175.

[9] Shaefer, C., (1987) "The ARGOT Strategy: Adaptive Representative Genetic Optimizer Technique." *Genetic Algorithms and their Applications: Proc of the Second International Conference*

[10] Starkweather, T., Whitley, D., Mathias, K., (1990) "Optimization Using Distributed Genetic Algorithms." *Parallel Problem Solving from Nature.* Springer/Verlag.

[11] Vose, M. and Liepins, G. (1991) "Punctuated Equilibria in Genetic Search." *Complex Systems*, 5:31-44.

[12] Whitley, D. (1989) "The GENITOR Algorithms and Selective Pressure: Why Rank Based Allocation of Reproductive Trials is Best." *Proceedings of the Third International Conference on Genetic Algorithms.* Morgan Kaufmann.

[13] Whitley, D., (1991) "Fundamental Principles of Deception in Genetic Search." *Foundation of Genetic Algorithms*. G. Rawlins, ed. Morgan Kaufmann.

[14] Whitley, D., (1992a) "Deception, Dominance and Implicit Parallelism in Genetic Search." *Annals of Mathematics and Artificial Intelligence*. 5:49-78.

[15] Whitley, D., (1992b) "An Executable Model of the Simple Genetic Algorithm." *Foundations of Genetic Algorithms-II*.

[16] Whitley, D. and Starkweather, T., (1990) "*GENITOR II*: A Distributed Genetic Algorithm." *J. Experimental and Theoretical Artificial Intelligence*. 2:189-214.

[17] Whitley, D., Mathias, K., Fitzhorn, P. (1991) "Delta Coding: An Iterative Search Strategy for Genetic Algorithms." *Proceeding of the Fourth International Conference on Genetic Algorithms*. Washington, D.C., Morgan Kaufmann.

PART 5

GENETIC OPERATORS AND THEIR ANALYSIS

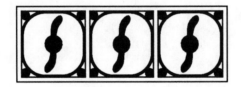

Real-Coded Genetic Algorithms and Interval-Schemata

Larry J. Eshelman and J. David Schaffer
Philips Laboratories
North American Philips Corporation
345 Scarborough Road
Briarcliff Manor, New York 10510

Abstract

In this paper we introduce *interval-schemata* as a tool for analyzing real-coded genetic algorithms (GAs). We show how interval-schemata are analogous to Holland's symbol-schemata and provide a key to understanding the implicit parallelism of real-valued GAs. We also show how they support the intuition that real-coded GAs should have an advantage over binary coded GAs in exploiting local continuities in function optimization. On the basis of our analysis we predict some failure modes for real-coded GAs using several different crossover operators and present some experimental results that support these predictions. We also introduce a crossover operator for real-coded GAs that is able to avoid some of these failure modes.

1 INTRODUCTION

A growing number of researchers in the genetic algorithm (GA) community have come to champion real-coded (or floating-point) genes as opposed to binary-coded genes, in spite of the fact that there are theoretical arguments purporting to show that small alphabets should be more effective than large alphabets. Although a few theorists have taken on this argument (Antonisse, 1989; Wright, 1991), the standard defense has been the practical one that experience shows that real-coded genes work better (Davis, 1991a). In this paper we take on the task of giving a theoretical defense of real-coded GAs. In the past such GA-heretics have been a small minority, who were largely ignored, but recently their numbers have been growing. Furthermore, in the last few years the GA community has begun to pay attention to the work of the Evolutionary Strategy approach in Europe which has from the beginning used real-coded genes (Bäck, Hoffmeister & Schwefel, 1991). Finally, descendents of the Evolutionary Programming approach have recently published a critique of GAs using binary coding and crossover, arguing that gaussian mutation on real-coded genes can be counted on to work well for a larger class of problems (Fogel & Atmar, 1990). (For more details on the history of real-coded GAs see Goldberg (1990).)

Although a number of advantages have been offered, we believe that three have been the primary motivation for real-coded GAs. First, real-coding of the genes eliminates the worry that there is adequate precision so that good values are representable in the search space. Whenever a parameter is binary coded, there is always the danger that one has not allowed enough precision to represent parameter values that produce the best solution values. Second, the range of a parameter does not have to be a power of two. Third, GAs operating on real-coded genes have the ability to exploit the gradualness of functions of continuous variables (where gradualness is taken to mean that small changes in the variables correspond to small changes in the function) (Wright, 1991).

In this paper we will concentrate on the third feature. Strictly speaking, the first point does not distinguish real-coded GAs from binary coded GAs. All computer solution methods require a discretization and hence a Nyquist-like assumption (i.e., no large variations between sample points). Given that computers have limited precision and "real-coded" (i.e., floating-point) values can be mapped onto integers, we will focus on what might be more properly called *integer-coded* GAs. (Several methods have been suggested for dealing with the precision problem without giving up a binary representation (Schraudolph & Belew, 1991; Shaefer, 1987).) We will not address the second alleged advantage, but will assume that the parameters of an integer-coded GA range over the same set of values as binary coded parameters—powers of two. In other words, it will be assumed that GAs that we will be comparing will use representations that have the same range and precision for any given function. The only difference is that the integer-coded GA will create new individuals by operating on strings of integers rather than bit strings. This approach allows us to use the same functions for comparing the performance of a GA that uses integer coding with a traditional GA that uses binary coding.

2   INTERVAL-SCHEMATA AND CROSSOVER

Most of the theoretical objections to real-coded GAs assume that the crossover operator operates at parameter boundaries (e.g., Goldberg, 1990, 1991). But many implementors of real-coded GAs use more vigorous forms of crossover. Davis combines parameter-bounded crossover with a crossover operator that averages (some of) the parameters (Davis, 1991a). Wright's linear crossover operator creates three offspring: Treating the two parents as two points, p_1 and p_2, one child is the midpoint of p_1 and p_2, and the other two lie on a line determined by p_1 and p_2: $1.5p_1 - 0.5p_2$ and $-0.5p_1 + 1.5p_2$ (Wright, 1991). Radcliffe's flat crossover chooses parameters for an offspring by uniformly picking parameter values between (inclusively) the two

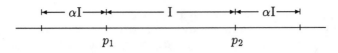

Figure 1: BLX-α

parents parameter values (Radcliffe, 1990). We use a crossover operator that is a generalization of Radcliffe's which we call blend crossover (*BLX-α*). It uniformly picks values that lie between two points that contain the two parents, but may extend equally on either side determined by a user specified GA-parameter α (see Figure 1). For example, BLX-0.5 picks parameter values from points that lie on an interval that extends $0.5I$ on either side of the interval I between the parents. (These are the extrema used by Wright.) BLX-0.0, on the other hand, is equivalent to Radcliffe's flat crossover. (Of course, there are many other possible versions of crossover that operate on integer or real-coded parameters.)

What all these crossover operators have in common is that they exploit the parameter intervals determined by the parents rather than the patterns of symbols they share. Holland's language of schemata was developed for strings of symbols. We feel that this terminology is too restrictive for analyzing real-coded GAs. Something analogous, yet distinct, is needed for GAs that manipulate interval values rather than bit values. We suggest that the relevant concept is an *interval-schema*.

Let $n = 2^L$ be the size of the range for integers that could be coded as L-bit strings. The number of interval-schemata (including all possible subranges) that can be defined over this range of integers is:

$$\sum_{i=1}^{n} i = \frac{n(n+1)}{2}$$

Thus, 36 interval-schemata can be defined for a parameter whose range is $[0,7]$. There are two interval-schemata of length 7, $[0,6]$ and $[1,7]$. At the other extreme are the 8 short interval-schemata, $[0,0]$, $[1,1]$, etc. Every specific parameter value is a member of at least n interval-schemata (points that lie at the extrema) and up to a maximum of $\lfloor (n+1)^2/4 \rfloor$ interval-schemata (points that lie in the center of the interval). In particular, a value of k for the parameter that ranges from $[0, n-1]$ is a member of $(k+1)(n-k)$ interval-schemata. Furthermore, the two points at positions k_1 and k_2 $k_1 < k_2$ have $(k_1+1)(n-k_2)$ interval-schemata in common.

It should be noted that interval-schemata are similar to Wright's connected schemata (Wright, 1991). Both interval-schemata and connected schemata are concerned with intervals. However, connected schemata are like ordinary schemata except the "don't-cares" are restricted to the lower order bits. Thus, 01### corresponds to the interval $[8,15]$ (assuming binary coding). The problem with this way of analyzing intervals is that some intervals are not representable with symbol-schemata—for example, no single symbol-schema describes the interval $[7,10]$. Thus, according to Wright's analysis for a parameter of range $[0, n-1]$ there are $2^{L+1} - 1$ connected schemata, far fewer than the $n(n+1)/2$ interval-schemata (where $n = 2^L$) identified by our analysis.

Of course, if this is to be more than an abstract counting exercise it must be related to how the algorithm samples interval-schemata. Clearly any selection mechanism that allocates exponentially increasing trials to the observed best individuals will also allocate exponentially increasing trials to the observed best interval-schemata. So the important question is how crossover preserves and explores interval-schemata.

Parameter-bounded crossover, Davis's averaging operator, and Radcliffe's flat crossover (BLX-0.0) all have the property that the offspring are members of the same interval-schemata of which the parents are common members. (In Radcliffe's terms, these operators are "respectful.") These algorithms differ, however, as to how many new interval-schemata are potentially reachable in a single crossover event. Parameter-bounded crossover and averaging crossover are both strongly biased toward certain interval-schemata over others. BLX-0.0, on the other hand, is much less biased in this respect, although it does have a bias toward points near the center of the interval. It should be noted that this difference in reachability has a parallel for symbol-schemata using either two-point crossover (2X) or uniform crossover (UX). Because UX eliminates the positional bias of 2X, many more schemata are potentially reachable via a single crossover event using UX than 2X (Eshelman, 1991).

The negative side of greater reachability is that repeated applications of such a crossover operator are also more likely to disrupt the schemata that make the two parents better than average. As we have argued elsewhere, however, for a GA to be effective, it must not simply preserve schemata, it must test them in new contexts, and this entails disrupting schemata (Schaffer, Eshelman, & Offutt, 1991). Furthermore, we have shown that by combining a disruptive crossover operator with a conservative selection mechanism that maintains a population of the best M individuals seen so far (where M is the population size) one often gets the best of both worlds—vigorously testing schemata in new contexts while preserving the best schemata discovered so far.

The way an interval-processing GA (IPGA) processes interval-schemata is analogous to the way a symbol-processing GA (SPGA) processes symbol-schemata. To understand the parallel, it is important to note that long interval-schemata correspond roughly to low order symbol-schemata. Both are characterized by not being very specific. As search progresses a SPGA will progressively focus its search on higher order schemata whereas an IPGA will progressively focus on shorter interval-schemata. In the former case, the SPGA has narrowed the search down to certain partitions, whereas in the latter case the IPGA has narrowed the search to certain contiguous regions. The interval-schemata that are being searched are those bounded by the parameter extrema contained in the population. As these values narrow, the search becomes more and more focused, taking its samples from a smaller and smaller region of the parameter range. In this way an IPGA exploits the local continuities of the function.

Finally, it should be noted that the above analysis is easily extendible to multi-parameter functions. If the function has two parameters, for example, then each instance (point in a two dimensional space) is a member of a set of rectangle-schemata. The number of rectangle-schemata that a particular instance is a member of will be the product of the number of intervals the first parameter is a member of times the number of intervals the second parameter is a member of.

3 FAILURE MODES OF AN IPGA

Every successful search algorithm exploits some biases allowing it to favor some samples over others. Every bias also has an Achilles' heal—a problem can always be devised that will mislead a search method depending on a special bias. An IPGA is no exception. If the problem has no local (Euclidean) continuities, or if they lead away from the optimal solution, then an IPGA is likely to have difficulties. In this section we are concerned, however, with failure modes that arise even for problems that on the surface appear as natural for an IPGA as a SPGA. We shall concentrate on two failure modes: failure to propagate good schemata and premature convergence.

3.1 FAILURE TO PROPAGATE GOOD SCHEMATA

There are a number of situations where an IPGA will have difficulty propagating good schemata, but it is instructive to consider an extreme case—a needle on a plateau. In other words, there is only a single value in the interval that is good, and all other values are equally bad. Of course, finding the optimum will be equally hard for a SPGA as an IPGA. However, suppose that the function consists of a number of independent needle-on-a-plateau genes, or suppose the genes have some structure, but there is a plateau in the region of the optimum. The successful algorithm requires a crossover operator that has a fairly high likelihood of passing on to the offspring those genes that are by chance the optimum allele. Clearly, 2X has this property, since it has a relatively high probability in a many-gene problem of passing on any single gene intact. Perhaps less obvious, UX will also be more successful than BLX-0.0 at propagating optimal values when surrounded by a plateau. If we look at the extreme case where an optimum value is crossed over with its complement, the cases appear the same. Suppose the optimum lies at one of the extrema of a parameter that ranges over 2^L values—e.g., it is coded as the integer 0 for an IPGA and a string of L zeros for a SPGA with binary coding. Then if the parameter is crossed over with its complement (i.e., $2^L - 1$ in the case of the IPGA and L ones in the case of a SPGA) to produce a single child, there is only a $1/2^L$ probability of the allele surviving in the child. Although the probability of propagating the allele is the same in the worst case, this is not the typical situation. To see this, the important thing to note is that if there is no structure around the optimum, then the mate is likely to be a random individual in this range.[1] In the case of BLX-0.0 the expected value of a randomly generated gene will differ from the optimum by one half the range, and in the case of UX one half the bits. In other words, the probability of propagating the optimum when mated with a randomly chosen individual is $2 \times 1/2^L = 1/2^{L-1}$ for BLX-0.0, whereas it is $1/2^{L/2}$ for UX. If the optimum lies in the center rather than an extrema, the probability of propagating the optimum doubles for BLX to $1/2^{L-2}$, since the expected distance of a randomly chosen value from the optimum is now 1/4 rather than 1/2, but for intervals coded

---

[1]This needs to be qualified. Initially, when the optimum is first discovered, the values on the plateau will tend to be random. But as this optimum value is repeatedly crossed over with other values, to the extent that some offspring survive these matings, there is a tendency for the nonoptimum values to pick up "fragments" from the optimum allele and to drift in the "direction" of the optimum for both a SPGA and an IPGA.

with more than four bits, UX will still have a higher likelihood of propagating the optimum than BLX-0.0.

3.2 PREMATURE CONVERGENCE

In order for a GA to make progress it must focus its search. Convergence of the population is the necessary consequence of this. Unfortunately, it is highly likely that some of the observed correlations will be spurious (due to sampling error) so that the population may converge to a suboptimal region, discarding schemata that contain the optimum. No search algorithm that uses its prior samples to bias its future samples is immune to sampling error. As we have argued elsewhere, however, the less disruptive a crossover operator is, the more susceptible the algorithm will be to sampling error (Schaffer, Eshelman & Offutt, 1991). Furthermore, some functions are more likely to present misleading samples (e.g., deceptive functions) than others. The strength of BLX-0.0 is that it produces its samples in the contiguous regions defined by the points contained in the population. This means that BLX-0.0 is less likely to prematurely converge to the values that correspond to the lower order bits. (We will make an important qualification to this below.) 2X, on the other hand, is much more likely to prematurely converge on the lower order bits. 2X is good at preserving contiguous chunks of the chromosome intact, whereas BLX-0.0 is good at testing small variations of contiguous chunks (at least for the lower order bits). Unlike 2X, UX has no positional bias. It will be better at searching the lower order bits than 2X, but not as good as BLX-0.0.

BLX-0.0 pays a price, however, for this ability to exploit local information. To understand this, we need to look at what it means for the population to converge for an IPGA. Again, it is helpful to first examine the parallel phenomenon in a SPGA. In discussing schemata it is typically noted that a population of size M consisting of bit strings of length L contains between 2^L and $M * 2^L$ schemata. As a counting exercise this is true, but what is critical is how many new schemata potentially can be sampled in the population by crossover. If all the bits are converged, then the population consists of identical members, each of which has the same 2^L schemata. However, since there is nothing to cross over, no new schemata are being sampled. More generally, the number of schemata that are relevant to search is between 2^D and $M * 2^D$ where D is the number of bits of L that are not completely converged to the same alleles. The same holds for interval-schemata. The number of interval-schemata being searched by a GA using BLX is limited by the maximum and minimum values of the parameters represented in the population. Just as 2X or UX cannot introduce new alleles, BLX-0.0 cannot extend the interval ranges.

IPGAs that operate on large intervals share with SPGAs that use a large cardinality alphabet a particularly serious failure mode—premature convergence in the first generation. If the range of the parameters (or cardinality of the alphabet) is large relative to the population size, then the algorithm is quite likely to start its search without some values represented. This is a fatal weakness for an IPGA if the optimal point is at one of the extrema of the interval, for a crossover operator bounded by the two points determined by the parents (as in the case of BLX-0.0) will never be able to find the optimum unless it is enveloped by the original population. More generally, unless the extrema in the initial population envelop the optimal point, it cannot be reached via BLX-0.0 (or parameter-bounded or average crossover). This fact is consistent with our analysis of interval-schemata. Of course, even if the initial population contains extrema that straddle the optimal point, this may be of no value if the points on one side of the optimal point are very poor so that they don't survive. (Imagine the optimum is at the bottom of an incline that abuts a cliff, so that it can be approached from only one side—any point sampled on the other side of the optimum won't survive.)

This problem of optimal-extrema can be overcome by letting the range from which an offspring is chosen extend on either side of the interval defined by the parents' parameter values (i.e., let $\alpha > 0$). We note that in the absence of selection pressure all values for $\alpha < 0.5$ will exhibit a tendency to population convergence toward allele values in the center of their ranges. Only when $\alpha = 0.5$ does the probability that an offspring will lie outside its parents become equal to the probability that it will lie between its parents. In other words, $\alpha = 0.5$ balances the convergent and divergent tendencies in the absence of selection pressure.

Although it is no longer the case that the child will be a member of the same interval-schemata that the parents share, the algorithm still narrows its focus as the search progresses, and thus differs from mutation.[2] But this process is no longer monotonic. Whenever a point is sampled that lies outside the population extrema and survives, the extrema are expanded. In the long run the extrema will narrow (assuming that there is local structure to the function), but in the process of narrowing and widening the algorithm now can also shift its focus laterally.

4 EMPIRICAL COMPARISONS

We have two goals in this section: (1) to test whether BLX-α is susceptible to the failure modes predicted in the previous section, and (2) to test how well a GA using BLX-α does on a range of functions in comparison to other crossover operators.

4.1 FAILURE MODE TESTS

We devised four functions to test BLX-α's predicted failure modes. Our purpose in choosing these functions was not to come up with challenging problems, but to choose a set of simple functions that will enable us to test the failure modes predicted in the previous section.

We tested each function using both a traditional GA and CHC (Eshelman, 1991). In order to place the emphasis on the effects of crossover, no mutation was used in either algorithm. CHC differs from the traditional GA in several respects: (1) *Cross generational elitist selection*: the parent and child populations are merged and the best M individuals are chosen, where M is the population size. (2) *Heterogeneous recombination* (incest prevention): only individuals who are sufficiently different (in terms of Hamming distance) are mated. (3) *Cataclysmic mutation* (restarts): only crossover is used to produce new offspring, but when the population converges, massive mutations are applied, preserving the best individual intact, and the search is resumed using only crossover. We tested four crossover operators: BLX-0.0, BLX-0.5, 2X, and HUX. (HUX is like UX, except exactly half the differing bits

---

[2]See section 5 for a discussion of BLX-α's relationship with mutation.

f-incline   f-V   f-cliff

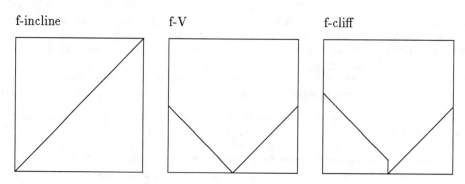

Figure 2: Failure mode test functions

Table 1: Failure Mode Tests

Number of times optimum found and trials to convergence							
	f-needles		f-incline		f-V		f-cliff	
	opt	trials	opt	trials	opt	trials	opt	trials
BLX-0.0	21	1597	0	1895	50	734	0	1926
BLX-0.5	20	1751	50	1418	50	968	49	1795
HUX	35	1409	47	1177	44	1153	47	1154
2X	29	831	9	1175	15	1175	8	1188
Tr-GA	11	1513	3	2111	0	1939	1	1563

are swapped at random.) Each of the four operators produces two children per mating. In each case, we used a population size of 50 and halted the search when either the minimum was found or the population converged (with no restarts). The traditional GA used proportional selection, the elitist strategy, a population size of 50, no mutation, and 2X with a crossover rate of 1.0.

The first function, f-needles, was devised to test the hypothesis that BLX-α would have difficulty in certain circumstances propagating good schemata. f-needles consists of five needles on five plateaus: for each of the five 6-bit genes a value of zero is given if the pattern is 111111 (i.e., 42 in gray code), and a value of one for all other patterns. The next three functions were devised to test the hypothesis concerning premature convergence (see Figure 2). The first, f-incline, is a simple incline problem with the minimum (the optimum) at one extreme: $f(x) = x$. The second, f-V, is a double incline or V function with the minimum at the center: $f(x) = |range/2 - x|$. The third, f-cliff, is similar to f-V except that the left incline has been raised so that there is a cliff on one side of the minimum: $f(x) = x - range/2 \ if (x \geq range/2)$ *otherwise* $f(x) = (6 \times range/10) - x$. For these three functions, x ranges from 0 to $2^{30} - 1$. For the SPGA tests the 30 bit string was interpreted as gray coded.

The results, averaged for 50 replications, are reported in Table 1. Our prediction that needles-on-plateaus would be relatively harder for BLX-α than HUX or 2X is confirmed by BLX-0.0's and BLX-0.5's worse performance on f-needles. Also as predicted, BLX-0.0 has difficulties on f-incline and f-cliff, but does quite well on f-V where the optimum lies in the center. The good performance of BLX-0.5 on these problems indicates that extending the interval outside the extrema determined by the parents overcomes a major shortcoming of BLX-0.0. Finally, the poor performance of 2X (for both a traditional GA and CHC) may come as somewhat of a surprise. It should be kept in mind, however, that no mutation was used with any of these runs, and a relatively small population size (50). The main failure mode for 2X on these functions is premature convergence—and in the case of the latter three functions, premature convergence on the lower order bits. Uniform crossover (HUX) is much less susceptible to this, and BLX-0.5 is even less so.

4.2 PERFORMANCE TESTS

We have seen that BLX-0.0 is susceptible to several failure modes as predicted, but that one of these shortcomings, the inability to sample parameter values that lie outside the extrema represented in the population can be overcome by using BLX-0.5. Our goal in this section is to discover how well BLX-0.5 performs relative to a crossover operator that operates on bit strings.

As our test suite we used a set of 11 functions for which we have extensive performance data (Schaffer, Caruana, Eshelman & Das, 1989; Eshelman & Schaffer, 1991; Eshelman, 1991). There are twelve problems in the test suite, but one of them (f10, a graph partitioning problem) is by nature a binary problem and so was not suitable for BLX-α. To these 11 we added two functions that other researchers have reported pose challenges to binary coded GAs—f13 and f14.

f13 is a function studied by Fogel and Atmar requiring the solving of a system of 10 linear equations with a 1.0 probability of the off-diagonal coefficient being non-zero (Fogel & Atmar, 1990). f14 is a 45 variable dynamic control problem studied by Janikow and Michalewicz (1991). For f13 we used 8-bit parameters and for f14, 10-bit parameters. (The binary representations of all functions were interpreted as gray coded.)

Table 2 summarizes some of the features of the thirteen functions. We used several variants of Davis's random bit-climber (i.e., an iterative, bit-wise hillclimber) as a measure of the difficulty of these functions (Davis, 1991b). Functions for which some version of the bit-climber significantly outperformed all versions of the GA were given an Easy rating, whereas functions for which some version of the GA significantly outperformed the bit-climber were given a Hard rating.[3] Functions for which the best version of the bit-climber and the GA performed about the same were given a Moderate rating.

We used CHC (with HUX as the crossover operator) as our benchmark since in two previous studies we showed that CHC outperformed a traditional GA for functions f1-f12 (Eshelman, 1991; Eshelman & Schaffer, 1991).[4] We ran CHC using BLX-0.5

---

[3] No version of the bit-climber was able to find the optimum for functions f8, f9, f11, f12 and f13. Furthermore, by shifting both axes by a small amount so that the optimum is no longer at the origin, functions f6 and f7 become hard for the bit-climber relative to the GA.

[4] For 10 of the 12 functions we were not able to find any parameter settings for a

Table 2: Function Summary

fnc	np	bpp	len	ep	bcr	description
f1	3	10	30	N	E	parabola
f2	2	12	24	Y	H	Rosenbrock's saddle
f3	5	10	50	N	E	stair steps
f4	30	8	240	N	H	quadratic with noise
f5	2	17	34	Y	E	Shekel's foxholes
f6	2	22	44	Y	M	sine envelope sine wave
f7	2	22	44	Y	M	stretched V sine wave
f8	16	4	64	Y	H	FIR filter
f9	30	5	150	Y	H	30-city TSP, sort representation
f11	20	5	100	N	H	needle on a plateau
f12	20	5	100	N	H	deceptive
f13	10	8	80	Y	H	10 linear equations
f14	45	10	450	Y	M	dynamic control problem

fnc   function
np   number of parameters
bpp   bits per parameter
len   string length
ep   epistasis among parameters (Yes, No)
bcr   bit-climber rating (Easy, Moderate, Hard)

and HUX on the 13 functions, halting each run when either the optimum was found or 50,000 evaluations had been completed. (Since f4 is noisy, it was required to be only "close" (two standard deviations) to the minimum.) Unlike the failure mode tests, restarts were enabled for these runs. The results are summarized in Table 3 based on 50 replications. BLX-0.5 did significantly better than HUX for f1, f2, f4, f13, and f14, whereas HUX did significantly better for functions f3, f5, f6, f11 and f12. For the remaining three functions there is no significant difference.

The five problems for which BLX-0.5 is the winner are the kind of functions that one might expect BLX-0.5 to do well on. They are all smooth, continuous functions (except for the discretization introduced by the representation). f1 and f4 are continuous and monotonic (in Euclidean space) with independent parameters. In the other three functions, f2, f13, and f14, there is interaction among the parameters. f2 is continuous and monotonic, but the discretization produces local minima in the region near the optimum. Based on the fact that a bit-climber can do quite well on f14, it seems that f14 is also monotonic. f13, on the other hand, seems to have many local minima. (We tried a variety of hillclimbers on f13, but none of them did very well.)

---

traditional GA (tr-GA) so that it outperformed CHC-HUX. The two exceptions were f1 (the easiest problem in the suite) and f12 (a deceptive problem). For seven of the functions, f5-f11, CHC performed significantly better. We tested the tr-GA using 840 different combinations of parameter settings for f1-f10, and 20 different combinations for f11 and f12, and picked the best settings for each function when comparing to CHC. CHC, on the other hand, was tested using its default parameter settings (population size of 50 and a restart mutation rate of 0.35).

Table 3: Performance tests

	BLX-0.5	sem	HUX	sem
Mean number of trials to find optimum				
f1	*874*	20	1089	25
f2	*4893*	357	9065	591
f3	2005	119	*1169*	27
f4	*933*	24	1948	97
f5	5561	588	*1396*	38
f6	14736	1998	*6496*	725
f7	3425	68	3634	291
f8	5822	522	7279	515
Mean performance				
f9	424.6	0.3	429.2	3.5
f11	1.5	0.2	*0.0*	0.0
f12	9.0	0.3	*1.2*	0.1
f13	*21.7*	2.4	61.8	11.2
f14	*16241.2*	30.1	38272.4	1039.0

The five cases for which BLX-0.5 lost are more instructive. BLX-0.5's poor performance on f11 and f12 is no surprise, since neither one has continuous variables with the sort of gradualness that BLX-0.5 can exploit. f11 consists of a 20 independent 5-bit genes, each of which is a needle on a plateau lying at one extrema. f12 consists of 20 independent 5-bit genes, each of which is deceptive. f3, like f11, also contains plateaus. Each of its 5 independent 10-bit genes has more structure than those of f11, but next to the optimum value there is a small plateau that gives BLX-0.5 some problems. (Each gene consists of 12 plateaus or steps, the optimum step being at one extrema, and only 10% as large as the previous step.) f5 consists of evenly spaced wells with sloped floors sunk in a plateau. Thus, the optimum value and the 24 suboptima are up against cliffs.

BLX-0.5's poor performance on f6 is harder to explain. It is a noiseless, continuous function without any plateaus or cliffs. On examination, it turns out that f6 illustrates not a defect in BLX-0.5 so much as a fortuitous advantage presented to HUX by the representation chosen. Since we believe that this is an important phenomenon, we will examine f6 in some detail.

Figure 3-a shows a 2D cross section through the origin of f6. f6 is cylindrically symmetric about the z axis. Figure 3-b shows a small region of f6 around the origin as seen from "above". The point in the center is the global optimum, and the concentric circles marked with dashed lines are the regions of the second, third, and fourth best local optima (counting from the center). The concentric circles marked with solid lines are the peaks of the ridges that separate the local optima.

Figure 3-c plots the points generated and evaluated during a single run of CHC using HUX and Figure 3-d plots the subset of points generated that are accepted into the parent population by replacing the worst members. Figures 3-e and 3-f show corresponding plots for BLX-0.5. Figures 3-c and 3-e dramatically illustrates the difference in how schemata are sampled via a SPGA and an IPGA—patterns

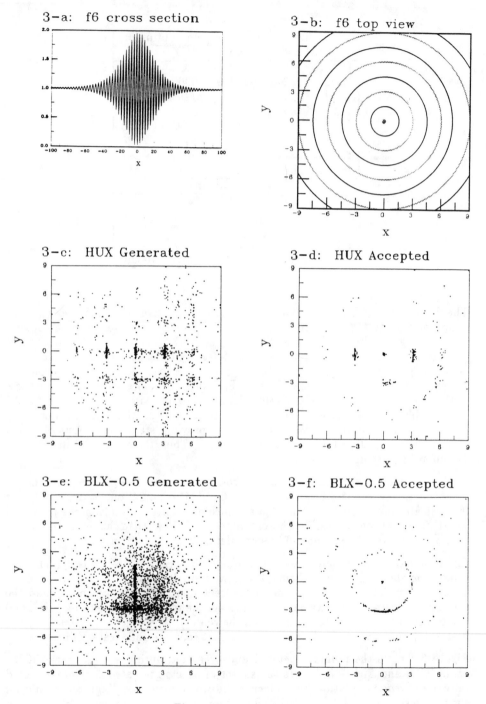

Figure 3: Function f6

vs intervals. One can see the outline of a grid-like structure filling much of Figure 3-c but not 3-e. (Keep in mind that the source (i.e., parents) of the points shown in Figures 3-c and 3-e are some of the points in figures 3-d and 3-f.)

To really understand the differences between the two search algorithms a dynamic dimension is needed. Both algorithms tend to get trapped in the best suboptimal region indicated by the inner, dashed circle in Figure 3-b. And both tend to favor points in this inner circle that intersect the x and y axes, although this tendency seems to be much stronger with HUX.[5] Given that good points tend to cluster in these areas, all crossover needs to do to put a point in the central region, where the global optimum resides, is to recombine a point in the (0, 3) region with a point in the (3, 0) region (for example). This would be easy to do with parameter-bounded crossover, but it turns out that in this case it isn't that hard with uniform crossover (HUX) either. The reason is that the function was fortuitously discretized so that the spacing between the concentric circles is nearly a power of 2. (The circle marking the best suboptima crosses the axes 65820 units from the center which is very close to $2^{16} = 65536$.) Since Gray coded values that are powers of two apart differ by only two bits (except when they are neighbors), the Gray coded values of the points where the inner suboptimum circle crosses either of the axes will differ by only two higher order bits from the optimum point in the center. However, some of the neighboring points in this suboptimal region will differ from the optimum by only one bit. By shifting both the axes by an amount that is not a multiple (or a near multiple) of the distance between concentric circles, e.g., 2^{14}, the points in the best suboptimal region will always differ from the optimum by at least two bits. This makes the problem somewhat harder for HUX. The grid-like pattern generated by HUX for the shifted problem is similar to that shown in figure 3-c except the spacing between "lines" is half as much as before (see Figure 4). BLX-0.5's performance is not affected, but HUX's performance deteriorates to that of BLX-0.5. We can make f6 even harder for HUX by rescaling the problem so that the distance between "valleys" is one-third of that of the original problem. This does not affect the performance of BLX-0.5, but HUX now does significantly worse than BLX-0.5.

In summary, the experimental data corresponds nicely with the theory. The functions for which BLX-0.5 does worse than HUX are either predicted by theory to be hard for BLX-0.5, or are cases where HUX has an advantage due to the periodicity (natural or fortuitous) of the problem.[6]

---

[5]The reason why points tend to cluster along the axes is that in these regions many new good points can be generated by changing one parameter value by a small amount whereas in regions that intersect lines at 45° to the axes good neighboring points can be generated only by changing both parameters by nearly equal and small amounts.

[6]L. Davis has discovered a similar phenomenon where a choice of representation (e.g., symmetrical about the origin) can make the problem easy for a bit-climber (Davis, 1991b). A SPGA, however, is much much more robust in this respect—there are many more ways to fortuitously introduce patterns which it can exploit.

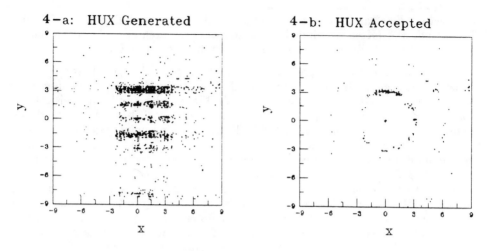

Figure 4: Function f6-shifted (output re-centered)

5 CROSSOVER VERSUS MUTATION

Although we are not making any claims about the general superiority of blend crossover, or BLX-0.5 in particular, the results of the empirical tests we have conducted have convinced us that BLX-0.5 is a powerful operator. It might be questioned, however, as to what kind of operator it really is. In particular, is BLX-0.5 really a crossover operator or is it more like a mutation operator? If one insists upon looking at BLX-0.5 from the symbol-schemata point of view it will not appear to qualify as a crossover operator. Even from the point of view of our interval-schemata framework, it fails to fully qualify as crossover—for unlike BLX-0.0, not all the interval-schemata commonly containing the parents are preserved in the offspring. In Radcliffe's terminology, BLX-0.5, unlike BLX-0.0, is not strictly a *respectful* recombination operator (Radcliffe, 1991). But before excommunicating BLX-0.5 from the growing communion of crossover operators, it is important to ask the converse question as to how it differs from mutation.

Unlike the typical mutation operator used with a real-coded GA, BLX-0.5's "step-size" is self-adjusting, and is a function of the extent to which the population is converged. If it is a mutation operator, it is a very special mutation operator that shares with crossover the property of increasingly focusing search. There are search algorithms in the literature that do use a dynamically adjusting step-size, so this won't necessary put BLX-0.5 in the crossover category (Beale & Bentley, 1984; Bäck, Hoffmeister & Schwefel, 1991). But unlike these other operators BLX-0.5 does not use aggregate information to determine the step size, but like crossover, uses the specific information contained in the parents being paired, i.e., in Radcliffe's terminology, it preserves some of the *locality formae* common to both parents (Radcliffe, 1991). This is an important feature that can easily be illustrated by an example. Suppose that the function has two parameters coded as two intervals. Furthermore, suppose that there are two good regions in the search space—one where both parameters have values at the low extreme and the other where both parameters have values at the high extreme. If the population is about evenly divided between instances from both these regions, then any mechanism that bases the mutation step-size on aggregate statistics is going to have a difficult time focusing in on the good regions. On the other hand, BLX-0.5 will have difficulty only when the parents are from different regions. This will only happen half the time. The other half the time parents will be chosen from the same region and search will progress.

The important point is that BLX-0.5, like all true crossover operators, but unlike mutation operators, including ones that are dynamically adjusted, implicitly exploits higher order correlations. Genes are not adjusted simply on the basis of the aggregate value of other instances of the same gene. Crossover implicitly takes into account the interaction among the genes when generating new instances. This is the source of crossover's power as a search operator.

6 CONCLUSION

With the new tool of interval-schemata, the reasons behind the empirical successes reported for real-valued GAs can now be better understood. Both IPGAs and SPGAs have the property of implicit parallelism. They differ in their biases. IPGAs exploit local continuities, whereas SPGAs exploit discrete similarities. It is natural to expect that for different problems different biases will provided a competitive advantage.

One future line of research is to explore an algorithm that uses both crossover operators, creating half the children using BLX-0.5 and half using HUX. Preliminary results suggest that the performance of an algorithm using both operators is better than the average of the performances of each operator used separately, and on some problems, e.g., f6-shifted, performs better than either alone.

References

H. J. Antonisse. (1989) A New Interpretation of Schema Notation that Overturns the Binary Encoding Constraint, *Proceedings of the Third International Conference on Genetic Algorithms*, Morgan Kaufmann, San Mateo, CA, 86-91.

T. Bäck, F. Hoffmeister & H. Schwefel. (1991) A Survey of Evolution Strategies, *Proceedings of the Fourth International Conference on Genetic Algorithms*, Morgan Kaufmann, San Mateo, CA, 2-9.

G. O. Beale & S. E. Bentley. (1984) Parameter Estimation Using Microprocessors and Adaptive Random Search Optimization, *IEEE Transactions on Industrial Electronics IE-31*, 1 85-89.

L. Davis. (1991a) Hybridization and Numerical Representation, in *The Handbook of Genetic Algorithms*, L. Davis (editor), Van Nostrand Reinhold, New York, 61-71.

L. Davis. (1991b) Bit-Climbing, Representational Bias, and Test Suite Design, *Proceedings of the Fourth International Conference on Genetic Algorithms*, Morgan Kaufmann, San Mateo, CA, 18-23.

L. J. Eshelman. (1991) The CHC Adaptive Search Algorithm: How to Have Safe Search When Engaging in Nontraditional Genetic Recombination, in *Foundations of Genetic Algorithms*, G. J. E. Rawlins (editor), Morgan Kaufmann, San Mateo, CA, 265-283.

L. J. Eshelman & J. D. Schaffer. (1991) Preventing Premature Convergence in Genetic Algorithms by Preventing Incest, *Proceedings of the Fourth International Conference on Genetic Algorithms*, Morgan Kaufmann, San Mateo, CA, 115-122.

D. B. Fogel & J. W. Atmar. (1990) Comparing Genetic Operators with Gaussian Mutations in Simulated Evolutionary Processes Using Linear Systems, *Biological Cybernetics 63*, 111-114.

D. E. Goldberg. (1990) Real-Coded Genetic Algorithms, Virtual Alphabets, and Blocking, IlliGAL Report 90001, Illinois Genetic Algorithms Laboratory Dept. of General Engineering University of Illinois at Urbana-Champaign, Urbana, IL.

D. E. Goldberg. (1991) The Theory of Virtual Alphabets, In *Parallel Problem Solving from Nature*, H. P. Schwefel and R. Männer (editors), Springer-Verlag, Berlin, 13-22.

C. Z. Janikow & Z. Michalewicz. (1991) An Experimental Comparison of Binary and Floating Point Representations in Genetic Algorithms, *Proceedings of the Fourth International Conference on Genetic Algorithms*, Morgan Kaufmann, San Mateo, CA, 31-36.

N. J. Radcliffe. (1990) Genetic Neural Networks on MIMD Computers, Ph.D. Dissertation, Dept. of Theoretical Physics, University of Edinburgh, Edinburgh, UK.

N. J. Radcliffe. (1991) Forma Analysis and Random Respectful Recombination, *Proceedings of the Fourth International Conference on Genetic Algorithms*, Morgan Kaufmann, San Mateo, CA, 222-229.

J. D. Schaffer, R. A. Caruana, L. J. Eshelman & R. Das. (1989) A Study of Control Parameters Affecting Online Performance of Genetic Algorithms for Function Optimization, *Proceedings of the Third International Conference on Genetic Algorithms*, Morgan Kaufmann, San Mateo, CA, 51-60.

J. D. Schaffer, L. J. Eshelman & D. Offutt. (1991) Spurious Correlations and Premature Convergence in Genetic Algorithms, in *Foundations of Genetic Algorithms*, G. J. E. Rawlins (editor), Morgan Kaufmann, San Mateo, CA, 102-112.

N. N. Schraudolph & R. K. Belew. (1991) Dynamic Parameter Encoding for Genetic Algorithms, Technical Report LAUR90-2795, Los Alamos National Laboratory, Los Alamos, NM.

C. G. Shaefer. (1987) The ARGOT Strategy: Adaptive Representation Genetic Optimizer Technique, *Genetic Algorithms and Their Applications: Proceedings of the Second International Conference on Genetic Algorithms*, Lawrence Erlbaum Associates, Hillsdale, NJ, 50-58.

A. Wright. (1991) Genetic Algorithms for Real Parameter Optimization, in *Foundations of Genetic Algorithms*, G. J. E. Rawlins (editor), Morgan Kaufmann, San Mateo, CA, 205-218.

Genetic Set Recombination

Nicholas J. Radcliffe

njr@epcc.ed.ac.uk

Edinburgh Parallel Computing Centre

University of Edinburgh

King's Buildings

EH9 3JZ

Scotland

Abstract

The application of genetic algorithms to optimisation problems for which the solution is a set or multiset (bag) is considered. A previous extension of schema analysis, known as forma analysis, is further developed and used to construct principled representations and operators for problems in this class. The extensions to forma analysis include the introduction of genes whose values cannot be assigned independently and a method for mediating between desirable but sometimes incompatible properties of recombination operators.

1 Introduction

This paper is concerned with optimisation problems for which the solution is a set or multiset (bag). Examples include selecting an investment portfolio, choosing a connectivity for a neural network and finding the best sites for a network of retail outlets given a choice of possible locations. Both the case in which the size of the set or multiset is fixed and the case in which it is subject to optimisation are considered.

The approach taken is based on *forma analysis,* an extension to schema analysis (Holland, 1975) developed previously (Radcliffe, 1991a, 1991b). Section 2 presents a brief review of forma analysis, but the reader unfamiliar with this approach may find it helpful to read Radcliffe (1991b), which provides a gentler introduction. Various formae (generalised schemata) for sets and multisets are introduced in sections 4–6, and suitable recombination operators for their manipulation are constructed. The key problems with set and multiset optimisation arise when their size is constrained. In this case, the natural formae are said to be *non-separable,* and the construction of satisfactory recombination operators is especially hard. This difficulty is tackled in section 5.2 by introducing genes whose values cannot always be assigned independently. This is achieved through the formalism of *non-orthogonal bases.* A new recombination operator, called *random assorting recombination,* is introduced to deal with this case in section 7.

2   Forma Analysis: Summary and Definitions

This section reviews forma analysis as developed in Radcliffe (1991a, 1991b). In later sections this will be used to analyse recombination of sets and multisets.

2.1   Equivalence Relations and Formae

Let S be a search space and let Ψ be a set of equivalence relations over S. Then the equivalence classes of all the equivalence relations in Ψ are referred to as *formae,* which are generalisations of schemata. The set of all formae induced by a set of equivalence relations Ψ will be written $\Xi(\Psi)$. Formae satisfy the "schema theorem" (Holland, 1975) provided that suitable disruption coefficients are chosen (Radcliffe, 1991a, Vose & Liepins, 1991). Disruption is discussed further in section 7.3.

It will be assumed throughout this paper that there are enough equivalence relations in Ψ to ensure that specifying all of the equivalence classes to which a solution in S belongs suffices to identify that solution uniquely.

2.2   Respect

A recombination operator X can be conveniently described by a function

$$X : S \times S \times \mathcal{K}_X \longrightarrow S \qquad (1)$$

which takes two parent solutions A and B from the search space, S together with a control parameter $k \in \mathcal{K}_X$, and produces a child in S. The control parameter, k, determines which of the (typically) many possible children is produced, so that for one point crossover k would be the cross point and for uniform crossover it would be a binary mask. A recombination operator X is said to *respect* the set $\Xi(\Psi)$ of formae induced by the equivalence relations in Ψ if and only if

$$\forall \xi \in \Xi \; \forall A \in \xi \; \forall B \in \xi \; \forall k \in \mathcal{K}_X : \; X(A, B, k) \in \xi. \qquad (2)$$

Thus respect requires that all children produced by recombination alone are members of all formae to which both their parents belong. For example, suppose there were equivalence relations for hair and eye colour in Ψ. Then if both parents had blue eyes and brown hair, respect would require that recombination only produce children with blue eyes and brown hair.

2.3   Similarity Set

It is convenient to introduce the notion of the *similarity set* of two parents, which is—loosely—the set of all solutions which share the characteristics that the two parent solutions

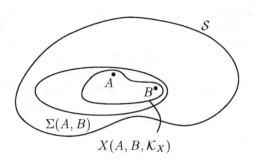

Figure 1: A recombination operator is said to *respect* a set of formae if given any pair of solutions A and B, all of their children under recombination are members of all the formae to which *both* parents belong. The *similarity set* of A and B, written $\Sigma(A, B)$, is the smallest forma which contains them both. This can be constructed as the intersection of all formae containing both parents, illustrated above. Respect amounts to the requirement that each child produced by recombination lies in the similarity set of its parents. Equivalently, the image $X(A, B, \mathcal{K}_X)$ of the control set $\mathcal{K}_X$ should be a subset of $\Sigma(A, B)$, as shown.

share. Formally, the similarity set of A and B, written $\Sigma(A, B)$, is defined by

$$\Sigma(A, B) = \bigcap \{\xi \in \Xi(\Psi) \mid A, B \in \xi\}. \tag{3}$$

This is the intersection of all formae to which both parents belong—the smallest forma to which they both belong provided that this intersection is itself an equivalence class induced by some equivalence relation in Ψ.

Respect can be seen to be identical to the requirement that every child lie in the similarity set of its parents, as is illustrated in figure 1. This allows respect to be re-expressed as

$$\forall A, B \in \mathcal{S} : \ X(A, B, \mathcal{K}_X) \subset \Sigma(A, B). \tag{4}$$

2.4 Assortment

A recombination operator X is said *properly to assort* the formae in $\Xi(\Psi)$ if and only if

$$\forall \xi_1, \xi_2 \in \Xi(\Psi) \ (\xi_1 \cap \xi_2 \neq \emptyset) \ \forall A_1 \in \xi_1 \ \forall A_2 \in \xi_2 \ \exists k \in \mathcal{K}_X : \ X(A_1, A_2, k) \in \xi_1 \cap \xi_2. \tag{5}$$

Thus proper assortment requires that, given parents $A_1 \in \xi_1$ and $A_2 \in \xi_2$, a single recombination can generate a child in $\xi_1 \cap \xi_2$ provided this intersection is non-empty. Continuing the same example used to illustrate respect, if one parent has blue eyes and the other has brown hair, then recombination must allow the construction of a child with blue eyes and brown hair, provided that these characteristics are compatible.

2.5 Separability and Random Respectful Recombination

A set $\Xi(\Psi)$ of formae is said to be *separable* if it is capable of being simultaneously respected and properly assorted. The *random respectful recombination* operator (R^3)

makes a uniform random selection from the similarity set of the two parents. This operator respects and properly assorts the formae if they are separable. To see this, recall that respect is the requirement that children lie in the similarity set of their parents (equation 4). R^3 guarantees this, and generation of any solution outside the similarity set would by definition violate respect. So if respect and assortment are to be compatible conditions, any solutions which the latter requires be capable of production must fall within the similarity set. Since every solution in the similarity set is generated with non-zero probability by the R^3 operator, it must properly assort as well as respect whenever these conditions are compatible.

2.6 Complete Orthogonal Basis

For present purposes an equivalence relation $\sim$ in Ψ is conveniently expressed as a binary function

$$\psi : \mathcal{S} \times \mathcal{S} \longrightarrow \{0, 1\} \tag{6}$$

which returns 1 if its arguments are equivalent and 0 if they are not:

$$\psi(A, B) = \begin{cases} 1, & \text{if } A \sim B, \\ 0, & \text{otherwise.} \end{cases} \tag{7}$$

The intersection of two equivalence relations $\psi_1, \psi_2 \in \Psi$ can then be defined by

$$(\psi_1 \cap \psi_2)(A, B) = \begin{cases} 1, & \text{if } \psi_1(A, B) = \psi_2(A, B) = 1, \\ 0, & \text{otherwise.} \end{cases} \tag{8}$$

Given this, a subset $E \subset \Psi$ is said to constitute a *complete orthogonal basis* for Ψ provided that

- (*Completeness*) Each relation $\psi \in \Psi$ can be constructed as the intersection of some subset of the basic relations:

$$\forall \psi \in \Psi \, \exists E_\psi \subset E : \bigcap E_\psi = \psi. \tag{9}$$

- (*Orthogonality*) Given any subset of the equivalence relations in E, the intersection of each possible combination of equivalence classes (basic formae) induced by these equivalence relations should be non-empty. Formally, let Ξ_ψ be the set of formae induced by the equivalence relation ψ and given a subset F of the equivalence relations in Ψ, let

$$\Xi_F = \prod_{\psi \in F} \Xi_\psi, \tag{10}$$

the space of vectors of equivalence classes induced by the various relations in F. Then orthogonality requires that

$$\forall F \subset E \, \forall (\xi_1, \xi_2, \ldots, \xi_{|F|}) \in \Xi_F : \bigcap_{i=1}^{|F|} \xi_i \neq \emptyset . \tag{11}$$

This means, in effect, that the basic equivalence class (basic forma) can be chosen independently for each basic equivalence relation in E without introducing incompatibilities. (It should be noted that the definition of orthogonality given here is different from that given in Radcliffe (1991b). The earlier definition, which required only pairwise compatibility between basic equivalence relations, is not strict enough for the purposes of this paper, and is altogether a less satisfactory definition.) If equation 11 is satisfied only for sets F up to some size N then the basis will be said to be *orthogonal to order N*.

2.7 Genes, Alleles and Gene Transmission

A recombination operator is said to be *strictly transmitting* if it is the case that for each basic equivalence relation $\psi \in \Psi$ the children produced by recombination always lie in the same basic forma as one or other of the parents. (Equivalently, every child should be equivalent to at least one of its parents under each basic equivalence relation.) Strict gene transmission trivially implies respect. Basic equivalence relations can be identified with *genes*, and basic equivalence classes (basic formae) with *alleles*. Thus if eye colour were one of the basic equivalence relations, eye colour would constitute a "gene" with alleles "blue", "green" and "brown".

2.8 Example: Schemata

In later sections formae which differ significantly from schemata will be introduced, but for illustration the characteristics of one-point crossover, uniform crossover[1] and random respectful recombination (R^3) are compared for the case of k-ary (base k) schemata in table 1.

Schemata can be identified as equivalence classes induced by certain equivalence relations as follows. Consider the set of equivalence relations $\{\psi_i\}$, defined by

$$\psi_i(A, B) = \begin{cases} 1, & \text{if } A_i = B_i, \\ 0, & \text{otherwise,} \end{cases} \tag{12}$$

which relate two solutions if they have the same value (allele) for the ith gene. The ith such equivalence relation can conveniently be denoted

$$\psi_i = \square\square \cdots \square\blacksquare\square\square \cdots \square \tag{13}$$

where $\square$ is the familiar "don't care" character used to describe schemata, and the "care" symbol $\blacksquare$ occurs at the ith position. The set of such equivalence relations over all positions i induce all first order schemata (basic formae, alleles) and form a complete orthogonal basis for equivalence relations constructed by the arbitrary intersection of members of the basis. Such intersected equivalence relations induce all higher order schemata.

One-point, two-point and uniform crossover can immediately be seen to transmit genes, because each of a child's genes comes from one or other parent. By virtue of their strict transmission, these operators also plainly respect schemata. While uniform crossover also properly assorts schemata, because a child can be constructed with any combination of its parents' genes in a single recombination, one-point crossover does not, because 1010 cannot be constructed from parents 0000 and 1111. It does, however, weakly assort in the sense that repeated incestuous recombination will allow an arbitrary admixture of parental genes.

R^3 is guaranteed to respect and properly assort schemata since schemata can be seen to be separable by virtue of their separation by uniform crossover. R^3 can be implemented by initially copying across all genes common to the two parents into the child and then filling remaining positions with random alleles. Except in the case $k = 2$, this random allocation of alleles to genes which take on different values in the parents prevents random respectful recombination from strictly transmitting genes.

---

[1] with parameter half, i.e. with each bit in the mask equally likely to be one or zero.

Table 1: Summary of the characteristics of various recombination operators.

Operator	Respect	Assortment	Gene Transmission
one-point	•	weak	•
two-point	•	weak	•
uniform	•	proper	•
R^3	•	proper	$k = 2$ only

3 Sets and Multiset Recombination

Recall that the distinction between a set and a multiset is that duplication of elements is not significant in sets, so that

$$\{ a, a, b \} \equiv \{ a, b \}, \tag{14}$$

whereas in multisets (also known as "bags") an element may appear more than once

$$\{\!\{a, a, b\}\!\} \neq \{\!\{a, b\}\!\}. \tag{15}$$

(The notation $\{\!\{ \cdots \}\!\}$ is used in this paper to indicate a multiset.) A number of different set and multiset optimisation problems may be distinguished. In general there will be a "universal set", $\mathcal{E}$, from which elements are drawn. The aim is to construct a set or multiset consisting of elements drawn from this universal set so as to optimise some property of the resulting set or multiset. Examples could include finding locations for bottle banks so as to maximise recycling in some area, selecting members of a committee to make a environmental impact assessment or choosing connections in a neural network to minimise its average learning time to some acceptable error (Radcliffe, 1992). Whitley (1987) has studied the use of genetic search over restricted poker hands using a multiset formulation; this is discussed in Radcliffe (1990).

4 Recombining Fixed-Size Sets

Given a universal set $\mathcal{E}$, the search space for sets of fixed size N is

$$\mathcal{S} = \{A \subset \mathbb{P}(\mathcal{E}) \mid |A| = N\}, \tag{16}$$

(figure 2), where $\mathbb{P}(\mathcal{E})$ is the *power set* (set of all subsets) of $\mathcal{E}$:

$$\mathbb{P}(\mathcal{E}) = \{B \subset \mathcal{E}\}. \tag{17}$$

4.1 Equivalence Relations and Formae

Given a universal set $\mathcal{E}$, with $a \in \mathcal{E}$, and solutions $A, B \in \mathcal{S}$, let

$$\psi_{\{a\}}(A, B) = \begin{cases} 1, & \text{if } a \in A \cap B \text{ or } a \notin A \cup B, \\ 0, & \text{otherwise.} \end{cases} \tag{18}$$

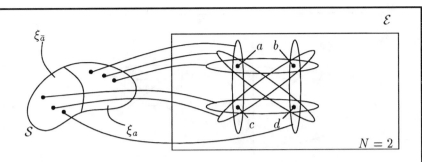

Figure 2: The equivalence relation $\psi_{\{a\}}$ for fixed size sets of size two drawn from a universe $\mathcal{E} = \{a, b, c, d\}$ partitions $\mathcal{S}$ into two formae, ξ_a, consisting of those sets which contain a and $\xi_{\bar{a}}$ consisting of those which do not.

This equivalence relation induces two equivalence classes, one comprising the solutions containing the element a and another comprising those which do not (figure 2). There is clearly an equivalence relation $\psi_{\{x\}}$ of the form described by equation 18 for *each* $x \in \mathcal{E}$. Moreover, these are intuitively natural candidates for a basis for a set Ψ of equivalence relations which might generate all formae specifying the presence or absence of any subset of the elements in $\mathcal{E}$. As will now be demonstrated, if the rule for intersection of equivalence relations described by equation 8 is followed, the set

$$E = \left\{ \psi_{\{x\}} \mid x \in \mathcal{E} \right\} \tag{19}$$

forms a basis, orthogonal to some order K, for a set of equivalence relations Ψ which induces a useful set of formae. Ψ can be defined by

$$\Psi = \left\{ \psi \mid \exists\, E_\psi \in E : \; \psi = \bigcap E_\psi \right\}. \tag{20}$$

To see this, consider the intersection of $\psi_{\{a\}}$ and $\psi_{\{b\}}$, which will be denoted $\psi_{\{a,b\}}$. According to the definition of intersection for equivalence relations (equation 8)

$$(\psi_{\{a\}} \cap \psi_{\{b\}})(A, B) = \begin{cases} 1, & \text{if } \psi_{\{a\}}(A, B) = \psi_{\{b\}}(A, B) = 1, \\ 0, & \text{otherwise.} \end{cases} \tag{21}$$

This equivalence relation, illustrated in figure 3, induces four equivalence classes, which might conveniently be written

$$\begin{aligned} \xi_{ab} &= \{A \in \mathcal{S} \mid a \in A, b \in A\}, \\ \xi_{a\bar{b}} &= \{A \in \mathcal{S} \mid a \in A, b \notin A\}, \\ \xi_{\bar{a}b} &= \{A \in \mathcal{S} \mid a \notin A, b \in A\}, \\ \xi_{\bar{a}\bar{b}} &= \{A \in \mathcal{S} \mid a \notin A, b \notin A\}. \end{aligned} \tag{22}$$

The generalisation of this is rather obvious. A general equivalence relation, $\psi \in \Psi$, has an associated *description set*, conveniently written $\langle \psi \rangle$, which is a subset of the universal set $\mathcal{E}$. Members of the search space (themselves subsets of $\mathcal{E}$) are then equivalent under ψ

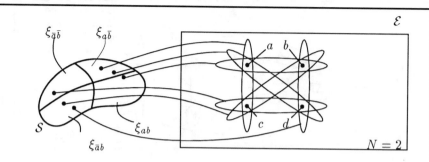

Figure 3: The equivalence relation $\psi_{\{a\}}$ for fixed size sets of size two drawn from a universe $\mathcal{E} = \{a, b, c, d\}$ partitions $\mathcal{S}$ into two formae, ξ_a, consisting of those sets which contain a and $\xi_{\bar{a}}$ consisting of those which do not.

precisely if they contain the same subset of the members of the description set $\langle\psi\rangle$ (figure 3). Formally,

$$\psi(A, B) = \begin{cases} 1, & \text{if } \langle\psi\rangle \cap A = \langle\psi\rangle \cap B, \\ 0, & \text{otherwise.} \end{cases} \tag{23}$$

It is clear that E (defined in equation 19) does indeed form a basis for Ψ, but that this basis is not fully orthogonal. Completeness follows because

$$\forall\psi \in \Psi : \psi = \bigcap_{x \in \langle\psi\rangle} \psi_{\{x\}}, \tag{24}$$

so that every relation can be expressed as an intersection of the basic relations defined in equation 19, as required for completeness (equation 9). Orthogonality to order K follows provided that for up to K basic equivalence relations it is possible independently to choose whether or not a set should include the element labelling the basic equivalence relation. K is given by

$$K = \max\{N, |\mathcal{E}| - N\}. \tag{25}$$

To see this, notice that if $|\mathcal{E}| \geq 2N$ then $K = N$, because after picking N elements to include it would be impossible to include an $N + 1$th. Similarly, if $|\mathcal{E}| < 2N$ then once $|\mathcal{E}| - N$ elements had been chosen for exclusion from the set it would be impossible to exclude any more. The lack of full orthogonality will prove to be problematical when recombination operators are constructed.

A general forma ξ induced by an equivalence relation $\psi \in \Psi$ can then be characterised by a partition of the description set $\langle\psi\rangle$. It thus becomes convenient to describe a forma by a 2-tuple

$$\langle\xi\rangle = (\xi^+, \xi^-) \tag{26}$$

where

$$\xi^+ \cap \xi^- = \emptyset \tag{27}$$

and

$$\xi^+ \cup \xi^- = \langle\psi\rangle \tag{28}$$

with the interpretation

$$A \in \xi \iff \left(A \cap \xi^+ = \xi^+ \text{ and } A \cap \xi^- = \emptyset \right). \tag{29}$$

This says that a set A is a member of the forma ξ if and only if it contains all those elements in ξ^+ and none of those in ξ^-. Having introduced this formalism, it is possible to identify the similarity set of two solutions with respect to the formae $\Xi(\Psi)$ induced by Ψ. This will allow the random respectful recombination operator R^3 to be constructed.

Recall that the similarity set of two solutions is defined by

$$\Sigma(A, B) = \bigcap \{\xi \in \Xi(\Psi) \mid A, B \in \xi\}. \tag{30}$$

Clearly A and B will share membership of a forma ξ if and only if

$$A \cap \xi^+ = B \cap \xi^+ = \xi^+ \tag{31}$$

and

$$A \cap \xi^- = B \cap \xi^- = \emptyset. \tag{32}$$

Equation 31 can be satisfied if and only if

$$\xi^+ \subset A \cap B \tag{33}$$

and equation 32 can be satisfied if and only if

$$\xi^- \subset \mathcal{E} - (A \cup B), \tag{34}$$

where the minus sign denotes set subtraction. It is thus clear that when performing an intersection of all formae satisfying these conditions the similarity set as specified in equation 30 must be described by

$$\langle \Sigma(A, B) \rangle = \left(A \cap B, \; \mathcal{E} - (A \cup B) \right). \tag{35}$$

The R^3 operator makes a random (uniform) selection from this similarity set. For example, with the universe

$$\mathcal{E} = \{a, b, c, d, e, f\} \tag{36}$$

and $N = 3$,

$$\langle \Sigma(\{a, b, c\}, \{a, d, e\}) \rangle = \left(\{a, b, c\} \cap \{a, d, e\}, \; \mathcal{E} - (\{a, b, c\} \cup \{a, d, e\}) \right)$$

$$= \left(\{a\}, \{f\} \right). \tag{37}$$

This describes the forma containing those sets which contain a and exclude f:

$$\Sigma(\{a, b, c\}, \{a, d, e\}) = \left\{ \{a, b, c\}, \{a, b, d\}, \{a, b, e, \}, \{a, c, d\}, \{a, c, e\}, \{a, d, e, \} \right\}. \tag{38}$$

Thus, R^3 for these formae can be understood as an operator which

1. copies all the elements which are common to the two parents into the child;
2. fills the remaining places in the child with a random selection of the unused elements from the two parents.

A child C of A and B thus has the natural properties

$$A \cap B \subset C \subset A \cup B. \tag{39}$$

It is clear, therefore, that in this case R^3 strictly transmits genes, a gene being labelled by an element of $\mathcal{E}$ and an allele corresponding to the presence or absence of that element (equation 18). Notice, however, that the formae are not separable, with the consequence that neither R^3 nor any other respectful operator can assort them. To see this, simply observe that $\{a, b, c\}$ is a member of the forma ξ_{bc} and $\{a, d, e\}$ is a member of the forma ξ_d but that R^3 cannot produce a member in the intersection $\xi_{bc} \cap \xi_d = \xi_{bcd}$ of these formae because respect restricts the choice of children to those in the similarity set given in equation 38. This arises directly from the restriction to fixed-size sets. A way of trading-off respect and assortment is discussed in section 7.

An alternative way of viewing this operator is to imagine a conventional linear chromosome in which every position represents an element from the universal set, and to imagine an operator like uniform crossover, but constrained so that the total number of 1's in the child is constant and equal to N, the fixed size of the set.

5 Recombination of Fixed-Size Multisets

5.1 Equivalence Relations and Formae

The extension of the previous case from sets to multisets is in essence simple, but involves one complication. The basic idea will be that rather than specify whether or not certain elements are in the multiset, a forma will specify the *multiplicities* of some elements. Again, assume that $\mathcal{E}$ is a universal set from which all elements are to be drawn, but that elements may now be taken more than once. Then let $\mathbb{P}_m(\mathcal{E})$ be the *multipower set* of $\mathcal{E}$, that is, the set of all multisets whose elements are drawn from $\mathcal{E}$. Then the *multiplicity* function

$$m : \mathcal{E} \times \mathbb{P}_m(\mathcal{E}) \longrightarrow \mathbb{Z}^+ \cup \{0\} \tag{40}$$

is defined so that $m(x, A)$ is the number of copies of x in the multiset A.

A forma for multisets could either specify exact multiplicities for certain elements or could give bounds on their multiplicities. Since the former is a special case of the latter, where the bounds are maximally tight, the more general case will be examined.

A forma is now conveniently described by a set of 3-tuples of the form $(x, N_x^!, N_x^!)$ each of which is understood to specify that the multiplicity $m(x, A)$ of the element x in the multiset A lies in the inclusive range $N_x^!$ to $N_x^!$. For example, a forma ξ with the description set

$$\langle \xi \rangle = \{(a, 0, 0), (b, 1, 3)\} \tag{41}$$

contains all those multisets over $\mathcal{E}$ of size N which contain no copies of a and contain between one and three copies of b (figure 4). Such formae are closely related to the "range formae" discussed in Radcliffe (1991a).

5.2 Non-Orthogonality

There are a number of sets of equivalence relations which could be constructed to generate these formae. An obvious starting point is equivalence relations which induce formae

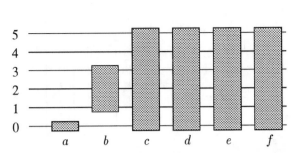

Figure 4: A visualisation of the forma ξ which has the description set $\langle \xi \rangle = \{(a, 0, 0), (b, 1, 3)\}$. The full ranges for elements c to f correspond to the "don't care" character familiar from conventional schemata.

defined with respect to a single element x from $\mathcal{E}$. The equivalence relation ψ which induces the forma described by $\langle \xi \rangle = \{(x, N_x^1, N_x^1)\}$ would have the same description set

$$\langle \psi \rangle = \{(x, N_x^1, N_x^1)\} \tag{42}$$

and would be defined by

$$\psi(A, B) = \begin{cases} 1, & \text{if } (m(x, A) < N_x^1 \text{ and } m(x, B) < N_x^1) \\ & \text{or } m(x, A), m(x, B) \in [N_x^1, N_x^1] \\ & \text{or } (m(x, A) > N_x^1 \text{ and } m(x, B) > N_x^1), \\ 0, & \text{otherwise,} \end{cases} \tag{43}$$

where

$$[N_x^1, N_x^1] = \{n \in \mathbb{Z} \mid N_x^1 \le n \le N_x^1\}. \tag{44}$$

As was the intention, formae can now specify a range of multiplicities for any element and a single equivalence relation will be seen to suffice to define up to three ranges simultaneously. The natural candidates for a basis are the equivalence relations which divide the range of multiplicities for a single element into a lower portion and an upper portion, as shown in figure 5,

$$E = \left\{ \psi \in \Psi \mid \langle \psi \rangle = \{(x, 0, N_x^1)\}, x \in \mathcal{E}, N_x^1 \in [0, N^\star] \right\} \tag{45}$$

where $N^\star$ is the maximum allowed multiplicity for an element. These equivalence relations can easily be seen to be complete, for any equivalence relation with a description set $\{(x, N_x^1, N_x^1)\}$ can be constructed as an intersection of the relations with description sets $\{(x, 0, N_x^1 - 1)\}$ and $\{(x, 0, N_x^1)\}$ (figure 6). Equivalence relations defined with respect to more than one member of $\mathcal{E}$ can then be constructed trivially by intersection.

It is easy, however, to see that the relations in E do not satisfy the condition of orthogonality specified in equation 11. To verify this, simply note that if a multiset is a member of the forma with description set $\{(x, 0, 1)\}$ (induced by ψ_1, as labelled in figure 5) it cannot also be a member of the forma with the description set $\{(x, 4, N^\star)\}$ (induced by ψ_3) as would be required if E were orthogonal (equation 11). This is because no multiset can have both fewer than two and more than three copies of the element x. Thus a multiset cannot

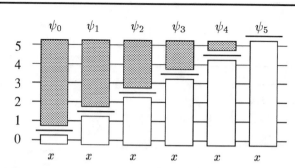

Figure 5: The set of equivalence relations with description sets of the form $\{(x, 0, N)\}$ divide the range $[0, N^\star]$ into a lower and and upper portion as shown: the equivalence relations may be thought of as simple dividing lines at integer-plus-half values.

be constructed with independent choice of its forma membership of for the equivalence relations ψ_1 and ψ_3, as shown in figure 5, violating orthogonality.

Rather than abandon this potential basis, it is instructive to return to the analogy with linear algebra which led to the original formulation of the conditions on a basis, namely completeness and orthogonality. In linear algebra there is a weaker notion than orthogonality known as *linear independence*: a set of vectors is said to be linearly independent if no one of them can be expressed as a linear combination of the others. Following this analogy, the following definition suggests itself:

- (*Independence*) A set E of equivalence relations will be said to be *independent* if no one of the relations $\psi \in E$ can be expressed as the intersection of some subset of the others, i.e.

$$\forall \psi \in E \nexists E_\psi \subset E - \{\psi\} : \bigcap E_\psi = \psi. \qquad (46)$$

The set E defined in equation 45 satisfies this condition of independence,[2] as well as completeness.

The purpose of introducing the notion of a complete orthogonal basis for a set of equivalence relations was to generalise the notion of a gene and allow a principle of strict gene transmission to be extended to more general formae. It will be demonstrated below that the weaker notion of a non-orthogonal *independent basis* suffices for the definition of genes, and thus is adequate for the original purpose. Using the same definition of genes and alleles for non-orthogonal bases as for orthogonal bases, (i.e. genes are the basic equivalence relations and alleles are the basic equivalence classes) it is now possible to construct the *random transmitting recombination* operator (RTR) induced by the basis E for Ψ, described by equation 45.

---

[2] A rather minor point which should nevertheless be made in passing is that the formae now being considered violate closure as discussed in Radcliffe (1991a, 1991b): this turns out to be unimportant.

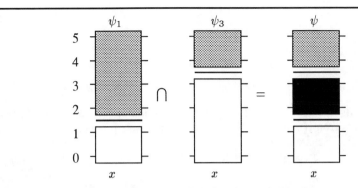

Figure 6: Any equivalence relation ψ defined on a single element can be constructed as an intersection of relations from the basis E, defined by equation 45. Here ψ has the description set $\langle \psi \rangle = \{(x, 2, 3)\}$ and is constructed as the intersection of ψ_1 with description set $\langle \psi_1 \rangle = \{(x, 0, 1)\}$ and $\langle \psi_3 \rangle = \{(x, 0, 3)\}$.

5.3   Random Transmitting Recombination

The random transmitting recombination operator[3] (RTR) can be defined in a way similar to random respectful recombination (R^3), the difference being that instead of selecting from the entire similarity set of the two parents, RTR selects from that subset of solutions which share every gene with at least one parent. This subset, which for parents A and B is written $\Gamma(A, B)$, is called their *dynastic potential,* and is defined by

$$\Gamma(A, B) = \left\{ C \in \Sigma(A, B) \mid \forall \psi \in E : \psi(C, A) = 1 \text{ or } \psi(C, B) = 1 \right\}. \qquad (47)$$

The RTR operator picks each element in the dynastic potential of the parents with equal probability, and both strictly transmits genes (which trivially implies respect) and properly assorts formae, provided that these conditions are compatible. (In the special case of k-ary schemata, RTR is identical to uniform crossover with parameter half.)

5.4   Application to Fixed Size Multisets

The formalism developed above can now be applied to the problem of recombining fixed-size multisets. The similarity set of two solutions (now multisets) is the forma with the description set

$$\langle \Sigma(A, B) \rangle = \left\{ (x, N_x^!, N_x^!) \mid N_x^! = \min \left\{ m(x, A), m(x, B) \right\}, \qquad (48) \right.$$
$$\left. N_x^! = \max \left\{ m(x, A), m(x, B) \right\} \right\}$$

This similarity set contains all those multisets of the given fixed size N which have at least as many copies of each element as the parent with fewer copies, and no more than the number held by the parent with more. For example, if the chosen fixed size for the multisets is five, and the universal set $\mathcal{E}$ is given by equation 36, then given

$$A = \{a, a, a, b, c\}, \qquad (49)$$

---

[3]previously called *inheritance crossover*

and

$$B = \{a, b, b, c, d\},\qquad(50)$$

the similarity set $\Sigma(A, B)$ is described by

$$\langle \Sigma(A, B) \rangle = \{(a, 1, 3), (b, 1, 2), (c, 1, 1), (d, 0, 1), (e, 0, 0), (f, 0, 0)\}.\qquad(51)$$

The similarity set itself thus contains those multisets containing $\{a, b, c\}$ together with exactly two elements from $\{a, a, b, d\}$. For fixed size multisets it happens that, given the equivalence relations discussed, the dynastic potential of any parents A and B is identical to their similarity set. To see this, consider any basic equivalence relation ψ with the description set

$$\langle \psi \rangle = \{(x, 0, n)\}.\qquad(52)$$

This has two equivalence classes, described by

$$\langle \xi_1 \rangle = \{(x, 0, n)\}\qquad(53)$$

and

$$\langle \xi_1 \rangle = \{(x, n + 1, N^\star)\}.\qquad(54)$$

If both parents belong to the same basic forma, then their similarity set is clearly a subset of this forma, so gene transmission imposes no extra requirement. If, however, they belong to different basic formae, then since there are only two of these, the requirement to lie in their union is no restriction at all, because

$$\xi_1 \cup \xi_1 = \mathcal{S}.\qquad(55)$$

Thus dynastic potentials for these equivalence relations are indeed identical to similarity sets and so it can be seen that strict gene transmission is in this case no stronger a requirement than respect. In this special case, RTR reduces to R^3. This is not, of course, true for general formae, as is shown by the difference for the k-ary schemata discussed in section 2.8.

Notice that the restriction to multisets of fixed size, as was the case with sets, ensures non-separability, so that RTR and R^3 are unable to assort the formae discussed. This can be seen by from the same example as was used in section 2.1 to demonstrate that formae for fixed size sets are non-separable, re-interpreting the sets as multisets. This weakness is discussed further in section 7.

6   Recombining Variable-Size Multisets

Variable-size multisets can be dealt with simply by relaxing the constraint of fixed size as discussed in the previous section. The formae then arrived at are separable and random transmitting recombination (which is in this special case again identical to R^3) not only properly assorts and respects the formae, but also strictly transmits genes.

In summary, R^3/RTR for variable-size multisets simply inserts a number of copies of each element from the universal set which is bounded by the number of copies in the two parents, and in doing so strictly transmits genes and properly assorts the formae induced by the equivalence relations generated by the independent basis of equation 45.

7 Assorting Non-Separable Formae

When there are non-separable formae, such as those discussed in this paper, the question arises as to which among respect and proper assortment should be given priority. R^3 and RTR both give priority to respect, but arguably proper assortment, which embodies the exploratory power of the search, is more important.

The present section shows how it is possible to trade off the degree of violation of respect with the "thoroughness" of assortment. (Here thoroughness may be taken to mean the likelihood of generating the solutions required by assortment.) Recall that proper assortment only requires that the probability of generating a child in the intersection of a given pair of formae to which the parents belong be finite. (Equation 5 says that there must be some control parameter which allows generation of a solution in the intersection, but does not specify any required density of such parameters.) It is therefore technically possible to guarantee proper assortment by defining an operator which respects the given formae with a very high probability $1 - \varepsilon$, but with low probability ε randomly selects a solution which violates respect. (Indeed, mutation could be viewed as performing this rôle in combination with a respectful recombination operator.) More narrowly, the choice of solution outside the similarity set (the set of children allowed by respect) could be restricted to those which are required to be capable of being generated by assortment. The parameter ε can be viewed as controlling the degree to which respect is violated. While any non-zero value for ε technically guarantees assortment, clearly the larger the value, the more thorough will be the assortment.

7.1 Random Assorting Recombination

The following *random assorting recombination* operator (RAR) uses these ideas to ensure proper assortment by sacrificing respect (and, by implication, gene transmission) in a controlled way. This operator takes a positive integer parameter w (for "weight") which is like an inverse of the parameter ε discussed above.

1. Place w copies of each allele present in both parents in a bag (multiset).

2. Add to the bag one copy of each allele present in only one parent.

3. Repeatedly draw alleles from the bag without replacement. Whenever it is possible to add the allele to the child being formed, do so; otherwise discard the allele. Continue until the bag is empty or the child is fully specified, i.e. until a basic forma (allele) has been chosen for every one of the basic equivalence relations (genes).

4. If the child is not fully specified at the end of this process, assign alleles to any remaining genes at random, from among the remaining legal values.

A number of observations should be made about this operator.

- The operator is general, and can be applied to any problem in which genes are properly specified in the sense used in this paper, provided that their number is finite.

- If the formae are induced by a set of equivalence relations with an orthogonal basis, this operator will separate them, and reduces to RTR. To see this, simply observe that the only circumstances in which it would not be possible to add an allele drawn from the bag to the child would be those in which a different allele for the same gene had already be chosen from the other parent.

- If the formae are separable, RAR will separate them. To see this, observe that if a gene is common to both parents, only one allele will be placed in the bag for that gene (albeit w times). In the initial phase of RAR's operation, only combinations of alleles which assortment requires will ever be included in the child. If, therefore, assortment is compatible with respect, this cannot compromise the ability to include the shared alleles in the child, as required by respect.

- The action of the operator is only unusual, therefore, when the formae are non-separable. In other cases, the value of w is irrelevant. With non-separable formae, the higher the parameter w is set, the greater will be the degree of respect which RAR achieves and the less thorough will be the assortment. This is because the higher w is set, the more likely is it that alleles common to the two parents will be drawn early, so that incompatibilities between respect and assortment will be more likely to be concluded in favour of respect. General guidance as to the appropriate value for w probably requires experimental evidence, though $w = 2$ has aesthetic appeal because filling the bag then amounts merely taking every allele from each parent.

- Finally, the account of the RAR given describes its theoretical application. In most circumstances it will be possible to find implementations which are very much more efficient than that described, but which replicate its behaviour identically.

7.2 Example of Random Transmitting Recombination

Consider again the example used in section 4.1 to show that fixed-size sets are non-separable. Recall that $\mathcal{E} = \{\, a, b, c, d, e, f \,\}$, the fixed size is 3 and the parents are $A = \{\, a, b, c \,\}$ and $B = \{\, a, b, d \,\}$. The full genetic description of A is now the singleton forma

$$\xi_{abc\bar{d}\bar{e}\bar{f}} = \{A\}. \tag{56}$$

Similarly, B is given by

$$\xi_{ab\bar{c}d\bar{e}\bar{f}} = \{B\}. \tag{57}$$

Taking $w = 3$, the bag will initially be filled as follows:

$$\{\!\{ a, a, a, b, b, b, c, \bar{c}, d, \bar{d}, \bar{e}, \bar{e}, \bar{e}, \bar{f}, \bar{f}, \bar{f} \}\!\}. \tag{58}$$

Clearly drawing out a solution containing $\{\, c, d \,\}$ is now possible in a number of ways, but is not particularly likely because a and b are each three times as likely to get drawn out at each stage as are c and d. Notice that, just as there is a small possibility of being unable to include a and b, there is a possibility of having to include, say, f, which is present in neither parent. This would happen, for example, if the first three alleles drawn happened to be $\bar{c}$, $\bar{d}$ and $\bar{e}$. In this case, when the $\bar{f}$ allele was chosen it would be discarded because it is impossible for a solution to this problem to omit c, d, e and f, and in the final stage an e allele would necessarily be introduced. This slightly counterintuitive behaviour is required by assortment, and again becomes ever more unlikely as w is increased.

7.3 Forma Disruption

Depending partly on the weight used, random assorting recombination can be a fairly disruptive operator. As with other recombination operators, the amount disruption can be reduced by biasing the operator to take genes preferentially from one parent (Syswerda, 1989, Spears & DeJong, 1991). This may well prove to be sensible in the current context. It may further be appropriate to introduce linkage and reordering operators, though these are out of vogue. Such extensions will not, however, be discussed in detail in this paper.

8 Summary

This paper has shown how forma analysis can be used to construct operators for principled recombination of sets and multisets. A single operator, defined in generic terms and called *random assorting recombination* (RAR), suffices for this. Previous work had introduced the notion of a complete orthogonal basis for a set of equivalence relations and shown that when such a basis exists a linear chromosome can be constructed which uniform crossover can manipulate effectively in the sense that it transmits and properly assorts genes. This paper has demonstrated that in cases where no orthogonal basis exists it may still be possible to construct a non-orthogonal basis. Such a basis suffices for the definition of genes, but these cannot be independently assigned because some combinations of alleles will be illegal. Moreover, in some cases it will be possible to transmit and properly assort genes forming a non-orthogonal basis, but conventional operators will be unable to do this. The RAR operator is guaranteed to transmit and properly assort genes whenever this is possible. The paper has further shown that when respect and assortment are incompatible, it is possible to parameterise the degree of violation of respect and to use this to control the thoroughness of assortment. The random assorting recombination operator provides a convenient and general mechanism for doing this.

References

(Holland, 1975) John H. Holland. *Adaptation in Natural and Artificial Systems*. University of Michigan Press (Ann Arbor), 1975.

(Radcliffe, 1990) Nicholas J. Radcliffe. *Genetic Neural Networks on MIMD Computers*. PhD thesis, University of Edinburgh, 1990.

(Radcliffe, 1991a) Nicholas J. Radcliffe. Equivalence class analysis of genetic algorithms. *Complex Systems*, 5(2):183–205, 1991.

(Radcliffe, 1991b) Nicholas J. Radcliffe. Forma analysis and random respectful recombination. In *Proceedings of the Fourth International Conference on Genetic Algorithms*, pages 222–229. Morgan Kaufmann (San Mateo), 1991.

(Radcliffe, 1992) Nicholas J. Radcliffe. Genetic set recombination and its application to neural network topology optimisation. *Neural Computing and Applications*, 1(1), 1992.

(Spears and De Jong, 1991) William M. Spears and Kenneth A. De Jong. On the virtues of parameterised uniform crossover. In *Proceedings of the Fourth International Conference on Genetic Algorithms*, pages 230–236. Morgan Kaufmann (San Mateo), 1991.

(Syswerda, 1989) Gilbert Syswerda. Uniform crossover in genetic algorithms. In *Proceedings of the Third International Conference on Genetic Algorithms*. Morgan Kaufmann (San Mateo), 1989.

(Vose and Liepins, 1991) Michael D. Vose and Gunar E. Liepins. Schema disruption. In *Proceedings of the Fourth International Conference on Genetic Algorithms*, pages 237–243. Morgan Kaufmann (San Mateo), 1991.

(Whitley, 1987) Darrell Whitley. Using reproductive evaluation to improve genetic search and heuristic discovery. In *Proceedings of the Second International Conference on Genetic Algorithms*. Lawrence Erlbaum Associates (Hillsdale), 1987.

Crossover or Mutation?

William M. Spears
Navy Center for Applied Research in Artificial Intelligence
Code 5510
Naval Research Laboratory
Washington, D.C. 20375-5320
E-mail: SPEARS@AIC.NRL.NAVY.MIL

Abstract

Genetic algorithms rely on two genetic operators - crossover and mutation. Although there exists a large body of conventional wisdom concerning the roles of crossover and mutation, these roles have not been captured in a theoretical fashion. For example, it has never been theoretically shown that mutation is in some sense "less powerful" than crossover or vice versa. This paper provides some answers to these questions by theoretically demonstrating that there are some important characteristics of each operator that are not captured by the other.

1 INTRODUCTION

One of the major issues in genetic algorithms (GAs) is the relative importance of two genetic operators: mutation and crossover. In the 1960's, L. Fogel *et al.* (1966) illustrated how mutation and selection can be used to evolve finite state automatons for a variety of tasks. Simultaneously, in Europe, Rechenberg (1973) investigated "evolution strategies" that again concentrate on mutation as the key genetic operator. Sophisticated versions of evolution strategies, with adaptive mutation rates, proved quite useful for function optimization tasks (Baeck *et al.*, 1991; Schwefel, 1977). Recent studies confirm this view, illustrating the power of mutation (Schaffer *et al.*, 1989). D. Fogel has continued the earlier work of L. Fogel and makes an even stronger claim - that crossover has no general advantage over mutation (Fogel & Atmar, 1990).

On the other hand, proponents of the Holland (1975) style of genetic algorithm believe that crossover is the more powerful of the two operators. Considerable effort has been spent in analyzing crossover and its effects on performance (e.g., De Jong, 1975; Spears & De Jong, 1991; Vose & Liepins, 1991). In most of these analyses mutation is considered to be a background operator and of secondary importance. To support these views, experimental results have been presented, illustrating the power of crossover (e.g., De Jong, 1975). Most recently, Schaffer & Eshelman (1991) empirically compare mutation and crossover, and conclude that mutation alone is not always sufficient.

Unfortunately, empirical comparisons can often be disputed or may be misleading. For example, Schaffer & Eshelman speculate that implementation and representation may explain Fogel's results. Similarly, it can be speculated that Schaffer & Eshelman did not implement mutation reasonably (e.g., with an adaptive rate). To date, there has been no theoretical justification to support either camp's beliefs. It has never been theoretically shown that crossover is in any sense more powerful than mutation, or that mutation is more powerful than crossover. Similarly, no theoretical basis exists for supposing that both operators are necessary and perform different roles within the GA.

In this paper we show that, in a general sense, both camps are correct, although we dispute the stronger claim of Fogel and Atmar. We define two potential roles of any genetic operator, disruption and construction, and consider how well mutation and crossover perform these roles. Our results show that in terms of disruption, mutation is more powerful than crossover, although it lacks crossover's ability to preserve alleles common to individuals. However, in terms of construction, crossover is more powerful than mutation.

2 DISRUPTION THEORY

Holland provided the initial formal analysis of the behavior of GAs by showing how they allocate trials in a near optimal way to competing low order hyperplanes if the disruptive effects of the genetic operators is not too severe (Holland, 1975). Considerable attention has been given to estimating the *disruption rate* of crossover, i.e., the probability that a particular application of crossover will be disruptive. As has been pointed out, however, another important consideration is not just how *often* a sample will be disrupted, but *how* it will be disrupted (Eshelman *et al.*, 1989). In this section we will first consider a theory of disruption rates for crossover, and show how we can compare this with a disruption rate theory for mutation. We then briefly review both mutation and crossover with respect to how they disrupt hyperplane samples.

2.1 DISRUPTION RATES

Holland's initial analysis of the sampling disruption of 1-point crossover (Holland, 1975) has been extended to n-point crossover and a parameterized (P_0) uniform crossover (De Jong, 1975; Syswerda, 1989; Spears & De Jong, 1991), where n is the number of crossover points and P_0 represents the probability of swapping alleles between two parents. These results estimate the likelihood that the sampling of a kth-order hyperplane (H_k) will be disrupted by a particular form of crossover.

For example, given a 3rd-order hyperplane (H_3), one can compute the probability that an application of n-point or uniform crossover will disrupt the sampling of that hyperplane.

It turns out to be easier mathematically to estimate the complement of disruption, the likelihood that a hyperplane sample will survive crossover, which we denote as P_s. We can also refer to the survival and disruption of individuals within a hyperplane H_k. If an individual survives with respect to H_k, it remains within H_k. If an individual is disrupted with respect to H_k, it is no longer within H_k. Finally, it should be noted that if each application of crossover is independent, we can interpret P_s as the probabilistic survival of an individual within a hyperplane H_k. For example, if N individuals are within some H_k, we expect roughly $N \cdot P_s$ individuals to remain (survive) in H_k after crossover.

Figure 1 illustrates P_s for 3rd-order hyperplanes. For n-point crossover the probability that a sample will survive depends on the order of the hyperplane, its defining length, and the number of crossover points n. For uniform crossover the probability of survival depends on the order of the hyperplane and the probability of swapping alleles, P_0.[1] The reader should note that uniform crossover is labelled as "P_0 uniform" in Figure 1. For example, ".1 uniform" indicates that $P_0 = 0.1$.

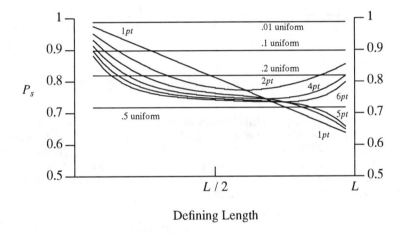

Figure 1: Crossover Survival, 3rd-Order, $P_{eq} = 0.5$

These results are time dependent in the sense that they are affected by the degree to which the population has converged. To see this, we need to define what we mean by convergence. In our theory we denote $P_{eq}(d)$ to be the probability that two hyperplanes will have the same allele at a particular defining position d (De Jong & Spears, 1992). As a useful simplification we also denote P_{eq} to be the average of the $P_{eq}(d)$'s over all d. Since we assume a bit level representation, $P_{eq} = 0.5$ represents the condition when the population is first randomly initialized and each allele has an equal probability of being a 1 or 0. When P_{eq} is close to 1, the population is nearly converged and lacks diversity. Figure 2 illustrates how crossover is affected by the convergence of the population. The horizontal axis represents the convergence of the population ($0.5 \le P_{eq} \le 0.9$). For the sake of simplicity we illustrate only uniform crossover, where P_0 ranges from 0.1 to 0.5 in increments of 0.1. These values are useful because the levels of disruption provided

---

[1] See Spears & De Jong (1991) for more precise details.

by n-point crossover are roughly bounded by the disruption levels of uniform crossover when $0.1 \le P_0 \le 0.5$ (see Figure 1). Note that disruption decreases as the diversity of the population decreases.

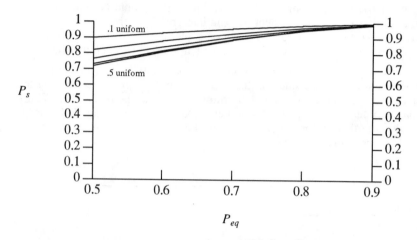

P_s

P_{eq}

Figure 2: Crossover Survival as a Function of Convergence, 3rd-Order

In the previous paragraphs we provided a review of disruption rate theory for n-point and P_0 uniform crossover. Can we now provide a similar theory for mutation? At first blush this would appear to be difficult, since crossover is a function of two individuals, while the mutation of one individual is not affected by the mutation of another. More precisely, how can P_s represent the probability of survival of one individual within a hyperplane H_k, given that two individuals are involved in the crossover operation?

The answer lies in the (often hidden) assumption that crossover is used to create two offspring, as opposed to one. Consider the situation where one parent individual is within H_k, while the other parent individual is not. Then, after crossover, at most one offspring will also be within H_k, and P_s represents the probability of that event. Equivalently, if there are N individuals within H_k, there will be roughly $N \cdot P_s$ individuals within H_k after crossover. Suppose, however, that both parents are in H_k. Then, after crossover, both offspring are guaranteed to be in H_k. In this case $P_s = 1$ and $N \cdot P_s = N$ individuals will survive crossover. If only one offspring were created, this analysis would not be correct. In summary, the assumption that both offspring are created is necessary to ensure that P_s correctly represents the independent survival of one individual within a hyperplane.

Since, for crossover, P_s really represents the probability of one individual within a hyperplane H_k surviving, we can compare this with a similar analysis for mutation. For mutation we want P_s to represent the probability that an individual in H_k will survive mutation. In this case, independence is trivial, since the mutation of one individual is not affected by the mutation of another. Again, if there are N individuals within some H_k, roughly $N \cdot P_s$ individuals will survive mutation. In this paper mutation is defined to be the operator that probabilistically selects a bit and flips that bit (recall that we are assuming a bit level representation). Again, suppose we have a 3rd-order hyperplane. Then the probability that an individual within that hyperplane will survive mutation is given by:

$$P_s(H_3) = (1 - P_m)^3$$

where P_m is the probability of mutating an allele. In general, we have:

$$P_s(H_k) = (1 - P_m)^k$$

for kth-order hyperplanes.

Given this analysis, we can now compare the disruptive effects of mutation with those of crossover. Figure 3 illustrates this with mutation rates of 0.01, 0.1, and 0.5. The curves for uniform crossover are the same as those in Figure 2.

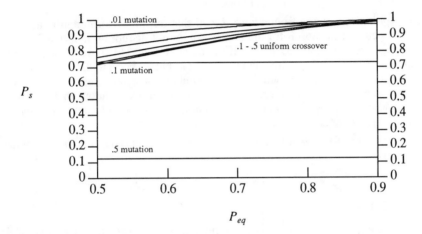

Figure 3: Survival Comparison, 3rd-Order

Figure 3 highlights several points. First, as expected, the disruptive effects of crossover are time dependent, while those of mutation are time independent. Second, the highest level of disruption for uniform crossover occurs when $P_0 = 0.5$. More interestingly, though, mutation can provide the same level of disruption as uniform crossover, if we allow the mutation rate to vary as a function of P_{eq}, k, and P_0.

In summary, then, we have introduced a disruption rate theory for mutation, and have compared this theory with our disruption rate theory for crossover. This comparison indicates that every level of disruption provided by crossover can be achieved with mutation alone. In fact, the comparison further indicates that crossover can not achieve the high levels of disruption that can be provided by mutation.

2.2   DISRUPTION DISTRIBUTIONS

Disruption rate theory estimates the likelihood that a genetic operator will disrupt a hyperplane sample. Again, we can also interpret this as the likelihood that individuals within a hyperplane will leave that hyperplane. It does not, however, indicate where those individuals will go. In other words, disruption rate theory does not indicate the distribution of disruptions, simply the likelihood that disruptions will occur.

Previous researchers (e.g., Eshelman *et al.*, 1989) have discussed the *exploratory power* of crossover operators, namely, the manner in which crossover disrupts individuals within hyperplanes. For example, suppose we consider the crossover of individuals from the two hyperplanes (the "#" denotes the "don't care" symbol):

> 1: 1####1
> 2: 0####0.

Uniform crossover will produce individuals from "######", while 1-point crossover will produce individuals from "1####0" and "0####1". In general, uniform crossover is more "explorative" than 1-point crossover. Eshelman *et al.* (1989) and Booker (1992) provide analyses of other biases in crossover operators.

What is the explorative power of mutation? Recall that our model of mutation assumes that a bit is flipped if it is chosen for mutation. We do not disrupt an individual within any hyperplane if the mutation rate is 0.0. If the mutation rate is 1.0, we always disrupt the individual, and produce the complement of the individual. For a mutation rate of 0.5, an individual will be disrupted with high probability, possibly creating any other individual. In summary, we can control the amount of exploration that mutation performs by adjusting the mutation rate. Mutation can provide any *amount* of exploration that crossover can provide.

Let us now compare the *type* of exploration that crossover and mutation provide. Suppose we consider individuals from the two hyperplanes:

> 1: 10####
> 2: 11####.

Crossover will only produce individuals from the hyperplane "1#####". The first "1" is guaranteed because it is common to the first defining position of both hyperplanes. Mutation, however, will *not* necessarily honor that guarantee, since it is a one individual operator and does not determine the commonality of alleles. Crossover, then, *preserves* alleles that are common to the individuals within the two hyperplanes (Radcliffe (1991) refers to this as "respect"). Preservation limits the type of exploration that crossover can perform. This limitation becomes more acute as the population loses diversity, since the number of common alleles will increase.

In summary, disruption analysis is the traditional analysis for describing the behavior of GAs in general, and crossover in particular. We have shown that mutation can provide any level of disruption that crossover can provide. We have also considered the form of disruption for both operators, by considering their exploratory power. Again, crossover has no advantage over mutation in terms of the amount of exploration that can be performed. They do differ, however, in the type of exploration. Crossover guarantees preservation of common alleles, while mutation does not. Given this evidence, then, we might suppose that there is some theoretical support for disputing Fogel's claim that crossover has no general advantage over mutation.[2] In the next section we will consider another potential difference between crossover and mutation.

---

[2] We are not implying that mutation has no advantage over crossover, however.

3 CONSTRUCTION THEORY

In the traditional theory, crossover is analyzed as a disruptive operator. However, more recently, Syswerda (1989) hypothesized that a more positive theory of crossover is constructive in nature.[3] For example, instead of calculating the probability that an existing hyperplane sample will be disrupted, we now calculate the probability that an individual within a hyperplane will be constructed from existing individuals within lower order hyperplanes. Syswerda's theory was extended by Spears & De Jong (1991) to include n-point and P_0 uniform crossover. This theory indicated that highly disruptive crossover operators are also highly constructive. Unfortunately, however, there was no theoretical evidence to indicate that mutation is not as constructive as crossover. In this section we will show that an analysis of the constructive abilities of mutation will provide us with that evidence.

Suppose, then, that we wish to create a theory of construction for mutation. More specifically, imagine that we wish to construct an individual within the 5th-order hyperplane "11111###" from an individual within another 5th-order hyperplane "11100###". This can be accomplished by mutating the 0's, while not mutating the 1's. In general, suppose we wish to construct an individual within a kth-order hyperplane H_k from an individual within another kth-order hyperplane H_s, when the two hyperplanes match on m alleles and do not match on n alleles (i.e., $k = n + m$). Then the probability of construction (denoted as P_{con}) is given by:

$$P_{con}(H_k, H_s) = (1 - P_m)^m (P_m)^n$$

In order to compare this with the constructive effects of crossover we again need to be careful about our assumptions. In this case we wish to compute the probability that crossover will construct an individual within a kth-order hyperplane from an individual in another hyperplane with m correct and n incorrect alleles, given an arbitrary mate. As an illustration, let us again imagine that we wish to construct an individual in the 5th-order hyperplane "11111###" from an individual in "11100###", using crossover. The individual in the hyperplane "11100###" will be crossed with an arbitrary individual from one of the four following hyperplanes:

> 1: ###00###
> 2: ###01###
> 3: ###10###
> 4: ###11###.

Of these four situations, only the last can result in the construction of an individual in the hyperplane "11111###". Each of these situations is not equally likely, unless $P_{eq} = 0.5$. For example, given the hyperplane "11100###" and the fact that $P_{eq} = 0.8$, we can compute that an individual from "11100###" will be crossed with an individual from hyperplane "###11##" with probability 0.04. In general, the probability that two hyperplanes differ in n defining positions is:

$$\prod_{d \in N} (1 - P_{eq}(d))$$

---

[3] In prior work we refer to this as "recombination" theory. Since we wish to extend this theory to mutation, the term "construction" seems more appropriate.

where N is the set of n defining positions.

We have now calculated the probability that a potentially successful recombination can occur. However, since this does not guarantee success, we also need to determine the probability that crossover will yield an individual within the hyperplane "11111###", given individuals from:

> 1: 11100###
> 2: ###11###.

This can be done in a straightforward fashion by using the earlier recombination theory of Spears & De Jong (1991), that deals with the construction of individuals within a hyperplane from individuals within two non-overlapping lower order hyperplanes (see the Appendix for details). In this theory, for example, it is possible to compute the probability that an individual within "11111###" will be constructed from the crossover of individuals from:

> 1: 111#####
> 2: ###11###.

This is a more general case of the above situation, in which an individual from "11100###" is crossed with an individual from "###11###" (i.e., because the "#" is more general than a "0"). Specifically, the two situations are identical if we state that $P_{eq}(d) = 0$ for the last two defining positions.[4]

In general, if we are interested in constructing an individual within a kth-order hyperplane from an individual within a hyperplane that has n incorrect alleles, we first need to compute the probability that some other individual contains the n correct alleles, to ensure that a potentially successful recombination can occur. Given these two individuals, we then compute the probability that construction will occur, by using a specific case of the earlier recombination theory, where $P_{eq}(d) = 0$ for those n defining positions. In summary, we can use the earlier recombination theory to create a construction theory for crossover that can be compared with the construction theory for mutation.

Using these theories, we can compare mutation and crossover from the viewpoint of construction. Figure 4 presents the comparison for 3rd-order hyperplanes, while Figure 5 presents the comparison for 8th-order hyperplanes. Mutation rates of 0.01, 0.1, and 0.5 are again compared with uniform crossover. Again, for the sake of simplicity, we illustrate uniform crossover where P_0 ranges from 0.1 to 0.5, because this roughly bounds the levels of construction provided by n-point crossover. It is important to note that a mutation rate of 0.5 yields the highest probability of construction. Due to symmetry, mutation rates above 0.5 yield lower probabilities.

Figures 4 and 5 illustrate several interesting points. First, when the population is diverse, mutation can not match the levels of construction that crossover can achieve. In fact, for 3rd-order hyperplanes, crossover has higher constructive levels until the population is 70% converged. Second, this advantage increases as the order of the hyperplane increases. For example, with 8th-order hyperplanes, crossover has higher

---

[4] Again, the theory allows one to define distinct $P_{eq}(d)$'s for each defining position.

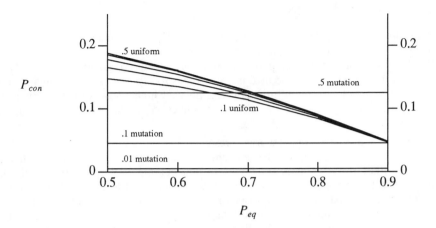

Figure 4: Construction Comparison, 3rd-Order

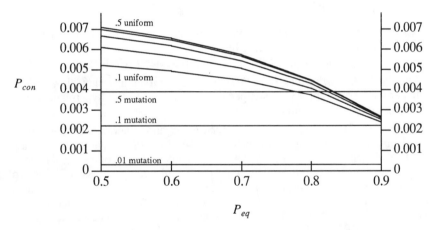

Figure 5: Construction Comparison, 8th-Order

constructive levels until the population is 80% converged.

Finally, note that it is impossible for mutation to simultaneously achieve high levels of construction and survival. This would appear to be important, since one without the other may not be extremely useful (i.e., it is nice if some of the constructions survive!). High construction levels with mutation are accomplished at the expense of survival (see 0.5 mutation), while good survival is at the expense of construction (see 0.01 mutation). In fact, crossover can simultaneously achieve higher levels of construction and survival than any particular amount of mutation.

4 SUMMARY AND DISCUSSION

These results provide a theoretical justification for Holland's belief that the role of crossover is to construct high order building blocks (hyperplanes) from low order building blocks. Mutation can not perform the role as well as crossover. Clearly, the role of crossover is construction and, in this case, crossover provides an advantage over mutation. In terms of disruption, mutation can provide higher levels of disruption and exploration, but at the expense of preserving alleles common to particular defining positions.

The disruption and construction theories presented here all concentrate on hyperplane (building block) analysis. Since the concept of a building block is often central to genetic algorithm research, it is important to connect our work with other relevant findings. First, this work does not assume a condition referred to as the *Static Building Block Hypothesis*. Next, we tie our work to the exploration and exploitation tradeoff, and indicate that our findings are consistent with experimental results. We conclude with the observation that our current distinction between crossover and mutation may not be necessary.

4.1 BUILDING BLOCKS

Since a role of crossover is construction, we would expect crossover to be useful on problems that have appropriate building blocks. What exactly is an appropriate building block? One possible answer lies in the following hypothesis (Grefenstette, 1992a):

> The *Static Building Block Hypothesis* (SBBH): Given any short, low order hyperplane partition, a GA is expected to converge to the hyperplane with the best static average fitness.

This hypothesis is often used as a base for theoretical and experimental work in genetic algorithms and implies that appropriate building blocks should have the highest average fitness. Unfortunately, as Grefenstette (1992a) indicates, the hypothesis is flawed in that a genetic algorithm is unlikely to determine the actual average fitness of a hyperplane, because the sampling of hyperplanes is biased. Although construction theory is concerned with the building of higher order hyperplanes from lower order hyperplanes, we do *not* make use of the SBBH. Rather, construction theory is consistent with what we will call the DBBH:

> The *Dynamic Building Block Hypothesis* (DBBH): Given any short [5], low order hyperplane partition, a GA is expected to converge to the hyperplane with the best dynamic (observed) average fitness.

In other words, a GA estimates the static average fitness from a dynamic biased sampling. As can be expected, the observed average fitness of a hyperplane can be quite different from its actual average fitness, implying that the GA may not converge to the hyperplane with the best static average fitness. Crossover, then, constructs higher order hyperplanes from lower order hyperplanes that have higher observed average fitness. These higher order hyperplanes may or may not bias search appropriately. Crossover works well with problems that have building blocks conducive to the creation of higher order building blocks that bias search correctly (see Vose & Liepins (1991) for a

---

[5] Actually, length is irrelevant for uniform crossover.

theoretical treatment of the relationship between crossover and building blocks). Although the appropriateness of building blocks is dynamic, and not well understood, some progress has been made in understanding the underlying issues. In the next section we outline some of the recent work. This work helps us to understand the roles of crossover and mutation in genetic algorithm search.

4.2 EXPLOITATION VS EXPLORATION

The issue concerning the relative importance of mutation and crossover can be viewed at a higher level. Mutation serves to create random diversity in the population, while crossover serves as an accelerator that promotes emergent behavior from components. The meta-issue, then, is the relative importance of diversity and construction. For the GA community, this is also related to the balance between exploration and exploitation. This meta-issue is the key to the difference in philosophy between Holland and Fogel. Specifically, Fogel *et al.* question the importance of recombination. They do not believe that natural selection selects individual traits (or, presumably, combinations of traits). Recombination is considered to be a third order factor, since it does not appear to occur frequently in nature (Atmar, 1992).

Of course, this does not necessarily imply that recombination is not useful for problems we wish to solve. Atmar is correct to remind us that "uncritical advocacy of a particular phenomenon" promotes "a blindness in perspective that is very difficult to dispel" (Atmar, 1992). Neither mutation nor crossover should be uncritically advocated or dismissed. Each operator plays a different role in the search process. *A priori*, it is difficult to specify the relative importance of each operator, for each problem. The appropriate balance of exploration and exploitation required for good performance depends on the amount of diversity in the population, the style of genetic algorithm used, and the purpose for which it is used.

For example, although GAs are often used as optimizers, our current understanding is that they attempt to maximize accumulated payoff (Holland, 1975). In this sense, they are greedy and should not necessarily be expected to find optimal solutions. Crossover can enhance this effect. Fogel and Atmar (1990) report that although the mean behavior of a GA with crossover outperformed the mean behavior of a GA without crossover (albeit insignificantly), they regard the winner to be the algorithm that found superior individual solutions. The GA without crossover had a much higher variance and found superior solutions in 6 of 10 trials. This effect has also been noted by Spears and Anand (1991), although they found that the results were dependent on population size. This indicates, then, that the GA practitioner should be clear about his or her goals. If optimality is sought, crossover may be deleterious. If the maximization of accumulated payoff is sought, mutation may be insufficient.

Similarly, greater disruption is more important for steady state genetic algorithms, since they suffer a higher allele loss than do their generational counterparts (De Jong & Sarma, 1992). It is also more important in non-stationary environments, where the optimal solution changes over time (Grefenstette, 1992b). We can conclude, then, that mutation will play an important role in these situations. Figures 4 and 5 support these ideas by suggesting that mutation becomes more important relative to crossover as the population loses diversity. Experiments with adaptive operator probabilities (Davis, 1989) support this analysis.

Crossover, however, will play an important role when construction and survival are required for good performance. This occurs when the population is diverse and problems consist of appropriate building blocks. Recent work suggests that fitness correlation (Lipsitch, 1991; Manderick *et al.*, 1991) and epistasis (Schaffer & Eshelman, 1991; Davidor, 1990) provide useful measures for determining the usefulness of crossover. For example, crossover appears to work well with functions that are highly correlated or have mild epistasis.

4.3   WHAT IS MUTATION?

Although our discussion of mutation and crossover stresses the differences between the two operators, it is also important to note that mutation can be greatly modified, minimizing those differences. The reason that crossover can exhibit high simultaneous levels of preservation, survival, and construction is that crossover shares information between fit individuals. Mutation, on the other hand, is often implemented with a parameter that is constant during genetic algorithm search. No information is shared when mutation is implemented in this fashion.

It is possible to implement mutation with a parameter that is adapted during genetic algorithm search. Population statistics, such as population convergence, are often used to adapt the mutation rate (Davis, 1989). The European community (e.g., Baeck *et al.*, 1991) go further, and explicitly adjust the mutation of each parameter, for every chromosome. One can easily imagine, then, a situation in which these mutation rates are based on finer grained population statistics, such as column convergence (De Jong, 1975). At this point, information can be communicated in a fashion similar to that of crossover. For example, we could measure the allele loss for each defining position and only mutate at defining positions with small allele loss, thus preserving common alleles. This would give mutation a disruption distribution more similar to that of crossover (see Section 2.2). At what point do we no longer call this mutation?

This leads us to the realization that standard mutation and crossover are simply two forms of a more general exploration operator, that can perturb alleles based on any available information (e.g., see Syswerda, 1992). It is not clear that the current distinction between crossover and mutation is necessary, or even desirable, although it may be convenient. The creation of more general operators, however, may lead to more robust biases. For example, it may be possible to implement one general operator that can specialize to mutation, crossover, or any variation in between. In our future work we intend to investigate this alternative.

Acknowledgements

I would like to thank Lawrence Fogel for suggesting this work in a dream. In this dream, Dr. Fogel approached me and said "You know, in terms of disruption theory, mutation can do everything crossover can. Crossover isn't necessary." I rushed to work and immediately began preparing this paper.

I would also like to thank Ken De Jong and John Grefenstette for many constructive discussions on methods for comparing crossover with mutation. Finally, I also wish to thank Diana Gordon, Alan Schultz, and four anonymous reviewers for helping to clarify earlier versions of this paper.

References

Atmar, W. (1992) The philosophical errors that plague both evolutionary theory and simulated evolutionary programming. In D. Fogel & J. Atmar (eds.), *Proceedings of the First Annual Conference on Evolutionary Programming*, 27-34. San Diego, CA: Evolutionary Programming Society.

Baeck, T., Hoffmeister, F., & Schwefel, H.-P. (1991) A survey of evolution strategies. *Proceedings of the Fourth International Conference on Genetic Algorithms*, 2-9. La Jolla, CA: Morgan Kaufmann.

Booker, L. B. (1992) Recombination distributions for genetic algorithms. *Proceedings of the Foundations of Genetic Algorithms Workshop*. Vail, CO: Morgan Kaufmann.

Davidor, Y. (1990) Epistasis variance: a viewpoint on GA-hardness. *Proceedings of the Foundations of Genetic Algorithms Workshop*, 23-35. Indiana University: Morgan Kaufmann.

Davis, L. (1989) Adapting operator probabilities in genetic algorithms. *Proceedings of the Third International Conference on Genetic Algorithms*, 60-69. La Jolla, CA: Morgan Kaufmann.

De Jong, K. A. (1975) *An analysis of the behavior of a class of genetic adaptive systems.* Doctoral Thesis, Department of Computer and Communication Sciences. University of Michigan, Ann Arbor.

De Jong, K. A. & Spears, W. M. (1992) A formal analysis of the role of multi-point crossover in genetic algorithms. *Annals of Mathematics and Artificial Intelligence Journal* 5, 1-26. J. C. Baltzer A. G. Scientific Publishing Company.

De Jong, K. A. & Sarma, J. (1992) Generation gaps revisited. *Proceedings of the Foundations of Genetic Algorithms Workshop*. Vail, CO: Morgan Kaufmann.

Eshelman, L. J., Caruana, R. A., & Schaffer, J. D. (1989) Biases in the crossover landscape. *Proceedings of the Third International Conference on Genetic Algorithms*, 10-19. La Jolla, CA: Morgan Kaufmann.

Fogel, L. J., Owens, A. J., & Walsh, M. J. (1966) *Artificial Intelligence Through Simulated Evolution.* New York: Wiley Publishing.

Fogel, D. B. & Atmar, J. W. (1990) Comparing genetic operators with gaussian mutations in simulated evolutionary processes using linear systems. *Biological Cybernetics* 63, 111-114.

Grefenstette, J. G. (1992a) Deception considered harmful. *Proceedings of the Foundations of Genetic Algorithms Workshop*. Vail, CO: Morgan Kaufmann.

Grefenstette, J. G. (1992b) Genetic algorithms for changing environments. Submitted to the *Parallel Problem Solving from Nature (PPSN) Workshop*. Brussels, Belgium.

Holland, J. H. (1975) *Adaptation in Natural and Artificial Systems.* Ann Arbor, Michigan: The University of Michigan Press.

Lipsitch, M. (1991) Adaptation on rugged landscaped generated by iterated local interactions of neighboring genes. *Proceedings of the Fourth International Conference on Genetic Algorithms,* 128-135. La Jolla, CA: Morgan Kaufmann.

Manderick, B., de Weger, M., & Spiessens, P. (1991) The genetic algorithm and the structure of the fitness landscape. *Proceedings of the Fourth International Conference on Genetic Algorithms,* 143-149. La Jolla, CA: Morgan Kaufmann.

Radcliffe, N. J. (1991) Forma analysis and random respectful recombination. *Proceedings of the Fourth International Conference on Genetic Algorithms,* 222-229. La Jolla, CA: Morgan Kaufmann.

Rechenberg, I. (1973) *Evolutionsstrategie: Optimierung Technischer Systeme nach Prinzipien der Biologischen Evolution.* Frommann-Holzboog, Stuttgart.

Schaffer, J. D., Caruana, R. A., Eshelman, L. J., & Das, R. (1989) A study of control parameters affecting online performance of genetic algorithms for function optimization. *Proceedings of the Third International Conference on Genetic Algorithms,* 51-60. La Jolla, CA: Morgan Kaufmann.

Schaffer, J. D., Eshelman, L. J. (1991) On crossover as an evolutionarily viable strategy. *Proceedings of the Fourth International Conference on Genetic Algorithms,* 61-68. La Jolla, CA: Morgan Kaufmann.

Schwefel, H.-P. (1977) *Numerical Optimization of Computer Models.* New York: John Wiley & Sons.

Spears, W. M. & Anand, V. (1991) A study of crossover operators in genetic programming. *Proceedings of the Sixth International Symposium on Methodologies for Intelligent Systems,* 409-418. Charlotte, NC: Springer-Verlag.

Spears, W. M. & De Jong, K. A. (1991) On the virtues of uniform crossover. *Proceedings of the Fourth International Conference on Genetic Algorithms,* 230-236. La Jolla, CA: Morgan Kaufmann.

Syswerda, G. (1989) Uniform crossover in genetic algorithms. *Proceedings of the Third International Conference on Genetic Algorithms,* 1-9. Fairfax, VA: Morgan Kaufmann.

Syswerda, G. (1992) Simulated crossover in genetic algorithms. *Proceedings of the Foundations of Genetic Algorithms Workshop.* Vail, CO: Morgan Kaufmann.

Vose, M. D., & Liepins, G. E. (1991) Schema disruption. *Proceedings of the Fourth International Conference on Genetic Algorithms,* 237-242. La Jolla, CA: Morgan Kaufmann.

Appendix[6]

Summary of the Survival Analysis

For n-point crossover, P_s is expressed in the order dependent form ($P_{k,s}$):

$$P_{2,s}(n, L, L_1) = \sum_{i=0}^{n} \binom{n}{i} \left[\frac{L_1}{L} \right]^i \left[\frac{L-L_1}{L} \right]^{n-i} C_s$$

and

$$P_{k,s}(n, L, L_1, \ldots, L_{k-1}) =$$
$$\sum_{i=0}^{n} \binom{n}{i} \left[\frac{L_1}{L} \right]^i \left[\frac{L-L_1}{L} \right]^{n-i} P_{k-1,s}(i, L_1, \ldots, L_{k-1})$$

Note that the survival of a kth-order hyperplane under n-point crossover is recursively defined in terms of the survival of lower order hyperplanes. L refers to the length of the individuals. The $L_1 \cdots L_{k-1}$ refer to the defining lengths between the defining positions of the kth-order hyperplane. The effect of the recursion and summation is to consider every possible placement of n crossover points within the kth-order hyperplane. The correction factor C_s computes the probability that the hyperplane will survive, based on that placement of crossover points.

For the ease of presentation we denote K to be the set of k defining positions, while X denotes a subset of K. Suppose that crossover results in a subset X of defining positions being exchanged. Then the hyperplane will survive if: 1) the parents match on the subset X, or 2) if they match on the subset $K - X$, or 3) they match on the set K. Hence, the general form of the correction is:

$$C_s = \prod_{d \in X} P_{eq}(d) + \prod_{d \in K-X} P_{eq}(d) - \prod_{d \in K} P_{eq}(d)$$

where $P_{eq}(d)$ is the probability that two parents have the same alleles on a particular defining position d.

For parameterized uniform crossover, P_s is also expressed in an order dependent form ($P_{k,s}$):

$$P_{k,s}(H_k) =$$
$$\sum_{I \subseteq K} (P_0)^{|I|} (1-P_0)^{|K-I|} \left[\prod_{d \in I} P_{eq}(d) + \prod_{d \in K-I} P_{eq}(d) - \prod_{d \in K} P_{eq}(d) \right]$$

where I is a subset of K, and P_0 is the probability of swapping two parents' alleles at each defining position. A graphical representation of these equations has been shown previously in Figure 1.

Recombination (Construction) Analysis for N-Point Crossover

In our definition of survival, it is possible for a hyperplane to survive in either child. Recombination can be considered a restricted form of survival, in which two lower order

---

6 This Appendix is a compendium of crossover theory from De Jong & Spears (1992).

hyperplanes survive to form a higher order hyperplane. The difference is that the two lower order hyperplanes (each of which exists in a different parent) must survive in the same individual, in order for recombination to occur.

In the remaining discussion we will consider the creation of a kth-order hyperplane from two hyperplanes of order m and n. We will restrict the situation such that the two lower order hyperplanes are non-overlapping, and $k = m + n$. Each lower order hyperplane is in a different parent. We denote the probability that the kth-order hyperplane will be recombined from the two hyperplanes as $P_{k,r}$.

An analysis of recombination under n-point crossover is simple if one considers the correction factor C_s defined earlier for the survival analysis. Recall that recombination will occur if both lower order hyperplanes survive in the same individual. If an n-point crossover results in a subset X of the k defining positions surviving in the same individual, then recombination will occur if: 1) the parents match on the subset X, or 2) if they match on the subset $K - X$, or 3) they match on the set K. Hence, the general form of the recombination correction C_r is:

$$C_r = \prod_{d \in X} P_{eq}(d) + \prod_{d \in K-X} P_{eq}(d) - \prod_{d \in K} P_{eq}(d)$$

Note the similarity in description with the survival correction factor C_s (the only difference is in how X is defined). In other words, given a kth-order hyperplane, and two hyperplanes of order n and m, $P_{k,r}$ is simply $P_{k,s}$ with the correction factor redefined as above.

Recombination (Construction) Analysis for Uniform Crossover

The analysis of recombination under uniform crossover also involves the analysis of the original survival equation. Note that, due to the independence of the operator (each allele is swapped with probability P_0), the survival equation can be divided into three parts. The first part expresses the probability that a hyperplane will survive in the original string:

$$P_{k,s,orig}(H_k) = \sum_{I \subseteq K} (P_0)^{|I|} \ (1 - P_0)^{|K-I|} \prod_{d \in K-I} P_{eq}(d)$$

The second part expresses the probability that a hyperplane will survive in the other string:

$$P_{k,s,other}(H_k) = \sum_{I \subseteq K} (P_0)^{|I|} \ (1 - P_0)^{|K-I|} \prod_{d \in I} P_{eq}(d)$$

The final part expresses the probability that a hyperplane will exist in both strings:

$$P_{k,s,both}(H_k) = \sum_{I \subseteq K} (P_0)^{|I|} \ (1 - P_0)^{|K-I|} \prod_{d \in K} P_{eq}(d) = \prod_{d \in K} P_{eq}(d)$$

Then:

$$P_{k,s}(H_k) = P_{k,s,orig}(H_k) + P_{k,s,other}(H_k) - P_{k,s,both}(H_k)$$

Note, however, that this formulation allows us to express recombination under uniform crossover. Again, assuming the recombination of two non-overlapping hyperplanes of order n and m into a hyperplane of order k:

$$P_{k,r}(H_k) = P_{m,s,orig}(H_m)\, P_{n,s,other}(H_n) \,+$$
$$P_{m,s,other}(H_m)\, P_{n,s,orig}(H_n) \,-$$
$$P_{m,s,both}(H_m)\, P_{n,s,both}(H_n)$$

This equation reflects the decomposition of recombination into two independent survival events. The first term is the probability that H_m will survive on the original string, while H_n switches (i.e., both hyperplanes survive on one parent). The second term is the probability that both hyperplanes survive on the other parent. The third term reflects the joint probability that both hyperplanes survive on both strings, and must be subtracted. It should also be noted that the third term is equivalent to $\prod_{d \in K} P_{eq}(d)$.

Simulated Crossover in Genetic Algorithms

Gilbert Syswerda

BBN Laboratories
BBN Systems and Technologies Corporation
10 Moulton Street
Cambridge, MA 02138

Abstract

A new class of crossover operator, simulated crossover, is presented. Simulated crossover treats the population of a genetic algorithm as a conditional variable to a probability density function that predicts the likelihood of generating samples in the problem space. A specific version of simulated crossover, bit-based simulated crossover, is explored. Its ability to perform schema recombination and reproduction is compared analytically against other crossover operators, and its performance is checked on several test problems.

1 INTRODUCTION

Suppose we set out to create an operator similar to crossover, but which does not perform recombination. Consider the following operator: select two binary population members according to some fitness criteria. Wherever these parents have a bit in common, copy that bit to the child; otherwise, randomly generate bits for the remaining positions of the child. Does this operator perform recombination? Does it do anything useful at all?

We should hope so, since this is just what crossover does. If a bit has the same value in both parents, crossover cannot change that bit and it is passed on to the child. If the values are different, then the bit will be copied from one or the other parent. For one-point, two-point, multi-point, and uniform crossover (assuming a uniform crossover operator with a 50% selection probability), each remaining bit has a 50% chance of coming from one parent or the other. Since the bits in each parent are different, this is the same as simply randomly assigning a 0 or 1 to the position.

Uniform crossover is exactly described by this procedure, since each bit is treated independently. With the other crossover operators, there are conditional dependencies among the bits, since crossover is done with contiguous groups of bits. In the average case however, where nothing is assumed about the location of the bits of schemata, the operators are equivalent.

Given two random parents of length l, how many bits in a child will be different from either parent after crossover? On average, half of the bits of the parents will match each other (both 0 or 1). Since the remaining bits are randomly assigned, half of these will also match. Thus, for random parents (e.g. chosen from a randomly initialized population), the average number of bits that are different from either parent is only $0.25l$. This number is the upper limit; it decreases as the population converges.

Since in the worst case only one fourth of the bits on average are changed, perhaps a bit mutation rate of 0.25 or less will accomplish what crossover does. Mutation alone, however, does not have the same effect, since crossover is changing only those bits that are different between two parents. The fact that some bits are the same is influenced by the state of the population and the selection criteria being used. If mutation could take into account these similar bits, then mutation would begin to behave like crossover.

We can take this one step further and ask ourselves how to generate from a population a child whose bits are like those, on average, that would be produced if two parents were selected and a child produced via crossover. One way to think of this is to treat the population as a variable to a conditional density function that computes the likelihood that any particular region of the problem space will be visited by a child produced via crossover from that population. This function can be treated as a probability distribution by normalizing the area under the function's curve to 1.0. We will call a mechanism that produces children in this manner *simulated crossover*.

For the purposes of this paper, crossover is defined as a syntactic operation on chromosomes (e.g. bitstrings). Recombination is defined as the combination of previously separate fit schemata or building blocks into the same individual. Simulated crossover concerns itself with bit patterns; *simulated recombination* would concern itself with combining building blocks, the values of which are defined in terms of the problem being solved.

2   BIT-BASED SIMULATED CROSSOVER (BSC)

In the steady-state model of genetic algorithms [Syswerda 1989, 1991], two members are chosen from a population according to some fitness criteria and combined via crossover to produce one or two new children. For this paper, it is assumed that only one child is produced. We require a mechanism that takes a population as a whole and produces a child with the same distribution as normal, or *explicit crossover*.

Consider a list of the values of a single bit position in a population (a bit column). It will contain some ratio of ones and zeroes. The owners of these bits, members of the population, have some probability of being selected, and these probabilities can be used to create a weighted average for the ones and zeroes in each bit column. This provides a probability for each bit as to whether it should be a zero or a one. These probabilities can be used to generate members that, on average, will have the same ratio of ones and zeroes for each bit position that crossover would have generated. We will call this procedure *bit-based simulated crossover* (BSC).

It is important to note that BSC will generate the same members as explicit crossover only in the average case; there are a number of reasons why there may be performance differences between BSC and explicit crossover. These differences are addressed throughout the remainder of this paper.

3 RECOMBINATION AND SIMULATED CROSSOVER

BSC ignores the bit patterns (or schemata) of individual population members. Instead, it uses the statistics of individual bits across an entire population. At first glance, it might seem that such an operator would not perform recombination. However, if we examine the probability of each operator performing recombination, we will see that BSC actually combines schemata with greater likelihood than explicit crossover in some cases.

To gain an intuitive understanding of this, consider an operator which simply generates random bitstrings to insert into a population. It "performs" recombination at some low rate, since it has a greater than zero chance of creating a population member with an appropriate combination of bits. We are interested in computing the average rate of production of new population members that contain combinations of existing schemata in the population.

Let S_i and S_j be two schemata in the population, P_i and P_j the probability that a population member will be chosen for crossover that contains S_i or S_j, and i and j the order of S_i and S_j. Let P_{EX} be the probability that explicit crossover will produce a child that contains both S_i and S_j, and P_{BSC} the probability that BSC will do the same. In the following analysis, it is assumed that no population member currently contains both S_i and S_j, explicit crossover produces only one child, any bits that do not belong to either S_i or S_j contain either a 0 or 1 with equal probability, and the bits belonging to S_i and S_j can be anywhere but do not overlap.

We are assuming that population members which do not contain some schema S_i have random values in those bit positions, except that those values do not specify schema S_i. Let R_i be the chance of crossover generating the schema by combining these "random" bits. To adjust for the not quite random conditions, the value of R_i is changed from $.5^i$ for purely random bits to $(.5(2^i - 1)/2^i)^i$. Likewise, let Q_i be the probability that crossover will reproduce a schema when crossing over S_i with random bits. Q_i changes from $.75^i$ to $(.5 + .5(.5(2^i - 1)/2^i))^i$.

To compute P_{EX} we must consider six possible combinations of parents with or without schemata i and j: $ii, ij, i\phi, jj, \phi j, \phi\phi$, where ϕ indicates a population member without either schema.

$$P_{ii} = P_i^2 R_i$$
$$P_{ij} = 2P_i P_j \, Q_i \, Q_j$$
$$P_{i\phi} = 2P_i(1 - P_i - P_j) \, Q_i \, R_j$$
$$P_{jj} = P_j^2 \, R_i$$
$$P_{\phi j} = 2P_j(1 - P_i - P_j) \, R_i \, Q_j$$
$$P_{\phi\phi} = (1 - P_i - P_j)^2 \, R_i \, R_j$$

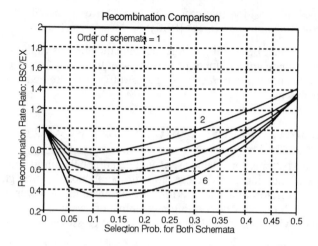

Figure 1: Schema Recombination. Comparison of the recombination rate of BSC against explicit crossover.

$$P_{EX} = P_{ii} + P_{ij} + P_{i\phi} + P_{jj} + P_{\phi j} + P_{\phi\phi} \tag{1}$$

The probability for BSC is given by:

$$P_{BSC} = (P_i + (1 - P_i) \,.5(2^i - 1)/2^i)^i \, (P_j + (1 - P_j) \,.5(2^j - 1)/2^j)^j . \tag{2}$$

Figure 1 depicts P_{BSC}/P_{EX} (the relative effectiveness of BSC compared to EX) for order 1 through order 6 schemata, plotted against the selection probability for each schema (the same for both). As we can see, the rate at which recombination occurs is generally lower for BSC than for explicit crossover, except when the selection probabilities approach 0.5, where BSC begins to perform recombination with greater likelihood. However, it is lower by only a factor of about 2 at worst for schemata of order 5. For single bits, BSC is twice as effective as explicit crossover, but this is somewhat an artifact of the assumption that both schemata do not exist in the same population member.

4   REPRODUCTION OF SCHEMATA

Another measure of the effectiveness of a crossover operator is the rate at which existing schemata survive the process. A related measure is the chance that a schema which exists in the population will exist in a child produced by crossover. Using the same terminology as above:

$$P_{EX} = P_i^2 + 2P_i(1 - P_i) \, Q_i + (1 - P_i)^2 \, R_i \tag{3}$$

$$P_{BSC} = (P_i + (1 - P_i) \,.5(2^i - 1)/2^i)^i \tag{4}$$

The ratio of P_{BSC}/P_{EX} for reproduction is presented in Figure 2. The rate at which population members are created with existing schemata is somewhat lower for BSC than

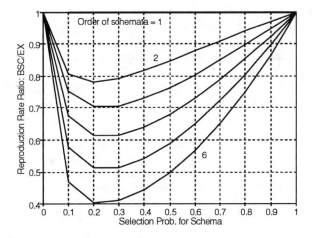

Figure 2: Schema Reproduction.

for explicit crossover, although not by a large margin. Also, the flip side of reproduction is exploration, which works well within steady-state genetic algorithms [Syswerda, 1989].

5   Simulating Simulated Crossover

Mutation must be taken into account when comparing the performance of BSC against explicit crossover. For explicit crossover, each bit of the child created is subjected to mutation according to some small probability.

Mutation can be included in the BSC procedure by adjusting the probabilities in the following manner.

Let P_m be the chance any bit will be changed by mutation, and P_1 be the chance that a bit will be a one. The actual chance of a bit being a one (P), including mutation, is given by

$$P = P_1 - P_1 P_m + (1 - P_1) P_m.$$

6   EMPIRICAL STUDIES

Testing the performance of BSC against various forms of explicit crossover requires a suite of test problems. This can be a challenge in and of itself, since few researchers agree on what constitutes an appropriate suite of problems.

Presented here are the results of testing on five problems of varying complexity and difficulty.

F6. F6 [Schaffer 1989] is a sinusoidal function in two parameters (22 bits each), scaled

```
f(1111) = 30   f(0100) = 22   f(0110) = 14   f(1110) = 6
f(0000) = 28   f(1000) = 20   f(1001) = 12   f(1101) = 4
f(0001) = 26   f(0011) = 18   f(1010) = 10   f(1011) = 2
f(0010) = 24   f(0101) = 16   f(1100) = 8    f(0111) = 0
```

Table 1: Function DF2. Given are the values associated with the binary patterns for each of 10 4-bit deceptive problem.

by a parabolic curve. Each parameter changes the period of the sine for the other parameter. At the ideal value for each parameter, the function is periodic in a power of 2 on the other parameter, potentially making it easy for mutation to search this function. Davis [Davis 1991] has reported that this function, as posed by Schaffer, can be solved successfully using his bit-climbing algorithm. F6 has been cast here as a maximization problem, with a maximum value of 1.0.

DF2. This is a GA-deceptive problem described in [Whitley 1991]. It consists of 10 fully deceptive 4-bit problems concatenated and summed together. (See Table 1). The deceptive peak for each subproblem is at 28 for bit pattern **0000**, providing a maximum deceptive value of 280. The true maximum has a value of 30 for bit pattern **1111**, providing a global maximum of 300.

DF2-300. This is similar to DF2, except that the value of the best bit pattern, **1111**, has been increased from 30 to 300.

TSP. This is a binary-coded traveling salesperson problem suggested by Smith [Shaefer, 1990]. Five bits were assigned to each of 30 cities, and the binary value of the bits was used to determine each city's place in the tour. This representation is a poor one for the TSP, making the problem difficult for a genetic algorithm. The problem was cast as a maximization problem.

RR4. This is the Royal Road function described in [Mitchell et al 1992]. The function consists of 8 groups of 8 contiguous bits, each worth 8 points for **11111111** and zero otherwise. Continuous pairs of groups of 8, when solved, result in an additional 16 points, contiguous pairs of pairs of 8 add an additional 32 points when solved, etc. This function seems easy at first glance, but turns out to be difficult for a GA, since the payoff given to higher-order schemata provides misleading information about the unsolved portions of the problem.

A steady-state population model was used for the empirical studies. Parent selection was by proportional fitness, where the fitness was scaled between the population best and worst. Deletion was by exponential ranking, with a 0.0001 chance for the population best being selected. The mutation rate was fixed at 0.003 per bit, except when using only mutation, in which case the rate was set to 0.03. No duplicate members were allowed in the population.

Four crossover operators were compared: one-point, two-point, uniform, and BSC. In addition, runs were made using just mutation (no crossover), with the mutation rate per bit set at 0.03 (ten times higher than when used with crossover).

Each operator was run with several population sizes: 5, 10, 25, 50, 100, 200, and 400. The graphs show the performance after various number of evaluations: typically 1000, 2000, ..., 5000 or 1000, 4000, 8000, ..., 20000. All data presented is the average of

30 independent runs.

6.1 F6

Figure 3 presents the performance graphs for F6. Notice first that BSC performed better than all other crossover operators. Also notice how well mutation alone did on this problem. This will be a recurring theme.

One and two-point crossover did relatively poorly, with the performance plummeting when the population size approached 10, but then rebounding a little with a population size of 5. This effect is likely due to the increased noise in the selection process simulating a higher mutation rate.

BSC performed well with a population size of 50; anything less severely degraded performance. Both one-point and two-point crossover have positive slopes at 5000 evaluations, suggesting that larger population sizes are best for them. Uniform crossover performed best at 200, while mutation did well with nearly any population size.

6.2 DF2

Figure 4 presents the performance graphs for DF2.

The performance of the various operators on DF2 is quite interesting. The best performer on this problem, by a considerable margin, was mutation, followed by two-point, one-point, uniform, and finally BSC.

Problems like DF2 have been devised specifically to fool genetic algorithms. A genetic algorithm that accurately follows the clues the population is providing will always fall into the trap. With a population size of 100 or more, BSC does just that with 100% effectiveness. With a population size of less than 100, the performance of BSC begins to increase (where on all the other problems, performance decreased). This is because with a population that is too small, single members can bias BSC, and a mistake on this problem can jump the population out of the deception and onto a local maximum. For BSC, performance continues to increase as the population size decreases to 5.

Since one and two-point crossover swap contiguous runs of bits, they will tend to accidentally preserve 1111 groupings even if their fitness is not apparent in the population. In contrast to the other test problems, one and two-point crossover performed best with a smaller population of about 50-100.

If making mistakes is a good strategy for this problem, then mutation should excel, and it does. With a population size of 50, it achieves the highest score.

Since an error-prone operator seems to be the best strategy for this problem, perhaps an even higher mutation rate would be better. When blindly mutating, however, there is a tradeoff between keeping what has already found and finding better solutions. Figure 5 presents a study of the performance of mutation at different rates in a population size of 100. The mutation rate that was used, 0.03, is best for this problem.

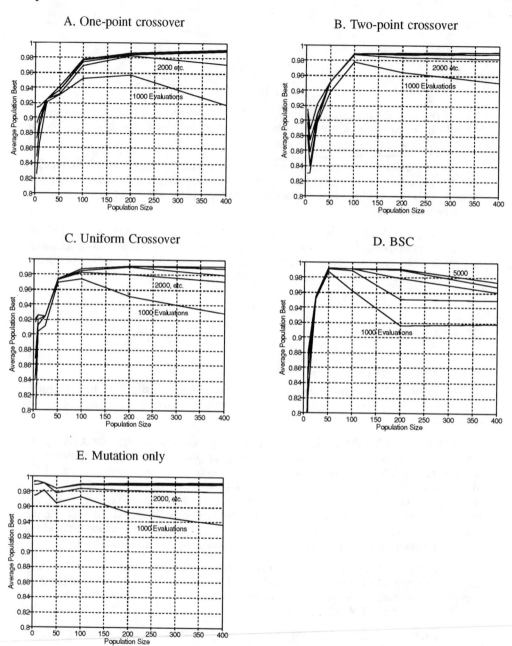

Figure 3: Function F6.

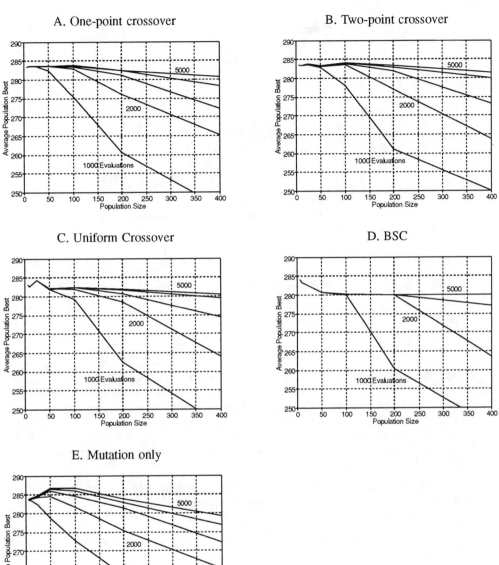

Figure 4: Function DF2.

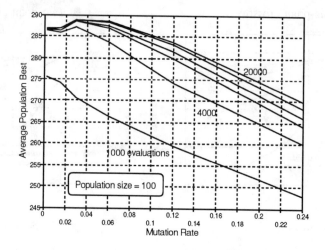

Figure 5: DF2 Mutation Rate Study.

6.3 DF2-300

To test the hypothesis that being deceived on DF2 is the correct thing to do from the viewpoint of a GA, a new function was devised that allows the value of the maximum string value, **1111**, to be seen above the noise generated by the rest of the representation. To accomplish this, the value associated with **1111** was increased from 30 to 300. There are still two peaks for each group of four bits, one at **1111** and the other at **0000**. What has changed is that the value of **1111** is now visible in the population above the background noise, removing the deception.

This seemingly small change in the function affects dramatically the performance of all operators. Each crossover operator, as seen in Figure 6, now solves this problem, with BSC being the best.

Mutation, in contrast to its relative performance on DF2, performed slowly, but would likely eventually solve the problem.

6.4 TSP

Mutation was a poor performer on this problem (Figure 7). BSC was best, followed by uniform, two-point and one-point crossover. All operators were continuing to improve the solution after 20000 evaluations.

At 20000 evaluations per run, one and two-point crossover require population sizes larger than 400, while uniform crossover does best with 200, BSC with 100, and mutation with 50.

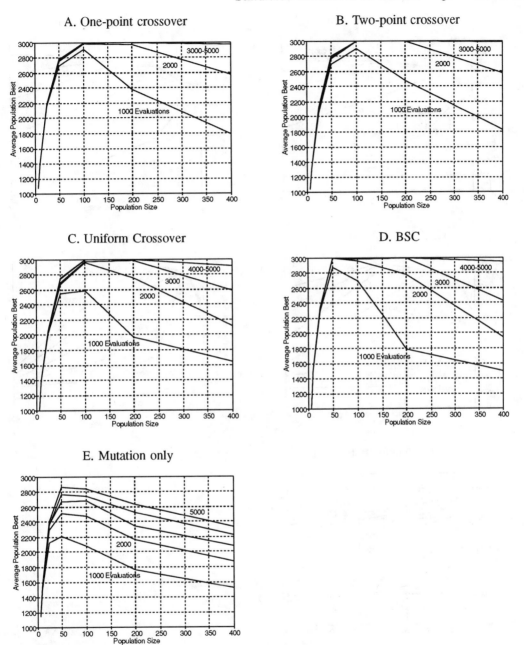

Figure 6: Function DF2-300, with the value of **1111** set to 300.

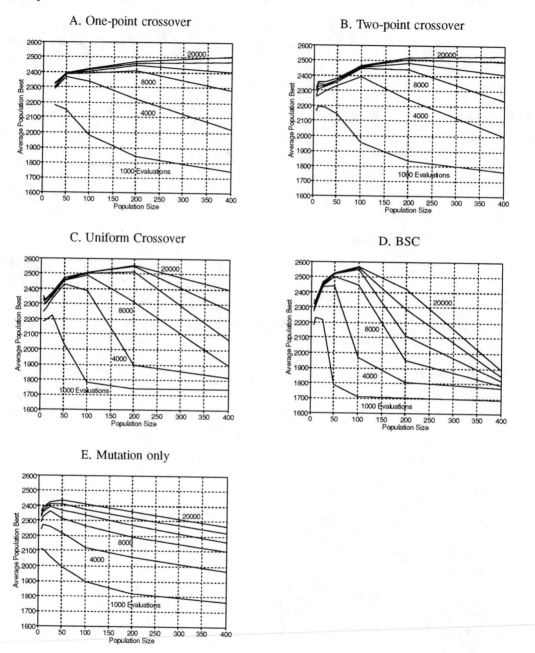

Figure 7: Function TSP.

6.5 RR4

The results in Figure 8 show how poorly the GA did on this problem. This problem is hard because the landscape has large plateaus due to no payoff being given until eight bits are correctly set, and because higher-order schemata, when discovered, provide a large evaluation increase. Both of these factors lead to premature convergence. One and two-point crossover did a little better than BSC, and uniform crossover and mutation did more poorly. All operators, except mutation, required a larger population than 400 given 10000 evaluations.

To reduce premature converge, this problem was run again with the parent selection procedure changed from proportional selection to random selection. Only the deletion selection procedure provided convergence. The results are presented in Figure 9. Performance is much better, with two-point crossover leading the pack. Both one and two-point still required larger population sizes. Uniform crossover and BSC were now happy with a population size of 100. The performance of mutation remained poor and essentially unchanged.

7 THE ROLE OF A POPULATION

Why does a genetic algorithm require a population? More specifically, why does a GA using BSC require a population? BSC replaces the explicit crossover operator (and mutation) of a genetic algorithm. Can the population be replaced as well?

To answer this, we must go back and examine the BSC algorithm more carefully. For each child generated, the value of each bit is selected based on the *selection probability* of the members of the population. An alternative might be to use the *expected value* of a bit. It is easy enough to compute; the expected value of a bit position with a particular value is simply the average fitness of population members with the bit position set to that value (ignoring the degenerate case of an entirely converged bit position).

For example, suppose the observed expected value for value **1** of some bit position is 2.0, and for value **0** the observed expected value is 1.0. The probability of generating a **1** is given by $2/(2 + 1) = 2/3$; the probability for **0** is given by $1/3$. A running tally could maintain statistics on the expected value of each bit position, and the population could thus be done away with.

This approach does not work, however, for reasons fundamental to why genetic algorithms work in the first place. Populations in a GA are not computing the expected value of schemata; they are computing the selection probabilities of schemata via an implicit control procedure which samples schemata that have better than average expected values with exponentially increasing frequency. Sampling according to expected value does not increase the sampling frequency of the observed best. If confidence is high that a **1** in some bit position has an expected value of 2.0 while a **0** has an expected value of 1.0, we would expect an optimal control strategy to nearly always sample using **1**. Previous work indicates that an optimal control strategy will sample the observed best at an exponentially increasing rate [Holland 1975, Goldberg 1989].

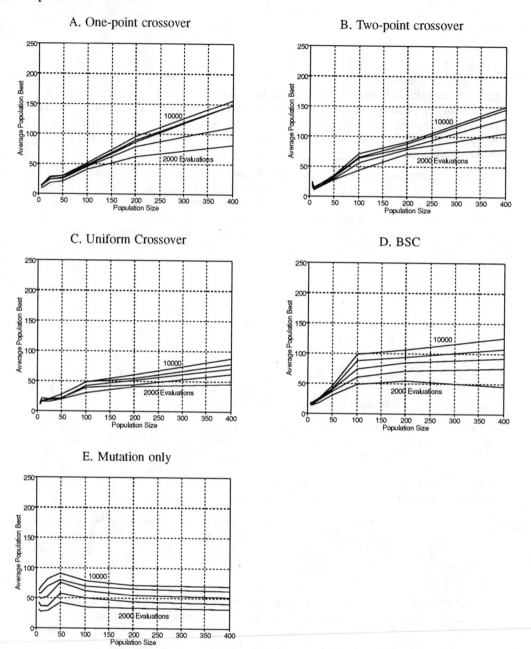

Figure 8: Function RR4.

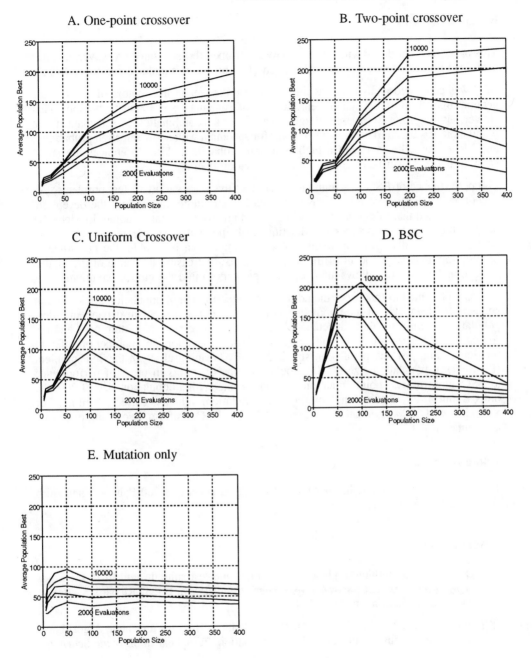

Figure 9: Function RR4, with random parent selection.

8 SUMMARY

This paper has introduced a new kind of crossover operator, simulated crossover, and has examined in some detail a particular version of it, bit-based simulated crossover (BSC).

BSC was compared analytically against explicit crossover in the context of steady-state genetic algorithms. The rate of reproduction of schemata was found to be somewhat slower, although not by a large margin. The rate of recombination of schemata in the population was found to be somewhat lower for schemata with a low selection probability, and higher for schemata with a high selection probability. The differences are small enough that BSC can be considered a recombination operator.

The performance of BSC was found to be competitive on a variety of test problems. An exception is DF2, where BSC is distinguished by unfailingly falling into the deception. It can be argued that this is precisely what a good crossover operator should do, since the deception is what the population is indicating. If the population size decreases to the point where fortuitous discovery of a subproblem maximum can be visible, the performance of BSC improves to the level of the other operators. Increasing the value of the hidden peak to make it visible allows all crossover operators to find the global maximum.

The data presented in this paper includes testing over a range of population sizes. As can be seen, the ideal size of the population is both operator and problem dependent. Ideal population size is also likely dependent on the version of genetic algorithm used and the settings of other parameters. Providing a comparison of operators on a number of different problems using just one population size is not very meaningful. With the TSP problem (Figure 7), a population size of 100 would rank the operators in order of performance as BSC, uniform, two-point, mutation, and one-point. With a population size of 200, the order is uniform, two-point, one-point, bsc, and mutation. Which is correct? Neither. For a maximum of 20000 evaluations, the true ranking is BSC, uniform, two-point, one-point and mutation.

Acknowledgments

The research presented here has benefitted from discussions with Jeff Palmucci and Herb Gish of BBN Laboratories.

References

[1] Davis, L. Bit-climbing, representational bias, and test suite design. In *Proceedings of the Fourth Int'l. Conference on Genetic Algorithms*, 18-23, Richard K. Belew, ed., Morgan Kaufmann, July 1991.

[2] Forrest S. and Mitchell M. What Makes a Problem Hard for a Genetic Algorithms? Some Anomalous Results and Their Explanation. To appear in *Machine Learning*.

[3] Goldberg, D.E. Simple genetic algorithms and the minimal deceptive problem. In Lawrence D. Davis, editor, *Genetic Algorithms and Simulated Annealing*, Research Notes in Artificial Intelligence, Los Altos, CA, 1987. Morgan Kaufmann.

[4] Goldberg, D.E. *Genetic Algorithms in Search, Optimization, and Machine Learning*. Addison-Wesley Publishing Company, 1989.

[5] Holland, J.H. *Adaptation in Natural and Artificial Systems.* University of Michigan Press, 1975.

[6] Mitchell, M, Forrest, S. and Holland, J. The Royal Road for Genetic Algorithms: Fitness Landscapes and GA Performance. In *Proceedings of the First European Conference on Artificial Life.* Cambridge, MA: MIT Press, 1992.

[7] Schaffer, J.D. A study of control parameters affecting online performance of genetic algorithms for function optimization. In *Proceedings of the Third Int'l. Conference on Genetic Algorithms and their Applications*, 2-9, J. D. Schaffer, ed., Morgan Kaufmann, June 1989.

[8] Shaefer, C., Smith, S. *The Argot Strategy II - Combinatorial Optimizations.* Thinking Machines Technical Report RL90-1 (1990).

[9] Syswerda, G. Uniform crossover in genetic algorithms. In *Proceedings of the Third Int'l. Conference on Genetic Algorithms and their Applications*, 2-9, J. D. Schaffer, ed., Morgan Kaufmann, June 1989.

[10] Syswerda, G. A study of reproduction in generational and steady-state genetic algorithms. In *Foundations of Genetic Algorithms*, 94-101, Gregory J.E. Rawlins, ed., Morgan Kaufmann, 1991.

[11] Whitley, L. D. Fundamental principles of deception in genetic search. In *Foundations of Genetic Algorithms*, 221-241, Gregory J.E. Rawlins, ed., Morgan Kaufmann, 1991.

PART 6

MACHINE LEARNING

Learning Boolean Functions with Genetic Algorithms: A PAC Analysis

Johannes P. Ros
Computer Science Department
University of Pittsburgh
Pittsburgh, PA 15260
ros@cs.pitt.edu

Abstract

This paper presents a genetic algorithm for learning the classes of Boolean conjuncts and disjuncts (which include the sets of kCNF and kDNF formulas) in the context of the distribution-free learning model (the *probably approximately correct* or PAC model). A PAC learner produces with high probability an accurate representation for an unknown concept from training examples that are chosen at random from a fixed but arbitrary probability distribution. The analysis of the genetic PAC learner is complete: For any reasonable crossover operator and any confidence and accuracy level, the result provides the number of generations and the size of the population to accomplish this learning task.

1 INTRODUCTION

The distribution-free model for concept learning was introduced by Valiant as an alternative to *exact* concept identification and to provide a general standard by which learning algorithms can be evaluated (Valiant, 1984; Kearns *et al.*, 1987). In this model the learner is presented with a collection of training examples which are labeled '+' or '−' depending on whether or not they belong to the unknown concept. These examples are chosen at random using a probability density function (pdf) on which we do not place any assumptions (hence distribution-free), except that it is time invariant. Informally, we say a concept class is PAC learnable if there is an efficient algorithm that produces with high probability an accurate description of an unknown concept from this class. The justification for this model can be found in Kearns *et al.* (1987).

The notion of a genetic algorithm was introduced by John Holland as a formal framework towards the construction of a unified theory of adaptation (Holland, 1975). Holland observed that many adaptive systems in economics, AI, nature and elsewhere, have the common goal of producing 'well-adapted' entities under uncertain circumstances. At the same time, he observed that progress towards a deeper understanding of such systems was prevented by domain-independent obstacles, which led him to believe that a study of formal frameworks for adaptive systems will not only enable future researchers to better understand the behavior of natural adaptive systems, but will also provide them with the fundamental tools necessary to build efficient artificial adaptive systems (e.g., genetic algorithms) in the context of many differents domains.

We investigate genetic algorithms (GAs) as learning algorithms in the context of the distribution-free learnability model (Ros, 1989). A class of GAs is introduced that are polynomial-time PAC learners for the classes of Boolean conjuncts and disjuncts, which include the set of kCNF and kDNF formulas. This is the first positive result and comprehensive analysis of genetic algorithms as PAC learners.

2 CONCEPTS, REPRESENTATIONS, PAC LEARNING

In this section we introduce the definitions for concept classes and representation classes, followed by a precise description of the Boolean functions considered in this paper. Second, we provide a short introduction to PAC learning. Finally, at the end of this section, we provide a proposition that forms the basis for the result in Section 3.

The following two definitions are essentially from Warmuth (1989).

Definition 2.1 An 'alphabet' Σ is any set, and the elements of alphabets are called 'symbols'. A 'word' w over an alphabet Σ is any finite length sequence of symbols of Σ, and $|w|$ denotes its 'length'. Σ^v denotes all words of length v, for any positive integer v. A 'concept' or 'language' c over an alphabet Σ is any subset of Σ^*. A 'concept class' C over an alphabet Σ is any set of concepts, that is, $C \subseteq 2^{\Sigma^*}$. For any concept $f \subseteq \Sigma^v$, the 'positive examples' of f are the elements inside f, the 'negative examples' of f are the elements outside f (or inside $\bar{f}$, where $\bar{f} \equiv \Sigma^v - f$). $\square$

Since we are only interested in Boolean functions, we have $\Sigma = \{0, 1\}$. It is important to make a distinction between concepts and *representations* for concepts. Hence the following definition (Warmuth, 1989).

Definition 2.2 A 'representation class' is a four-tuple (R, Γ, c, Σ) where Γ is an alphabet; R is a language over Γ, called the 'representation language'; Σ is an alphabet for the concepts represented by R; and c is a mapping from R to concepts over Σ^*. For each representation $r \in R$, the concept $c(r) \subseteq \Sigma^*$ is the concept represented by r. The concept class represented by (R, Γ, c, Σ) is the set $\{c(r) : r \in R\}$. For each $r \in R$, the length of r (i.e., the number of symbols in r) is $|r|$. $\square$

We represent the Boolean functions either as *conjuncts* or as *disjuncts* of at most n *attributes*, where an attribute is an Boolean function over v variables, as specified in the next definition. (Note: $\mathcal{N}$ denotes the set of natural numbers, $\mathcal{N}^+$ is $\mathcal{N} - \{0\}$, $\mathcal{R}^{(0,1)}$ is the set of real numbers within the interval 0 and 1 (0 and 1 excluded), $\mathcal{R}^{[0,1]}$ is the set of real numbers $\mathcal{R}^{(0,1)} \cup \{0, 1\}$, etc. The phrase 'if and only if' is abbreviated as 'iff'.)

Definition 2.3 For all $v, n \in \mathcal{N}^+$, let A^n denote the set of n attributes where each attribute is a Boolean function over the variables $x_1, \ldots, x_v$, that is, $A^n = \{a_1, \ldots, a_n \mid a_i : \Sigma^v \to \Sigma\}$. Since every attribute $a_i \in A^n$ is a Boolean function, for every example $e \in \Sigma^v$: $a_i(e) = 1$ or $a_i(e) = 0$, depending on whether e satisfies (does not satisfy, respectively) the function a_i. $\square$

The conjuncts and disjuncts are conveniently represented by strings of n bits, where each bit denotes the presence ('1') or absence ('0') of the corresponding attribute. Formally, we have the following two definitions.

Definition 2.4 Given the set of n attributes $A^n = \{a_1, \ldots, a_n\}$ as defined in Def. 2.3, the set of Boolean conjuncts of at most n attributes is represented by the representation class $(R_{(A^n, \wedge)}, \Gamma, c, \Sigma)$, where $\Gamma = \{0, 1\}$, $\Sigma = \{0, 1\}$, and $R_{(A^n, \wedge)} = \Gamma^n$ such that for all $r \in R_{(A^n, \wedge)}$ and all examples $e \in \Sigma^v$: $e \in c(r)$ iff $\left(\bigwedge_{a_i \in r} a_i(e) \right) = 1$. $\square$

Definition 2.5 Given a set of n attributes $A^n = \{a_1, \ldots, a_n\}$ as defined in Def. 2.3, the set of Boolean disjuncts of at most n attributes is represented by the representation class $(R_{(A^n, \vee)}, \Gamma, c, \Sigma)$, where $\Gamma = \{0, 1\}$, $\Sigma = \{0, 1\}$, and $R_{(A^n, \vee)} = \Gamma^n$ such that for all $r \in R_{(A^n, \vee)}$ and all examples $e \in \Sigma^v$: $e \in c(r)$ iff $\left(\bigvee_{a_i \in r} a_i(e) \right) = 1$. $\square$

We will use the notation $R_{(A^n, \oplus)}$ where $\oplus \in \{\wedge, \vee\}$ to denote either representation class (c, Γ, and Σ are omitted since they are implicitly understood).

Since every $r \in R_{(A^n, \oplus)}$ represents a Boolean function, it is convenient to write $r(e)$ as the value of the Boolean function described by r on input e, where e is a string from the set Σ^v. Formally, $r(e) = 1$ iff $e \in c(r)$; otherwise, $r(e) = 0$.

The number of attributes n must be polynomial in the number of Boolean variables v to obtain learning algorithms with run-time polynomial in v, which means that only certain Boolean concept classes can be considered. For any constant $k \in \mathcal{N}^+$, the set of Boolean functions represented by kCNF and kDNF formulas are examples of such Boolean classes. A kCNF formula is a conjunct of *clauses*, where a clause is a disjunct of at most k (out of $2v$) literals, where a literal is a propositional variable or its negation. For example, $(x_1 + x_3)(x_1 + \overline{x_2})(x_4)$ is a 2CNF formula. Similarly, a kDNF formula is a disjunct of *terms*, where a term is a conjunct of at most $2k$ literals. For example, $x_1 + x_2\overline{x_3} + x_4$ is a 2DNF formula.

For any $v \in \mathcal{N}^+$, consider the set of attributes $A^{2v} = \{x_1, \ldots, x_v, \overline{x_1}, \ldots, \overline{x_v}\}$. Then, the set $R_{(A^n, \wedge)}$ represents the Boolean functions computed by 1CNF formulas (*monomials, simple conjuncts*), and the set $R_{(A^n, \vee)}$ represents the set of Boolean functions that can be computed by 1DNF formulas (*simple disjuncts*). For instance, take $r = 110001$ ($v = 3$). Clearly, r belongs to $R_{(A^n, \wedge)}$ as well as $R_{(A^n, \vee)}$, but is interpreted differently. In the first case it stands for the conjunct $x_1 x_2 \overline{x_3}$, and in the second case for the disjunct $x_1 + x_2 + \overline{x_3}$. In general, let l be the number of clauses (or terms) with at most k literals. Clearly, $l < (2v)^{k+1}$, which is polynomial in v if k is constant. If A^l contains all possible clauses of at most k literals, then $R_{(A^n, \wedge)}$ represents the set of kCNF formulas. Similarly, $R_{(A^n, \vee)}$ represents the set of kDNF formulas if A^l contains all possible terms of at most k literals.

Under the PAC learning model, two representation classes are usually necessary to specify the learning problem: the *target class* C and the *hypothesis class* H, where C and H may or may not be identical. For any concept f whose representation belongs to C, the task of the learner is to construct a representation $h \in H$ within time polynomial in the parameters of the model such that the concept described by h is likely to be a good approximation for f. By a 'good approximation' we mean that the chances are small that h will misclassify future (positive or negative) examples (Def. 2.7).

The training examples are obtained from two *oracles*: *POS* and *NEG*. Upon request, the *POS* oracle returns a randomly selected positive example from the unknown concept f. Similarly, the *NEG* oracle returns a negative example. It is assumed that the pdfs over the positive and negative examples (defined as T_f and $T_{\overline{f}}$ respectively) are time invariant and unknown to the learner.

Definition 2.6 Let Δ be any finite set. Define $\mathbf{D}_\Delta$ as the set of all probability distribution functions (pdfs) with domain Δ. $\square$

Definition 2.7 For all $v \in \mathcal{N}^+$, let (H, Γ, c, Σ) denote a representation class such that for all the representations $h \in H$, $c(h) \subseteq \Sigma^v$. For all concepts $f \subseteq \Sigma^v$, all pdfs $T_f, T_{\overline{f}} \in \mathbf{D}_{\Sigma^v}$, all $\epsilon \in \mathcal{R}^{(0,1)}$, and all representations $h \in H$, the concept $c(h)$ is an 'ϵ-approximation for the concept f' iff $\sum_{e \in \overline{c(h)}} T_f(e) < \epsilon$ and $\sum_{e \in c(h)} T_{\overline{f}}(e) < \epsilon$. $\square$

It follows from this definition that if $c(h)$ is an ϵ-approximation for f, then with probability at least $1 - \epsilon$ the concept $c(h)$ agrees with f on the classification of a randomly selected positive (or negative) example. For example, take $v = 3$, $\oplus = \wedge$, and take the target function $f = x_1 x_2$. Clearly, the positive examples are $\{110, 111\}$, and the negative examples are $\{000, 001, 010, 011, 100, 101\}$. Let $h = x_1$ be the hypothesis. Observe that h is more general than f since $c(h)$ contains not only all the elements in f but also contains the negative examples 100 and 101. Despite their differences, however, $c(h)$ is an ϵ-approximation for f if $T_{\overline{f}}(100) + T_{\overline{f}}(101) < \epsilon$.

The class $R_{(A^n, \wedge)}$ is learnable from positive examples, and $R_{(A^n, \vee)}$ is learnable from negative examples (Valiant, 1984), which means that although the learner receives only positive (or negative) training examples, it is able to compute a hypothesis that is an ϵ-approximation to the target concept: In the future, the hypothesis will correctly classify positive *and* negative examples despite the fact the learner has only seen only one of the two. In the following paragraphs we will provide a simple and efficient (deterministic) learning technique that accomplishes this task. First we define the phrase 'algorithm A is a PAC learner for the class C' (Natarajan, 1991).

Definition 2.8 Let (C, Γ, c, Σ) denote the target and the hypothesis representation class. An algorithm A is a 'PAC learner for the class C' if for all $v \in \mathcal{N}^+$, all target concepts f whose representation(s) belong to C, all pdfs $T_f, T_{\overline{f}} \in \mathbf{D}_{\Sigma^v}$, all accuracy parameters $\epsilon \in \mathcal{R}^{(0,1)}$, and all confidence parameters $\delta \in \mathcal{R}^{(0,1)}$, A produces hypothesis $h \in C$ in time polynomial in v, $1/\epsilon$, and $1/\delta$, such that with probability at least $1 - \delta$, the concept $c(h)$ is an ϵ-approximation for f. $\square$

There is a simple and efficient *elimination* technique (Valiant, 1984) to PAC learn the concepts described by $R_{(A^n, \wedge)}$ from positive examples, and the concepts described by $R_{(A^n, \vee)}$ from negative examples. During the learning process the algorithm maintains a hypothesis h that is updated to assure consistency with the training examples. Initially, h contains all the attributes from the set A^n. The learner requests a positive example (from the oracle *POS*) if $\oplus = \wedge$, otherwise it requests a negative example (from *NEG*). Upon receiving the training example the learner evaluates the current hypothesis on this example. The algorithm proceeds immediately to the next stage if the classification as computed by the hypothesis is identical to the classification of the training example, otherwise it eliminates only those attributes from the hypothesis that are responsible for the erroneous classification.

For example, consider the target class $R_{(A^n, \wedge)}$. If the learner receives a (positive) example e that is rejected by the current hypothesis h (i.e., $h(e) = 0$), there are one or more attributes in h that causes h to produce '0' since h takes the conjunction of the attributes. By removing these attributes, h will (correctly) accept e. By duality, we obtain a similar process when we consider the target class $R_{(A^n, \vee)}$. If the learner receives a (negative) example e that is accepted by the current hypothesis h (i.e., $h(e) = 1$), there are one or more attributes in h that causes h to produce 1 since h takes the disjunction of the attributes. By removing these attributes, h will (correctly) reject e.

The duality of the classes $R_{(A^n, \wedge)}$ and $R_{(A^n, \vee)}$ allows us to analyze the performance of the algorithm at a slightly more abstract level. For that purpose, we introduce the function μ that can be used by the learner to determine whether or not a_i must be eliminated from the current hypothesis, given a training example e and its classification. In the next definition, μ will be defined such that $\mu = 0$ if attribute a_i must be eliminated from the current hypothesis, otherwise $\mu = 1$.

Definition 2.9 For all $v, n \in \mathcal{N}^+$, all Boolean functions $a_1 : \Sigma^v \to \Sigma$, $\cdots$, $a_n : \Sigma^v \to \Sigma$, all $\oplus \in \{\wedge, \vee\}$, all target concepts f whose representation(s) belong to $R_{(A^n, \oplus)}$ (where $A^n = \{a_1, \ldots, a_n\}$), all examples $e \in \Sigma^v$, and all classifications $\copyright \in \{+, -\}$, define the following 'elimination function' for all $i \in \mathcal{N}^{[1,n]}$:

$$\mu_f(i, e, \copyright) = \begin{cases} a_i(e), & \text{if } \copyright = \text{'}+\text{'} \\ \overline{a_i(e)}, & \text{otherwise.} \end{cases}$$

In the cases where the target concept f is understood, we will abbreviate $\mu_f(i, e, \copyright)$ by $\mu(i, e, \copyright)$. $\square$

Since GAs are randomized algorithms, we cannot expect the above elimination technique to work as it does with deterministic learners. In fact, GAs do not eliminate attributes but rather increase or decrease the *frequency* of the attributes in the population. Since hypothesis h is a structure selected at random from the final population, attributes are present in (or absent from) h with a certain probability. The next proposition shows these probability levels sufficient for the GA to PAC learn the class $R_{(A^n, \oplus)}$. This result forms the basis of Theorem 3.1.

Proposition 2.1 Assume L is an algorithm such that for all $v, n \in \mathcal{N}^+$, all Boolean functions $a_1 : \Sigma^v \to \Sigma$, $\cdots$, $a_n : \Sigma^v \to \Sigma$, all $\oplus \in \{\wedge, \vee\}$, all target concepts f (such that there exists a representation $r \in R_{(A^n, \oplus)}$ such that $c(r) = f$, where $A^n = \{a_1, \ldots, a_n\}$), all pdfs $T_f, T_{\bar{f}} \in \mathbf{D}_{\Sigma^v}$, all accuracy parameters $\epsilon \in \mathcal{R}^{(0,1)}$, all confidence parameters $\delta \in \mathcal{R}^{(0,1)}$, L receives randomly selected training examples (denoted by e) only from oracle *POS* (in which case $\copyright = $ '+') or only from *NEG* (in which case $\copyright = $ '–') depending on whether $\oplus = \wedge$ or $\oplus = \vee$, and produces hypothesis $h \in R_{(A^n, \oplus)}$ in time polynomial in v, $1/\epsilon$, and $1/\delta$ such that the following two properties hold for each position $i \in \mathcal{N}^{[1,n]}$:

- If $\mu(i, e, \copyright) = 0$ with probability at least ϵ/n, then attribute a_i is not in h (i.e., $a_i \notin h$) with probability at least $1 - \delta/n$;

- If $\mu(i, e, \copyright) = 1$ with probability 1, then attribute a_i is in h (i.e., $a_i \in h$) with probability at least $1 - \delta/n$.

Then, algorithm L is a PAC learner for the class $R_{(A^n, \oplus)}$. $\square$

The proof is omitted because of space limitations (Ros, 1992).

3   THE GENETIC PAC LEARNER

In this section, we describe the fitness function, the crossover operator and the genetic plan of the genetic PAC learner, which is followed by the result of this paper. (Mutation is not considered here.)

The GA maintains a population of bit strings of length n which are elements (words) from the set $R_{(A^n, \oplus)}$. The size of the population is denoted by N, and is fixed during the entire learning process.

Definition 3.1 For all $N \in \mathcal{N}^+$, let $\mathbf{P}_N$ denote the set of 'populations' B, where $B = \{r_1, \ldots, r_N\}$, and $r_k \in \Gamma^n$ for all $1 \leq k \leq N$. $\square$

The fitness function is a function of three arguments: a representation (member, structure, formula), a training example and the classification of this example. At this point we consider only *additive* fitness functions: the total credit of an individual is the sum of the (sub-) credit received by its components. Other fitness functions such as multiplicative functions are not considered here since we have strong evidence that these fitness functions do not lead to polynomial time learning algorithms (Ros, 1992). The sub-credits are limited to integers, resulting in a integral (total) credit value for each structure.

Definition 3.2 Take $\Sigma = \{0, 1\}$ as the concept alphabet, take $\Gamma = \{0, 1\}$ as the representation alphabet, and let μ be the elimination function as defined in Def. 2.9. Define $FIT_{\Sigma, \Gamma, \mu}$ as the set of 'fitness functions' g, such that for all representations $r \in R_{(A^n, \oplus)}$, all examples $e \in \Sigma^v$, and all classifications $© \in \{+, -\}$,

$$g(r, e, ©) = \sum_{j=1}^{n} G_{r(j) \mu(j, e, ©)},$$

where G_{xy} is a positive integer for all $x, y \in \Gamma$. We have abused the notation by using '$r(j)$' to denote the value of the j-th bit in r (which denotes the presence or absence of attribute a_j in r, that is, $r(j) = 1$ iff $a_j \in r$). However, there should be no confusion, since j denotes a position, rather than a word from Σ^v (see second paragraph after Def. 2.5). We shall use this notation only in the above context. □

Recall from Section 2 that the event '$\mu(j, e, ©) = 0$' indicates that attribute a_j must be eliminated, while the event '$\mu(j, e, ©) = 1$' means that attribute a_j should not be eliminated (yet). Then, we can interpret the different types of sub-credit as follows. For all the positions j (in representation r), all examples e and all classifications $©$, attribute a_j receives the credit

- G_{00}, if it is absent from r, and e shows that it should be absent.
- G_{01}, if it is absent from r, but e shows that it should be present.
- G_{10}, if it is present in r, but e shows that it should be absent.
- G_{11}, if it is present in r, and e shows that it should be present.

Naturally, not every fitness function from $FIT_{\Sigma, \Gamma, \mu}$ gives us a genetic PAC learner. In fact, one must carefully select the values for G_{00}, G_{01}, G_{10}, and G_{11} to obtain the desired behavior (Def. 3.3). If good performing structures receive too much credit with respect to an average structure, the GA may fail due to premature convergence. If such structures receive too little credit, the GA may fail due to the background noise from the selection stage (this stage will be discussed later in this section).

Definition 3.3 For all integers $a > 1$ and all rationals $b > 1$, let $x, y \in \mathcal{N}^+$ such that $b = (x + y)/y$, and let $z_1, z_2 \in \mathcal{N}^+$ such that $z_1 a = z_2(x + y)$. Then, the the set of 'productive fitness functions' $PFIT_{\Sigma, \Gamma, \mu} \subseteq FIT_{\Sigma, \Gamma, \mu}$ are the fitness functions $g \in FIT_{\Sigma, \Gamma, \mu}$ for which the four sub-credits are positive integers such that

$$G_{11} = z_1 a, \quad G_{01} = z_1(a - 1), \quad G_{00} = z_2(x + y), \quad \text{and} \quad G_{10} = z_2 x.$$

Since $G_{00} = G_{11}$, it follows that the sub-credit values of any productive fitness function satisfy the following relations:

$$1 \leq G_{10} < G_{01} < G_{11} = G_{00}, \quad \text{iff } a > b.$$

The symbols a and b are called 'fitness ratios' and it follows from the above that $a = \frac{G_{11}}{G_{11}-G_{01}}$, and $b = \frac{G_{00}}{G_{00}-G_{10}}$. $\square$

Notice that z_1 and z_2 can be chosen arbitrarily provided that the equality $z_1 a = z_2(x + y)$ is satisfied. Since the performance of the algorithm depends on the *ratio* between the various sub-credits (i.e., the fitness ratios a & b) and not their individual values, one typically chooses the smallest values for z_1 and z_2 such that the above equality holds.

A typical crossover operator produces two structures, called *offspring*, from two arguments, called *parents*. For simplicity, we use only one structure from the offspring and ignore the other, leaving the results essentially unaffected.

Definition 3.4 We define the 'crossover operator' as a (possibly randomized) function from $(\Gamma^n)^k$ to Γ^n, for any $k \geq 2$. For all $w_1, \ldots, w_k \in \Gamma^n$ (where $n, k \in \mathcal{N}^{[2,\infty)}$), let $w \in \Gamma^n$ such that $w \leftarrow \chi(w_1, \ldots, w_k)$, where χ is any crossover operator. Then, define the following events for all $i, j \in \mathcal{N}^{[1,n]}$ and all $s \in \mathcal{N}^{[1,k]}$:

$$copy(s, i) \equiv (w(i) \leftarrow w_s(i)),$$

$$single(i, j) \equiv \bigvee_{s \in \mathcal{N}^{[1,k]}} (copy(s, i) \wedge copy(s, j)).$$

In words, the event '$copy(s, i)$' occurs when the i-th symbol of the parent structure w_s is donated (copied, swapped) to the corresponding location in the offspring structure w. The event '$single(i, j)$' occurs when both the positions i and j receive their symbols from the same parent.

We define the 'shuffle factor' of χ, $SF(\chi)$, as the smallest probability that any two positions in the offspring structure receive their symbols from two *different* parental structures. More precisely,

$$SF(\chi) \equiv \min_{i \neq j}\{\Pr[\overline{single(i, j)}]\}. \quad \square$$

One may view the shuffle factor of a crossover operator as a characterization of its 'disruptiveness': the higher the shuffle factor, the more disruptive. For $k = 2$, the uniform crossover operator has a shuffle factor of 0.5 since for any two positions in the offspring structure, the probability that they receive their symbol from two different parents is 0.5. Similarly, the one-point crossover operator has a shuffle factor of only $\Theta(1/n)$ (where n is the length of the structure), since for any two positions in the final structure, the probability that the cut will be made between these two positions can be as small as $1/n$ (or of that magnitude, depending on the implementation). A shuffle factor of the value 1 can be obtained with a (deterministic) crossover operator which, on input of n parental structures, produces an offspring structure of length n in which each bit is copied from a different parent. This is the most powerful crossover operator in our framework.

We will only consider crossover operators that in some sense are 'well-behaved', as shown in the next definition.

Definition 3.5 The set $CROSS_{\Gamma,k}$ denotes the set of well-behaved crossover operators χ with the following properties. For all $n, k \in \mathcal{N}^{[2,\infty)}$, for all parental structures $h_1, \ldots, h_k \in \Gamma^n$, and for all crossover operators $\chi \in CROSS_{\Gamma,k}$, let $h \in \Gamma^n$ such that $h \leftarrow \chi(h_1, \ldots, h_k)$. Then, the following conditions hold for all $i, j \in \mathcal{N}^{[1,n]}$ such that $i \neq j$:

1. $copy(1, i) \vee \cdots \vee copy(k, i) = TRUE$ (always occurs).

2. $(\forall s, t : 1 \leq s \neq t \leq k)\ copy(s, i) \wedge copy(t, i) = FALSE$ (never occurs).

3. $(\forall s, t : 1 \leq s, t \leq k)\ copy(s, i) \wedge copy(t, j)$ is independent of r_s and r_t.

4. $0 < SF(\chi) \leq 1$. $\square$

It follows from the conditions (1) and (2) that every position in the offspring structure receives its symbol from one and only one parental structure. Condition (3) ensures that the probabilities of the copy events are not affected by any of the involved structures.

We are now ready to describe the genetic plan for our model. The initial population is a set of structures that are uniformly selected at random from the set $R_{(A^n, \oplus)}$. Then, the following steps are repeated for m generations. First, a training example is requested from the oracle. This is followed by the *credit stage* where every member in the population receives credit from the fitness function depending on how well the structure predicts the correct classification of the example. In particular, the structure's credit value depends on the number of attributes that are responsible for accepting (rejecting) the example. The credit stage is followed by the *selection stage* in which the GA forms the next population with the offspring obtained from crossing over individuals that are *proportionally* selected at random from the current population. The proportional selection scheme ensures that the individual's chance of being selected equals its credit value divided by the total credit collected by the members in the current population. After the last generation, the GA selects the hypothesis at random from the final population.

During the selection stage, a new population is created by repeating the following steps until the new population is filled. The GA selects $k \geq 2$ individuals from the current population at random according to the distribution of the credit values. The crossover operator is applied to these k parental structures and only one offspring structure is put into the new population.

One may view the new population as a sample of size N from the set of structures $R_{(A^n,\oplus)}$ according to some pdf. This pdf depends, of course, on the current population (denoted by B_{t-1}), the credit values received by the members of this population (which, in turn, depend on the fitness function g, the training example e and its classification $\copyright$), and the crossover operator χ. Hence, this pdf is denoted by $D_{B_{t-1},g,e,\copyright,\chi}$ (Def. 3.7). It is not difficult to see that $D_{B_{t-1},g,e,\copyright,\chi}$ is computable in polynomial time, which allows us to implement the genetic function as defined below (Def. 3.8).

Definition 3.6 For any pdf F defined on a finite domain Δ, and for every element $x \in \Delta$, the random variable (random function, experiment) $RAND(F)$ equals x with probability $F(x)$. $\square$

Definition 3.7 Let $B \in \mathbf{P}_N$ be any population. Take $R \equiv R_{(A^n,\oplus)}$. For any representation $r \in R$, define $M_B(r)$ as the set of positions $k \in \mathcal{N}^{[1,n]}$ for which $r = r_k \in B$ holds. Clearly, $0 \leq |M_B(r)| \leq N$. Define the 'population pdf' $D_B : R \to \mathcal{R}^{[0,1]}$ such that for every representation $r \in R$, the function $D_B(r)$ computes the proportion of the structures r in population B. Let $g \in FIT_{\Sigma,\Gamma,\mu}$ be any fitness function, $e \in \Sigma^v$ any training example, and $\copyright \in \{+, -\}$ any classification. Define the 'proportion pdf' $D_{B,g,e,\copyright} : R \to \mathcal{R}^{[0,1]}$ such that for every representation $r \in R$, the function $D_{B,g,e,\copyright}(r)$ computes the proportion of the structures r in population B with respect to the distribution of the credit values. Let $\chi \in CROSS_{\Gamma,k}$ be any crossover operator (for any $k \in \mathcal{N}^{[2,\infty)}$). Define the 'crossover pdf' $D_{B,g,e,\copyright,\chi} : R \to \mathcal{R}^{[0,1]}$ such that for any representation $r \in R$, the function $D_{B,g,e,\copyright,\chi}(r)$ computes the probability that structure r is the result of crossing over k structures from population B that are selected at random from the proportion pdf $D_{B,g,e,\copyright}$. Clearly, $D_B, D_{B,g,e,\copyright}, D_{B,g,e,\copyright,\chi} \in \mathbf{D}_{\Gamma^n}$. The above definitions are formally stated as follows.

$$M_B(r) \equiv \{k : r_k \in B, r_k = r\},$$

$$D_B(r) \equiv \frac{|M_B(r)|}{|B|}, \quad D_{B,g,e,\copyright}(r) \equiv \frac{|M_B(r)|g(r,e,\copyright)}{\sum_{s \in R} |M_B(s)|g(s,e,\copyright)},$$

$$D_{B,g,e,\copyright,\chi}(r) \equiv \mathbf{Pr}[\chi(RAND_1(D_{B,g,e,\copyright}), \ldots, RAND_k(D_{B,g,e,\copyright})) = r]. \quad \square$$

Definition 3.8 Take $\Sigma = \{0,1\}$ as the concept alphabet, take $\Gamma = \{0,1\}$ as the representation alphabet, let $\chi \in CROSS_{\Gamma,k}$ denote the crossover operator (for any $k \in \mathcal{N}^{[2,\infty)}$), let μ denote the elimination function, let $g \in FIT_{\Sigma,\Gamma,\mu}$ denote the fitness function, let $m \in \mathcal{N}^+$ denote the number of generations, and let $N \in \mathcal{N}^+$ denote the size of the population. The (random) function $GF_{\Sigma,\Gamma,\chi,m,g,N}$ is defined such that for all $v, n \in \mathcal{N}^+$ (such that $n \geq 2$), all Boolean functions $a_1 : \Sigma^v \to \Sigma$, $\ldots$, $a_n : \Sigma^v \to \Sigma$, all $\oplus \in \{\wedge, \vee\}$, all target concepts f (such that there exists a representation $r \in R_{(A^n, \oplus)}$ such that $c(r) = f$, where $A^n = \{a_1, \ldots, a_n\}$), all pdfs $T_f, T_{\bar{f}} \in \mathbf{D}_{\Sigma^v}$, all accuracy parameters $\epsilon \in \mathcal{R}^{(0,1)}$, and all confidence parameters $\delta \in \mathcal{R}^{(0,1)}$, $GF_{\Sigma,\Gamma,\chi,m,g,N}(v, n, A^n, \oplus, T_f, T_{\bar{f}}, \epsilon, \delta)$ computes the following function.

> **function** $GF_{\Sigma,\Gamma,\chi,m,g,N}(v, n, A^n, \oplus, T_f, T_{\bar{f}}, \epsilon, \delta)$
> /* $B_t \in \mathbf{P}_N$ denotes the t-th population,
> $\quad$ $B_t(j)$ denotes the j-th member of the t-th population, and
> $\quad$ $D_{uniform} \in \mathbf{D}_{\Gamma^n}$ is uniform distribution over elements in Γ^n; */
>
> **begin**
> $\quad$ **for** $j = 1 \cdots N$ **do** $\qquad$ /* initialize the first population */
> $\qquad$ $B_0(j) \leftarrow RAND(D_{uniform})$;
> $\quad$ **done**;
>
> $\quad$ **if** $\oplus = $ '$\wedge$' **then** $\qquad$ /* use positive examples */
> $\quad$ **begin** $T \leftarrow T_f$; $\copyright \leftarrow$ '+'; **end**;
> $\quad$ **else** $\qquad\qquad\qquad$ /* or use negative examples */
> $\quad$ **begin** $T \leftarrow T_{\bar{f}}$; $\copyright \leftarrow$ '−'; **end**;
>
> $\quad$ **for** $t = 1 \cdots m$ **do**
> $\qquad$ $e \leftarrow RAND(T)$; $\qquad$ /* obtain random training example */
> $\qquad$ **for** $j = 1 \cdots N$ **do** $\qquad$ /* execute credit & selection stage */
> $\qquad\quad$ $B_t(j) \leftarrow RAND(D_{B_{t-1},g,e,\copyright,\chi})$; $\quad$ /* see Def. 3.7 */
> $\qquad$ **done**;
> $\quad$ **done**;
> $\quad$ **output** $(RAND(D_{B_m}))$; $\quad$ /* select final hypothesis at random */
> **end**. $\square$

The result of this paper is the following.

Theorem 3.1 Let $R_{(A^n, \oplus)}$ be the target and hypothesis class such that the number of attributes $n \geq 2$ is polynomial in the number of Boolean variables v. Let ϵ be the accuracy parameter (s.t. $\epsilon/n < 0.25$), let δ be the confidence parameter and let G be a genetic algorithm that computes the function $GF_{\Sigma,\Gamma,\chi,m,g,N}$, where Σ and Γ are binary (concept & representation) alphabets, $\chi \in CROSS_{\Gamma,k}$ is a crossover operator (for any $k \in \mathcal{N}^{[2,\infty)}$), m is the number of generations, $g \in PFIT_{\Sigma,\Gamma,\mu}$ is a (productive) fitness function with the fitness ratios $a \equiv G_{11}/(G_{11} - G_{01})$, and $b \equiv G_{00}/(G_{00} - G_{10})$, and N is the size of the population. Let $\omega \equiv 1 - SF(\chi)$.

(a) If $\omega \geq \frac{\epsilon}{n^2 \ln(4n/\delta)}$ (s.t. $\frac{1}{1-\omega}$ is polynomial in v, $1/\epsilon$ and $1/\delta$), and if

$$m = \Omega\left(\frac{n^4 \omega \ln(n/\delta)}{\epsilon^2(1-\omega)}\left(\ln\frac{n^5 \omega \ln(n/\delta)}{\epsilon^2 \delta(1-\omega)}\right)^2\right), \quad a = \Theta\left(\frac{n^3 \omega(\ln(nm/\delta))^2}{\epsilon^2(1-\omega)}\right),$$

$$b = \Theta\left(\frac{n^2 \omega \ln(nm/\delta)}{\epsilon(1-\omega)}\right), \quad \text{and} \quad N = \Omega\left(\frac{n^{11}\omega^2(\ln(nm/\delta))^5}{\epsilon^4 \delta^3(1-\omega)^2}\right),$$

then G is a PAC learner for the class $R_{(A^n,\oplus)}$.

(b) If $0 < \omega < \frac{\epsilon}{n^2 \ln(4n/\delta)}$ (s.t. $1/\omega$ is polynomial in v, $1/\epsilon$ and $1/\delta$), and if

$$m = \Omega\left(\frac{1}{\omega \ln(n/\delta)}\left(\ln\frac{n}{\delta \omega \ln(n/\delta)}\right)^2\right), \quad a = \Theta\left(\frac{(\ln(nm/\delta))^2}{n\omega(\ln(n/\delta))^2}\right),$$

$$b = \Theta\left(\frac{\ln(nm/\delta)}{\ln(n/\delta)}\right), \quad \text{and} \quad N = \Omega\left(\frac{n^3(\ln(nm/\delta))^5}{\delta^3 \omega^2(\ln(n/\delta))^4}\right),$$

then G is a PAC learner for the class $R_{(A^n,\oplus)}$.

(c) For all rational numbers $b > 1$, if

$$m = \Omega\left(\frac{n^2 b^2 \ln(n/\delta)}{\epsilon(b-1+1/n)}\ln\frac{n^3 b^2 \ln(n/\delta)}{\epsilon \delta(b-1+1/n)}\right), \quad a = \Theta\left(\frac{nb^2\ln(nm/\delta)}{\epsilon(b-1+1/n)}\right),$$

$$0 \leq \omega \leq \frac{b(b-1)}{b(b-1)+4na}, \quad N = \Omega\left(\frac{n^7 b^4(\ln(nm/\delta))^3(b/(b-1))^{3(n+1)/n}}{\epsilon^2 \delta^3(b-1+1/n)^2}\right),$$

then G is a PAC learner for the class $R_{(A^n,\oplus)}$. $\square$

Corollary 3.1 If $\omega = 0$, and if

$$m = \Omega\left(\frac{n^2\ln(n/\delta)}{\epsilon}\ln\frac{n^3\ln(n/\delta)}{\epsilon\delta}\right), \quad a = \Theta\left(\frac{n\ln(nm/\delta)}{\epsilon}\right),$$

$$b = 1 + \Theta(1), \quad \text{and} \quad N = \Omega\left(\frac{n^7(\ln(nm/\delta))^3}{\epsilon^2\delta^3}\right),$$

then G is a PAC learner for the class $R_{(A^n,\oplus)}$. $\square$

The upper and lower bounds as stated in the theorem and corollary are formulated in orders of magnitude. The constants can be obtained from the Lemmas A.16 and A.17 in the author's dissertation (Ros, 1992).

The corollary shows the most optimal genetic learner for the class $R_{(A^n,\oplus)}$ in this framework. The corollary follows from the theorem (c) by taking $\omega = 0$ and $b = 1 + \Theta(1)$.

The key difference between (a), (b) and (c) is the condition under which each of them can be applied. According to the theorem, (a) can be applied if $\omega \geq \epsilon/(n^2\ln(4n/\delta))$, (b) can be applied if $0 < \omega < \epsilon/(n^2\ln(4n/\delta))$, and (c) can be applied if $0 \leq \omega \leq b(b-1)/(b(b-1) + 4na)$, where a and b are the fitness ratios as defined in Def. 3.3. Recall that $\omega \equiv 1 - SF(\chi)$ is a real number between 0 and 1 (i.e., $\omega \in \mathcal{R}^{[0,1)}$) that describes the effectiveness (disruptiveness) of the crossover operator: the more effective, the closer ω approaches the value 0. It is clear from (a) and the corollary that the crossover significantly affects the efficiency of the algorithm.

To understand the effectiveness of a crossover operator, and more generally, to gain insight into the expressions in (a), (b) and (c), it is important to understand the fundamental relationships between the crossover operator, the fitness function and the finite population. These relationships will surface upon examining one of the biggest problems for any genetic algorithm: to avoid premature convergence (PC).

The main problem for any genetic learner is to prevent the population from being dominated too early by strong performing individuals. This is especially relevant in an environment where the training examples are selected at random (as is the case in the PAC model), because those training examples that will point out the fallacy of dominating individuals, may arrive too late, in which case the learner may find itself stuck with a population that mainly consists of individuals that do not quite have the desired properties: We say the learner has 'converged prematurely'.

Strictly speaking however, one is never able to *completely* avoid PC since GAs are randomized algorithms: There is always a non-zero chance that PC occurs. However, stating this explicitly would be too tiresome, so that we will continue to write phrases like 'to avoid PC' when in fact we mean 'to reduce the chances of PC to sufficiently low levels'.

The crossover operator is one of the main tools available to the genetic algorithm to avoid PC. The effectiveness of the crossover is characterized by its shuffle factor. By definition, a larger shuffle factor increases the chances for every gene of being exchanged. By using a stronger crossover operator, it becomes more difficult for any structure to quickly dominate the population because there is more chance that the crossover breaks these individuals up and exchanges their parts with other (possibly weaker) structures, which prevents PC.

Premature convergence can also be avoided by reducing the rewards of good performing individuals compared to their weaker peers, thereby slowing the growth of dominating individuals. This is achieved by *increasing* the fitness ratios a and b (Def. 3.3). However, this results in more generations and larger population sizes. More generations are necessary because the growth of strong performing individuals have slowed down. A larger population size is necessary to reduce the selection noise (statistical fluctuation from the selection stage), since selection noise becomes more prominent when the difference in payoff between good and bad individuals decreases.

For this reason, it is better to combat PC by using stronger crossover operators than by increasing the fitness ratios. In fact, whenever the algorithm uses a more disruptive crossover, it can reduce the fitness ratios (which decreases the number of generations and the population size), while still maintaining its reliability.

This strategy is can be found in (a) and (c). According to (a), the fitness ratios, the number of generations and the population size decrease if the crossover operator is made more disruptive, that is, if ω is reduced (all other things being equal). We see a similar behavior in (c), except for the fact that ω does not appear explicitly in the expressions. Instead, one is able to manipulate the upper bound on the range of ω by the fitness ratio b: If $b \to \infty$, the upper bound increases and, consequently, the fitness ratio a, the number of generations and the population size increase, as expected. On the other hand, if b decreases, the upper bound on the range of ω decreases, and so do the other variables. Notice that the $(b - 1 + 1/n)$ factor in several denominators indicates that it is not necessary to push the $b - 1$ term lower than $1/n$. This is important because the size of the population goes to infinity (i.e., $N \to \infty$) when $b \to 1$.

As pointed out in the previous paragraph, it is not necessary to bring the fitness ratio b in (c) arbitrary close to 1. In fact, it is easy to see from

(c) that the optimal value for $b - 1$ is a constant (i.e., $\Theta(1)$). If $b - 1$ is a constant, we immediately obtain the following upper bound on the range for ω: $0 \leq \omega \leq \epsilon/(n^2\ln(nm/\delta))$. Clearly, there is a gap between the lower bound on ω for (a) to apply, and the upper bound on ω for (c) to apply. In other words, there is a gap between these two expressions which creates a situation where neither (a) nor (c) can be used. For this purpose, (b) has been provided. Although (b) behaves differently from (a) and (c) with regard to ω (i.e., all the variables *increase* when ω decreases), it is quite useful when $\epsilon/(n^2\ln(nm/\delta)) < \omega < \epsilon/(n^2\ln(4n/\delta))$, for $m > 4$.

The algorithm operates with a confidence level of $1 - \delta$. The population size N in (a), (b) and (c) is the only variable that depends on a factor that is polynomial in $1/\delta$ instead of $\ln(1/\delta)$. This is directly related to the method of selecting the final hypothesis. The genetic PAC learner selects the final hypothesis at random from the last population. On the basis of Proposition 2.1, the genetic PAC learner aims at a final population in which at least $N(1 - \delta)$ representations are syntactically identical and of high-quality.

It is not difficult to see that the $(1/\delta)^3$ factor in N can be avoided by using a majority vote as a method for selecting the final hypothesis from the last population. For example, one may aim for a final population in which a large - but *constant* - proportion of the representations are syntactically identical and of high-quality. Assuming that the majority-vote procedure selects the most frequently appearing representation in the last population, it will certainly produce a high-quality hypothesis when such a large proportion exists. This method eliminates the $(1/\delta)^3$ factor from N because the proportion is constant, hence independent of δ.

With regard to δ, it remains to argue why the variables in (a), (b) and (c) depend on the factor $\ln(1/\delta)$. The presence of this factor follows from a fundamental probability rule which - in general - says that the probability of an error decreases *exponentially* in the number of independent trials. In other words, $\ln(1/\delta)$ independent trials are sufficient to reach a confidence level of at least $1 - \delta$.

The algorithm operates with an accuracy level of $1 - \epsilon$, which essentially means that the important training examples appear with probability at least ϵ. In other words, if ϵ is reduced, the algorithm may need more time before it receives the important training examples due to a (possibly) lower probability of receiving them. This increases the changes of PC to occur, since the data necessary to reduce the proportion of prominent - but undesired - individuals in the population may arrive too late (or too infrequent). Therefore, to avoid PC when ϵ is reduced (all other things being equal), the algorithm has to increase the fitness ratios a and b, and hence the number of generations and the size of the population.

It is not surprising to have the number of generations, the fitness ratios and the size of the population in (a), (b) and (c) depend on a factor polynomial in the number of attributes n, since n is tightly related with the accuracy and confidence parameters ϵ and δ as follows: Generally speaking, for any fixed ϵ and δ, if n is increased, the algorithm must increase the accuracy and confidence levels for each *individual* attribute in order to maintain the *overall* accuracy and confidence levels. This follows from Boole's inequality which essentially says that, given an object of n components, the probability that the object won't fail is in the worst case inversely related to the number of components n.

4 SUMMARY AND DISCUSSION

In this paper we have introduced a class of genetic algorithms that are PAC learners for the classes of Boolean conjuncts and disjuncts. A complete analysis is provided: Given any (well-behaved) crossover operator, any confidence level $1 - \delta$, and any accuracy level $1 - \epsilon$, we provide (polynomial bounds on) the number of generations, the fitness function, the size of the population, and the number of (randomly selected) training examples such that for any (unknown) target concept and any probability distribution over the training examples, the GA produces with probability at least $1 - \delta$ a hypothesis with an accuracy of at least $1 - \epsilon$. (Mutation is not considered here.)

The work in this paper complements the various experimental studies of using GAs for learning Boolean concepts (e.g., Wilson, 1987; Bonelli *et al.*, 1990; McCallum *et al.*, 1990; Liepins and Wang, 1991), and other (theoretical) work that focus only on particular issues (in more general settings, however), such as representation (Liepins and Vose, 1990), stability of the population (Vose and Liepins, 1991a), deception (Liepins and Vose, 1990b; Liepins and Vose, 1991), and schema disruption (Vose and Liepins, 1991b).

Future work includes obtaining lower bounds on the number of generations and population size, thereby proving or disproving that GA's are inherently slow PAC learners (Liepins and Wang, 1991). It is well-known that the class $R_{(A^n, \oplus)}$ is PAC learnable by using $\Theta(1/\epsilon(n + \ln(1/\delta)))$ training examples (Ehrenfeucht *et al.*, 1988).

Other work may focus on using mutation to obtain genetic PAC learners for *drifting* concepts (Liepins and Wang, 1991).

Finally, we expect this work to apply outside the PAC domain (e.g., function optimization) by fixing the probability distributions over the training examples.

Acknowledgements

I would like to thank professor Robert Daley for many helpful comments about the material in this paper. This research was supported in part by the NSF grant CCR-8901795.

References

1. Blumer, Anselm, Andrzej Ehrenfeucht, David Haussler, and Manfred K. Warmuth, "Learnability and the Vapnik-Chervonenkis dimension", Journal of the ACM, vol. 36, no. 4, pp. 929-965, October 1989.

2. Bonelli, Pierre, Alexandre Parodi, Sandip Sen, and Steward Wilson, "NEWBOOLE: A fast GBML System", Proceedings of the 7th International Conference on Machine Learning, pp. 153-159, Austin, Texas, June 1990.

3. Davis, Thomas and Jose Principe, "A simulated annealing like convergence theory for the simple genetic algorithm", Proceedings of the 4th International Conference on Genetic Algorithms, pp. 174-181, San Diego, CA, July 1991.

4. Ehrenfeucht, Andrzej, David Haussler, Michael Kearns, and Leslie Valiant, "A general lower bound on the number of examples needed for learning", First Workshop on Computational Learning Theory, pp. 110-120, Cambridge, MA, August 1988.

5. Holland, John H., Adaptation in natural and artificial systems: an introductory analysis with applications to biology, control, and artificial intelligence, University of Michigan Press, Ann Arbor, MI, 1975.

6. Kearns, Michael, Ming Li, Leonard Pitt, and Leslie G. Valiant, "Recent results on Boolean concept learning", Proceedings of the 4th International Workshop on Machine Learning, pp. 337-352, Irvine, CA, 1987.

7. Liepins, Gunar E. and Michael D. Vose, "Representational issues in genetic optimization", Journal of Experimental and Theoretical Artificial Intelligence, vol. 2, no. 2, pp. 4-30, 1990.

8. Liepins, Gunar E. and Michael D. Vose, "Deceptiveness and genetic algorithm dynamics", Workshop on the Foundations of Genetic Algorithms, pp. 36-50, Bloomington, Indiana, July 1990.

9. Liepins, Gunar E. and Michael D. Vose, "Polynomials, basis sets, and deceptiveness in genetic algorithms", Complex Systems, vol. 5, pp. 45-61, 1991.

10. Liepins, Gunar E. and Lori A. Wang, "Classifier system learning of Boolean concepts", Proceedings of the 4th International Conference on Genetic Algorithms, pp. 318-323, San Diego, CA, July 1991.

11. McCallum, R. Andrew and Kent A. Spackman, "Using genetic algorithms to learn disjunctive rules from examples", Proceedings of the 7th International Conference on Machine Learning, pp. 149-152, Austin, Texas, 1990.

12. Natarajan, Balas K., Machine learning: a theoretical approach, Morgan Kaufmann Publishers, Inc., San Mateo, CA, 1991.

13. Rabinovich, Yuri and Avi Wigderson, "An analysis of a simple genetic algorithm", Proceedings of the 4th International Conference on Genetic Algorithms, San Diego, CA, July 1991.

14. Ros, Hans, "Some results on Boolean concept learning by genetic algorithms", Proceedings of the 3rd International Conference on Genetic Algorithms, pp. 28-33, Washington, DC, June 1989.

15. Ros, Johannes P., "Learning Boolean functions with genetic algorithms: a PAC analysis", Doctoral dissertation, Department of Computer Science, University of Pittsburgh, PA, July 1992.

16. Valiant, L.G., "A theory of the learnable", Communications of the ACM, vol. 27, no. 11, pp. 1134-1142, November 1984.

17. Vose, Michael D. and Gunar E. Liepins, "Punctuated equilibria in genetic search", Complex Systems, vol. 5, pp. 31-44, 1991.

18. Vose, Michael D. and Gunar E. Liepins, "Schema disruption", Proceedings of the 4th International Conference on Genetic Algorithms, San Diego, CA, July 1991.

19. Warmuth, Manfred K., "Towards representation independence in PAC learning", Analogical and Inductive Inference: International Workshop AII 1989, pp. 78-103, 1989.

20. Wilson, Steward W., "Classifier systems and the animat problem", Machine Learning, vol. 2, pp. 199-228, 1987.

Is the Genetic Algorithm a Cooperative Learner?

Helen G. Cobb
Navy Center for Applied Research in Artificial Intelligence
Naval Research Laboratory, Code 5514
Washington, D. C. 20375-5320
Internet: cobb@aic.nrl.navy.mil

Abstract

This paper begins to explore an analogy between the usual competitive learning metaphor presented in the genetic algorithm (GA) literature and the cooperative learning metaphor discussed by Clearwater, Huberman, and Hogg. In a blackboard cooperative learning paradigm, agents share partial results with one another through a common blackboard. By occasionally accessing the blackboard for a partial solution, an agent can dramatically increase its speed in finding the overall solution to a problem. The study of Clearwater et al. shows that the resulting speed distribution among the agents is lognormal. The GA can also be described in terms of an analogous cooperative learning paradigm. Unlike the blackboard learner, the GA shares information by copying and recombining the solutions of the agents. This method of communication slows down the propagation of useful information to agents. The slower propagation of information is necessary because the GA cannot directly evaluate parts of a solution or "partial solutions." The extent to which the GA is cooperative also depends on the choice of heuristics used to modify the canonical GA. The few test cases presented in this paper suggest that the GA may at times yield an approximately lognormal distribution or a mixture of lognormal distributions. While the results look promising, more analysis of the algorithm's overall process is required.

1 INTRODUCTION

Cooperation is generally recognized as an important underlying principle of our social systems. A natural extension of this principle is the application of cooperation in computational systems. Along with the increasing interest in developing parallel architectures, there has also been a corresponding increase in studying distributed problem solving (e.g., Bond and Gasser, 1988). Many artificial intelligence systems illustrate the benefits of cooperation. For example, the HEARSAY-II speech understanding system is one of the more noted systems (Fennell and Lesser, 1975).

The recent work by Clearwater et al. (Clearwater, Huberman, and Hogg, 1991; Huberman, 1990; Huberman and Hogg, 1987) examines the quantitative value of cooperative learning from both a theoretical and an empirical point of view. One of their papers emphasizes the importance and generality of their theoretical results (Clearwater, Huberman, and Hogg, 1991, pp. 2):

> It showed that cooperative searches, when sufficiently large, can display universal characteristics, independent of the detailed nature of either the individual processes or the particular problem being tackled. This universality manifests itself in two separate ways. First, the existence of a sharp transition from exponential to polynomial time required to find the solution as heuristic effectiveness improved [Huberman and Hogg, 1987]. Second, the appearance of a lognormal distribution in [an] individual agent's problem solving effectiveness. The enhanced tail of this distribution guarantees the existence of some agents with superior performance. This can bring about a combinatorial implosion [Kornfeld, 1982] with superlinear speed-up in the time to find the answer with respect to the number of processes.

The classical framework for analyzing generational genetic algorithm (GA) (Holland, 1975; De Jong, 1975) tends to emphasize the role of competition in the "survival of the fittest." Competing structures within the population explore various parts of the search space in parallel. The important ability of the GA to evaluate several schemata at once within the GA's population without "explicit bookkeeping" is called *implicit parallelism* (Holland, 1975; Goldberg, 1989). During each generation, the algorithm selects parts of the search space to explore: first, by applying the selection and reproduction operators to generate clones of relatively more fit structures, then by applying the crossover operators to generate new structures from substructures of the clones (i.e., possible *building blocks*), and finally, by mutating these new structures by a small amount to ensure some variety in the population. For function optimization problems, the competition over the generations evolves a population of relatively homogeneous structures. These structures provide an optimal and/or a near-optimal solution to the problem.

What if, instead of emphasizing the notion of competition, we describe the GA in a complementary way? Suppose we regard the GA population as a set of *cooperating agents*, where each agent has a trial solution to the problem. The selection function, based on environmental evaluations, effectively determines which solutions are to be shared among the agents. During reproduction, better performing agents provide their solutions to poorly performing agents. Agents also communicate parts of their solutions, i.e., "partial solutions," to other agents during crossover. Agents survive, but their trial solutions may change with each generation, depending on the information received from other agents and the amount of individual search each agent performs. In the standard GA, which uses haploid structures, each agent stores only one solution at a time.

If the generational GA does fall under the umbrella of the general framework called cooperative learning, especially when applied to problem solving and function optimization, then this additional perspective should enhance our understanding of the algorithm. Clearly, the GA is a cooperative learner in that it shares information among population members during reproduction and crossover, but perhaps a more important question is: "To what extent is the genetic algorithm a cooperative learner?"

The discussion above suggests a parallel between the competitive learning framework usually adopted when analyzing GAs and the cooperative learning framework discussed by Clearwater et al. The GA can be viewed from two complementary perspectives: competitive learning and cooperative learning. This paper further explores the relationship between the two perspectives within the standard GA. The paper then provides a brief analysis of the GA in terms of cooperative learning. A few empirical examples of GA performance illustrate the degree to which the GA is a cooperative learner. The empirical results indicates that although the GA is essentially cooperative, there are some significant differences between the cooperative learner described by Clearwater et al. and the GA. The Conclusions section addresses possible explanations for these differences.

2   OVERVIEW OF THE PROBLEM

This section first explains the assumptions behind blackboard cooperative learning and then compares these assumptions to the GA. Next, the section discusses the blackboard model of cooperative learning as a multiplicative process. A similar analysis is then applied to the Schema Theorem in GAs. This analysis shows that the GA is multiplicative with respect to the growth and decay of a schema; however, a schema may initially alternate in its growth and decay from one generation to the next, depending on the context of the current set of solutions in the population.

2.1 BLACKBOARD COOPERATIVE LEARNING

In the framework of cooperation in distributed problem solving, the blackboard learning system described by Clearwater et al. represents a simple form of cooperation called *results-sharing* (Smith and Davis, 1988). Smith and Davis describe results sharing among *knowledge sources* (KSs) as follows (Smith and Davis, 1988, pp. 66): "Results-sharing is a form of cooperation in which individual nodes assist each other by sharing partial results, based on somewhat different perspectives on the overall problem." In the blackboard learner, a KS is an agent that either searches independently for a solution, or posts and/or reads *hints* from a shared blackboard. If on the average, the hints are beneficial, then the synergetic effect of the interactions among the agents can have a substantial effect in reducing the search time of an agent. In effect, an agent enhances its performance by treating other agents as explicit (but fallible) teachers. If the search heuristics of the agents are effective, i.e., there is a high rate of posting good hints on the blackboard, then the reduction in an agent's search time can be dramatic.[1]

A hint does not have to be part of the final solution; a hint may only be correct in a limited context. In fact, if an agent were to occasionally receive a bad or misleading hint, the overall performance of the system would be unaffected so long as, on the average, hints are good. The value of a hint is not fixed. The same hint may be valued differently by different agents, depending on the processing context of the particular agent. Also, as the search progresses, the value of a partial solution tends to lessen. As more partial solutions are placed on the blackboard, an agent is more likely to have already tested a posted partial solution in a number of contexts. The rate of placing hints on the blackboard is determined by an agent's ability to generate partial solutions (and in realistic systems, the willingness of the agent to communicate them). In the blackboard cooperative learning framework, the agent is allowed to reconsider hints; however, an agent does not reexamine a possible overall solution more than once.

2.2 BLACKBOARD COOPERATIVE LEARNING AND THE GA

If we adopt the cooperative learning framework in thinking about the GA, then a generation is one of many repeated encounters among agents sharing solutions, and the population structures are the trial solutions of the agents that are formed by accepting, to a greater or lesser degree, the hints of other agents. Substructures within the structures, or schemata, represent partial solutions. The overall effect of applying the selection, reproduction and crossover operators to the population is analogous to selectively trading hints on a finite capacity blackboard. During each generation, an agent can respond in four different ways. An agent can: (1) maintain the same solution from the last generation, (2) adopt the solution of another agent, (3) modify its solution from the last generation by accepting hints from another agent, and (4) adopt another agent's solution and then modify that solution with hints from yet another agent. The initial diversity in the population aids individuals in their search; the mutation operator provides each agent with individual learning capability.[2] Each agent performs independent search to the extent that the pseudo-random number generator is uncorrelated.

When we examine the performance of both types of learners, there appears to be a number of similarities. The analysis provided by Clearwater et al. considers the trade-off faced by each agent between *cooperative learning* and the exploration of the search space. In the GA literature, this trade-off is usually expressed as *exploitation* versus exploration (Holland, 1975). Regardless of the terminology, placing extreme emphasis on cooperation (or

---

1. Depending on the quality of the heuristics, cooperation has a beneficial effect even when there are only a few agents involved. For example, a recent study of Clouse and Utgoff (Clouse and Utgoff, 1992) illustrates how a human expert's knowledge (i.e., a teacher having a "perfect heuristic") can accelerate the learning of a reinforcement learner. In the study, the expert (the first agent) periodically supplies actions to be taken (partial solutions) to the reinforcement learner (the second agent). Even with infrequently supplied additional expert knowledge, the reinforcement learner can learn in up to two orders of magnitude fewer trials.

2. Clearly, simple mutation represents the least directed of possible search strategies that an agent can pursue as an individual. The advantages of using hybrid approaches that combine a GA with gradient search strategies is currently a topic of ongoing debate (e.g., Ackley, 1991).

similarly exploitation) to the exclusion of exploration, or *vice versa*, is suboptimal. Let us consider the two cases. First, let us suppose that all of the agents dedicate their time to only seeking the partial solutions of other agents so that there is very little investment in individual search. This means that the agents can only form new hypotheses based on the combination of partial solutions already present in the population. If the number of agents is small, and if there is too much sharing of information among too few agents, it may not be possible for the agents to find a solution. More information can be exchanged by increasing the size of the population and/or by having agents perform more individual search. The same problem arises in the GA when using a high crossover rate with a harsh selection policy, a small population size, and a small mutation rate. These parameter settings emphasize exploitation in the GA's search while limiting the amount of exploration. In the worst case, the GA would not find the solution due to premature convergence. As the other extreme case, let us suppose that all of the agents in the blackboard cooperative learner work separately, without sharing information. In this case, the time required for an agent to find a solution may be very long, depending on the talent of the individual agent. The same problem occurs in the GA when using a large mutation rate, a weak selection policy, and no crossover. In the worst case, the GA would not converge.

Despite these overall similarities in behavior, there are some significant operational differences between the models as outlined below:

- In the blackboard model, each agent knows the correctness of a partial solution (though it may not be part of the optimal solution), whereas in the GA, the agent only knows the value of an overall solution, not partial solutions.

- The agent of the blackboard model knows when it has discovered an optimal solution so that the agent can halt its processing. In the GA, the goodness of a solution is relative to other solutions in the population. It is possible for an agent to destroy its current optimal solution to investigate possibly better, but actually worse, solutions.

- Hints remain on the blackboard for easy access, so that in the blackboard model partial solutions are not destroyed. In the GA, memory is local to agents, and the amount of that local memory is limited by the population size. A partial solution can be lost due to selection (when there is epistasis) or to mutation.

- In the blackboard model, all of the agents have access to all of the hints. (However, the search for new information may in reality cost time.) In the GA, the copying of solutions is usually performed among subsets of agents as a result of the selection process. The sharing of partial solutions is performed among pairs of agents during crossover.

2.3 MULTIPLICATIVE PROCESSES AND THE LOGNORMAL DISTRIBUTION

The lognormal distribution has been used to describe processes in many fields, including: industry, economics, biology, ecology, and geology (Aitchison and Brown, 1976; Shimizu, 1988). For example, Mosimann and Campbell (Mosimann and Campbell, 1988) describe an active tissue growth model in which the tissue produced at prior time steps actively participates in the production of tissue at the current time step. Thus, if W_k is the amount of tissue at time k, then the growth ratio, $X_{k+1} = W_{k+1}/W_k$. Given W_0 is the initial amount of tissue, the tissue at time k is

$$W_k = W_0 \prod_{i=1}^{k} X_i, \quad k = 1, ..., n. \tag{1}$$

Unlike the inert tissue model, which is additive, the active growth model allows for both the growth and the decay of tissue at different times. Values of X_i greater than one indicate growth; values of X_i between 0 and 1 indicate decay.

The asymptotic distribution for a product of *random variables* is generally lognormal. The name of the lognormal distribution is derived from the fact that a product can be transformed into a sum by taking the natural log of the terms. Because the sum of the log terms asymptotically converges to a Gaussian (Normal) distribution (as we know from the Central Limit Theorem), the limiting probability distribution of a product of random variables is usually described using the lognormal distribution. Given $x > 0$ in Eq. (1), the simple lognormal distribution is a function of a scaling parameter, μ, and a shape parameter, σ. For $-\infty < \mu < \infty$, and $\sigma > 0$, (Hahn and Shapiro, 1967, pp. 99) the two parameter lognormal is

$$f(x) = \frac{1}{\sigma x \sqrt{2\pi}} exp\left[-\frac{1}{2\sigma^2} (ln(x) - \mu)^2 \right] . \tag{2}$$

The three parameter version includes a threshold parameter, ε, which indicates the minimum value of x. Given $-\infty < \varepsilon < \infty$, $x > \varepsilon$, $-\infty < \gamma < \infty$, and $\eta > 0$ (Hahn and Shapiro, 1967, pp. 199), and

$$\eta = 1/\sigma \text{ and } \gamma = -\mu/\sigma$$

from the two parameter lognormal, the density of the three parameter lognormal is

$$f(x) = \frac{\eta}{(x - \varepsilon) \sqrt{2\pi}} exp\left[-\frac{\eta^2}{2} (\frac{\gamma}{\eta} + ln(x - \varepsilon))^2 \right] . \tag{3}$$

The three parameter lognormal can be transformed to the Gaussian distribution as follows:

$$z = \gamma + \eta ln(x - \varepsilon) . \tag{4}$$

If $\varepsilon = 0$, then the three parameter version of Eqn. (3) reduces to the two parameter version of Eq. (2).

2.4 MULTIPLICATIVE PROCESS IN COOPERATIVE LEARNING

The analysis of Clearwater et al. shows that the speedup in blackboard cooperative learning is due to the multiplicative aspect of the process. Their analysis is performed in two ways as briefly summarized below. The shape of the lognormal distribution shows that the characteristic acceleration of cooperative learning over individual learning. The long tail of the lognormal distribution represents an increase in the number of "superior performers" having fast search speeds.

2.4.1 Speed Distribution of Agents

Before examining multiplicative processes, let us start by examining the probability distribution of the time required for an agent to solve a problem individually using random search. Let us assume that an agent can look at a possible solution more than once (i.e., search with replacement). The probability distribution of an agent finding a solution at time t is

$$P(t) = \prod_{k=0}^{t-1} P_{failure}(k) P_{success}(t).$$ (5)

As Clearwater, et al. point out, if the probability of failure remains constant, then the resulting time distribution is a simple geometric. If the size of the search space is S, and the number of solutions is s, then the distribution becomes

$$f(t) = (1.0 - \frac{s}{S})^{t-1} (\frac{s}{S}).$$ (6)

If we change the replacement assumption so that the agent cannot reexamine a possible solution, then the fraction of failure s/S in Eq. (6) decreases as the search progresses due to a decrease in S. However, this decrease is initially small if the agent eliminates only one state in the search space at each step. Since there are still a large number of remaining potential solutions for the agent to examine, the distribution is still approximately geometric.

If we again change our assumptions so that each of the agents examines a changing blackboard for partial solutions, then $P_{failure}(t)$ in Eq. (5) becomes a random variable. The amount of reduction in the search space varies at each step as the product of the reduction at the previous time steps. Because Eq. (5) has the same form as Eq. (1), we can conclude that the time to solve the problem is lognormally distributed. If the time to solve the problem is lognormally distributed, then the speed of finding a solution (i.e., speed being the reciprocal of the time) is also lognormally distributed.

2.5 Examining the Distribution of Hint Values

Clearwater et al. (Clearwater, Huberman, and Hogg, 1991) also perform an analysis showing the multiplicative effect of hints in reducing the size of the search space. A lognormal distribution of hint values over a time interval results in a lognormal distribution in the speed of the agents finding a solution. The reduction in the size of the search space is expressed in terms of a recursive relationship as shown in Eq. (7). The equation includes the rate of individual working effort and the rate of cooperative sharing. The multiplicative effect of an additional j^{th} hint on the remaining size of the search space for the ith agent is

$$d_i(j) = \left[d_i(j-1) - \frac{w_i}{q}\right] h_j,$$ (7)

where $d_i(j)$ is the resulting number of solutions that the agent has left to consider, h_j is the effectiveness of a hint, w_i is the rate at which the agent works on exploring possible solutions (i.e., states in the search space), and q is the rate of examining posted hints of other agents. When h_j has a value less than one (and greater than zero), the jth hint reduces the number of possible solutions that an agent must consider. The closer the hint value is to zero, the more beneficial is the hint. Hints given during the beginning of the search are usually more effective than those received later on. Based on Eq. (7), the reduction in the effective search space from the initial time zero after the $(j - 1)^{th}$ hint is

$$d_i(j) = d_i(0) \prod_{k=1}^{j} h_j - \frac{w_i}{q} \sum_{l=1}^{j} \prod_{k=l}^{j} h_l. \tag{8}$$

If there is very little individual search (i.e., $w_i \ll q$), then the hint values in Eq. (8) are nearly multiplicative. As a result, the distribution of hint values for a fixed amount of time (or number of hints) is lognormally distributed. This distribution can be easily transformed into a lognormal distribution of solution times. Since the distribution of hints effectively gives a distribution of the search space remaining, this distribution is proportional to the distribution of time remaining for the agents to find the solution. Thus, Clearwater et al. again demonstrate that the asymptotic probability distribution of the times for discovering a solution (as well as the speed) is lognormally distributed among the agents.

2.6 SCHEMA ANALYSIS AND MULTIPLICATIVE PROCESSES

The Schema Theorem also expresses a recursive relationship: the number m of schema H in the population from one generation to the next. If $p_H(s, t-1)$ is the probability that schema H survives after the disruptive effects of crossover and mutation, and if $f_H(t-1)/\bar{f}(t-1)$ is the proportion of schema H generated, then according to the Schema Theorem, the number of schema H in the next generation, $m_H(t)$, is bounded as shown in Eq. (9) (Holland, 1975):

$$m_H(t) \geq m_H(t-1) \, [p_H(s, t-1)] \left(\frac{f_H(t-1)}{\bar{f}(t-1)} \right). \tag{9}$$

After $t - 1$ generations, the number of schema H is

$$m_H(t) \cong m_H(0) \prod_{k=0}^{t} \left(\frac{f_H(k)}{\bar{f}(k)} \right) [p_H(s, k)] . \tag{10}$$

GAs, when modified to perform function optimization, use fairly involved heuristics that go beyond the assumptions behind the Schema Theorem (e.g., single-point crossover, proportional selection).[3] These extended heuristics include the scaling of observed fitnesses, elitist strategies, ranked-based selection, various forms of crossover, adaptively changing crossover and mutation probabilities, etc. When performing function optimization, Eq.

---

3. De Jong points out the inappropriateness of using the Schema Theorem to analyze non-canonical GAs performing function optimization (De Jong, 1992).

(10) should probably be expressed in terms of some process-dependent heuristic, $\xi_H(k)$, which acts more like a random variable. Given

$$\xi_H(k) = \left(\frac{f_H(k)}{\bar{f}(k)}\right)[p_H(s, k)], \tag{11}$$

then

$$m_H(t) \equiv m_H(0) \prod_{k=0}^{t} (\xi_H(k)) . \tag{12}$$

If the heuristic $\xi_H(t)$ increases the effectiveness of the evaluation of schema H over time, then the GA is multiplicative in terms of increasing or decreasing *a particular schema H*. If schema H is considered to be the optimal solution to the problem, and if the heuristic $\xi_H(t)$ is effective, then the GA's distribution of solution speeds over the population members might be described using some variation of the lognormal distribution.

Notice that Eq. (12) has a form similar to Eqn. (1) above. Schema H in this context refers to a partial solution. In effect, when $\xi_H(t)$ is between 0 and 1, it is analogous to the *value* of a hint in Eq. (8). As $\xi_H(t)$ gets closer to zero, the effectiveness of the "hint" in reducing schema H improves. Alternatively, a product greater than one is analogous to increasing a desirable schema H in the population. The individual agent's work effort is implemented through the mutation operator. If the individual work effort is small, then the effect on $\xi_H(t)$ is small.

The GA implements hint sharing through reproduction and crossover. The combined effect of reproduction and crossover can be thought of as regulating the rate of access to a hint on the blackboard. However, unlike the blackboard learner, this access rate is inseparable from the value of the hints in Eqn. (12). The GA must use its population not only to investigate solutions, but it must also use this finite memory capacity to reflect the quality of hints within the population. The GA increases its memory allocation to better performing schema in order to increase the rate of access to better hint values. This dual use of memory means that a desirable schema H may initially go through several short cycles of growth and decay.

3 AN EMPIRICAL LOOK AT THE PROBLEM

The next phase of this study is to explore GA behavior empirically using a few test cases to see if the GA's speed distribution for finding a solution compares favorably to the lognormal distribution. In order to obtain a general estimate of the speed of finding solutions, the GA runs until all of the agents have found the optimal solution at least once. During the run, the time that an agent first discovers the optimum is recorded, even though the agent's answer may change subsequently due to crossover and mutation. Since the information is still within the population, for the most part, an agent is credited with success the first time it encounters the solution.

All of the experiments reported here use modifications of GENESIS-4.5 (Grefenstette, 1983). The modified GA treats solutions and agents separately during reproduction and crossover. The shuffling of population members to assure random mating means that agents, and not their solutions, are rearranged. In this way, it is possible to track when an agent first finds an optimal solution.[4]

This study examines three optimization problems: functions F3 and F5 from De Jong's test suite (De Jong, 1975), and the test function F7, studied by Schaffer et al. (Schaffer, Caruana, Eshelman, and Das, 1989). Clearwater et al. solve examples of cryptarithmetic problems using a blackboard cooperative learning system to test their theory empirically and to provide a quantitative evaluation of the value of cooperation in problem solving. The cryptarithmetic problem is an example of optimization under constraints. For purposes of comparison, this study also applies a GA to solve one of the more difficult of the cryptarithmetic problems reported by Clearwater et al.

This study represents the beginning a larger endeavor. The work starts by examining a few functions using informal graphical methods to determine the reasonableness of applying the lognormal model. When appropriate, the empirical data are fitted to the three parameter lognormal by using an inverse transformation technique in which distribution parameters are estimated from percentiles of the actual data. The data are transformed to the Gaussian distribution using Eq. (4), and then a Kolmogorov-Smirnoff goodness-of-fit test is performed between the data and a three parameter lognormal distribution having the same parameter values.[5]

---

4. Normally, in GENESIS-4.5, each generation is treated separately, since new structures represent the children of parent structures and not new solutions.

5. Goodness-of-fit techniques differ from usual statistical techniques (D'Agostino and Stephens, 1986). Usually, statistical experiments are designed to disprove a null hypothesis so that an alternative hypothesis can be accepted. In goodness-of-fit testing, the opposite approach is used: an effort is made to confirm the null hypothesis. The alternative hypothesis is often a composite hypothesis, giving little or no information about the distribution. As a result, proving the goodness of a fit is not usually definitive. Based on some understanding of the underlying process, a set of null hypothesis are tested; in turn, each of these null hypotheses are examined using several goodness-of-fit techniques.

4   EXAMINING SOME GA OPTIMIZATION PROBLEMS

In this study we are interested in emphasizing convergence or improvement in off-line performance, because we would like all of the agents to find the optimal solution. Very often, in order to achieve convergence, the GA needs a large population and a small mutation rate. So all of the optimization runs use a population size of 100 or 200 and mutation rate of 0.001. If we were more interested in on-line performance, a smaller population and a higher mutation rate would be desirable (Schaffer, Caruana, Eshelman, and Das, 1989). For F7, the most difficult problem of the three optimization problems, an elitist strategy is used instead of a higher mutation rate. All of the optimization runs use a scaling window of five generations.

Figure 1 (a) shows a graphic plot of the speed points for a run of F3. If a transformation of the log of the data to the Gaussian distribution is correct, the data points should lie along a straight line. Figure 1 (b) shows the corresponding fit to the empirical speed distribution among the agents. This fit passes the Kolmogorov-Smirnoff test at the 95% level of confidence.[6] Smoothing bin counts with neighboring bins provides an even better fit (not shown). Figure 1(c) show the growth rate of all schemata mapping into the optimal solution. During the beginning of the search, the optimal schemata exhibit short bursts of growth and then appear to decay (a growth ratio less than one) in a subsequent generation. The pattern of growth burst followed by decay, proceeds in a cyclic fashion until the growth rate begins to level out. Optimal schemata temporarily disappear (intact) while the GA tests other non-optimal schemata. When the optimal schemata reappear, the large burst of growth compensates for the temporary hiatus of the schemata.

Figures 2(a) and 2(b) shows the results of a comparable run using function F5. The non-linearity in Figure 2(a) indicates a shorter tail than what would be expected for the lognormal. The fit shown in Figure 2(b) passes the Kolmogorov-Smirnoff test at the 95% confidence level.

Figure 3 shows similar plots for function F7. By incorporating the elitist strategy, a large percentage of the agents in the population discover the optimal solution fairly quickly. The empirical distribution in this case is actually negatively skewed. The elitist strategy dominates the other learning heuristics. The result is an unidentifiable mixed distribution.

---

6. The Kolmogorov-Smirnoff test examines the general shape of the distribution and not expected bin counts. The bin counts in these experiments are too small to perform Chi-Square tests.

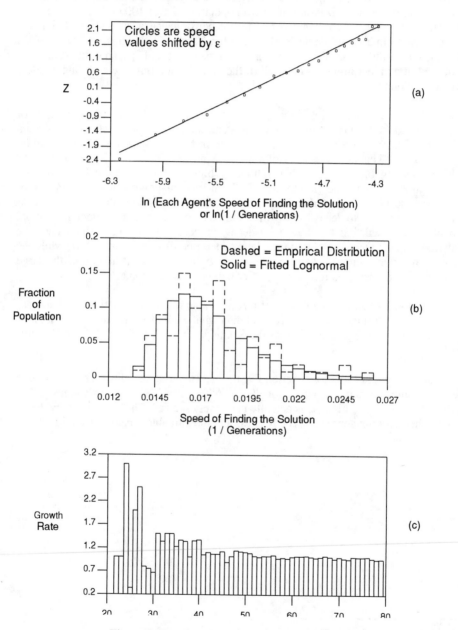

De Jong's F3: AGENTS' SPEED DATA
THREE PARAMETER LOG-NORMAL FIT
$Z = \gamma + \eta \times log\,(\,speed - \varepsilon\,)$, $\gamma = 11.9161$, $\eta = 2.0845$, and $\varepsilon = 0.01438$
Population = 100; $P_M = 0.001$, $P_C = 0.6$, Scaling Window = 5

Figure 1: Test Results for De Jong's F3 Function

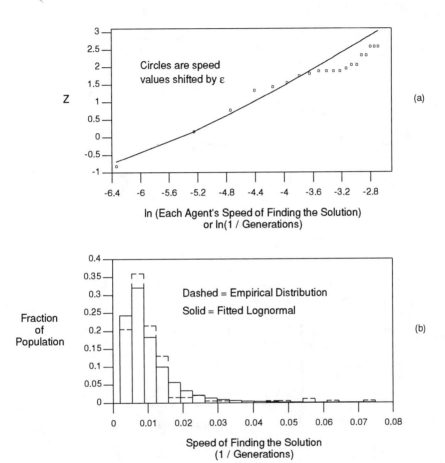

De Jong's F5: AGENTS' SPEED DATA
THREE PARAMETER LOG-NORMAL FIT
$Z = \gamma + \eta \times log\,(\,speed - \varepsilon\,)$, $\gamma = 6.3407$, $\eta = 1.2449$, and $\varepsilon = 0.002$
Population $= 200$, $P_M = 0.001$, $P_C = 0.6$, Scaling Window $= 5$

(a)

Circles are speed
values shifted by ε

Z

ln (Each Agent's Speed of Finding the Solution)
or ln(1 / Generations)

(b)

Fraction
of
Population

Dashed = Empirical Distribution

Solid = Fitted Lognormal

Speed of Finding the Solution
(1 / Generations)

Figure 2: Test Results for De Jong's F5 Function

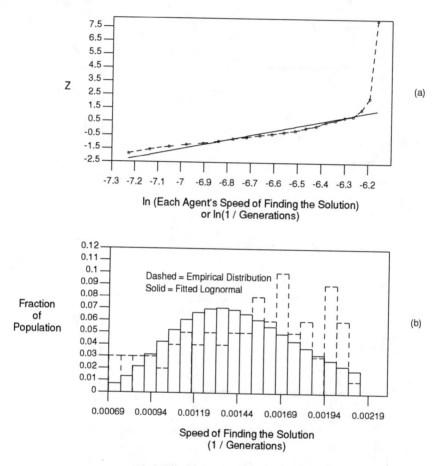

F7 OF SCHAFFER ET AL.: AGENTS' SPEED DATA
THREE PARAMETER LOG-NORMAL FIT
$Z = \gamma + \eta \times log\,(speed - \varepsilon)$, $\gamma = 23.2952$, $\eta = 3.5618$, and $\varepsilon = 0.0$
Population = 100; $P_M = 0.001$, $P_C = 0.6$, Scaling Window = 5, Elitist **Strategy**

Figure 3: Test Results for Function F7

4.1 THE CRYPARITHMETIC PROBLEM

In cryptarithmetic, an addition problem is expressed using letters. The objective of problem is to assign a *unique* numeral to each letter used so that the addition works correctly. For example, one problem explored by Clearwater et al. is DONALD + GERALD = ROBERT. In this problem, there is one solution and 3,628,800 legal combinations of decimal digits in the initial search space. The solution is A = 4, B = 3, D = 5, E = 9, G = 1, L = 8, N = 6, O = 2, R = 7, and T = 0.

A GA population structure consists of the assigned value for each of the ten letters; i.e., an allele is a decimal digit. For example, given the fixed letter sequence {A B D E G L O R T}, the correct assignment of the digits is {4 3 5 9 1 8 6 2 7 0}. In order to ensure that only legal representations are maintained in the population, the GA employs Syswerda's swap mutation, and a slight modification of his position-based crossover (Syswerda, 1991), which is called here, for convenience, *assignment crossover*. Assignment crossover ensures that the absolute allele assignments of the parents are maintained in the children as much as possible (not just relative positions). The assignment crossover proceeds in three steps: (1) exchange allele information at selected positions; (2) copy the alleles of a parent into the empty positions of the corresponding child (e.g., from parent 1 to child 1), unless the child just inherited the value from the other parent, and (3) place the crossed over values of a parent (e.g., alleles from parent 1 to child 2) into the remaining empty slots of its own child, if possible. Four positions are randomly selected for application of the operator.

Parent 1:	0	1	3	7	9	2	5	6	8	4
Parent 2:	7	3	5	9	8	4	0	2	6	1
Selected Positions:	*		*				*	*		
Child 1:	7	1	5	3	9	6	0	2	8	4
Child 2:	0	7	3	9	8	4	5	6	2	1

As a result of the assignment crossover above, parents and their corresponding children have the values 1, 9, 8, and 4 in the same positions. This crossover operator minimizes the disruption in the assignment of the digits.

The evaluation function in the GA computes the sum of DONALD + GERALD = ROBERT based on the current assignment of the letter values. For each column added correctly, the agent receives partial credit depending on the number of letters participating in the sum. For example, if in adding the right-most column, i.e., D + D = T, the agent were to get the correct answer, then the partial evaluation would be 2; if the agent were also to add the left most column correctly (after taking care of any carry, of course), then the agent would receive an additional 3 points. The totally correct answer is worth 14 points. Unlike most traditional GA problems, this problem explicitly uses partial evaluations to compute the overall fitness of an agent's trial solution. Nevertheless, population members only receive evaluations of entire solutions.

The mutation rate used for this problem is higher than the rate used in the preceding optimization problems ($p_m = 0.1$ instead of $p_m = 0.001$) because each letter can take on ten values instead of two. Without the use of a high mutation rate, the GA prematurely converges to a suboptimal solution. A small crossover rate (around 0.2) is used in conjunction with the high mutation rate. The GA also tends to prematurely converge when the crossover rate exceeds 0.21.

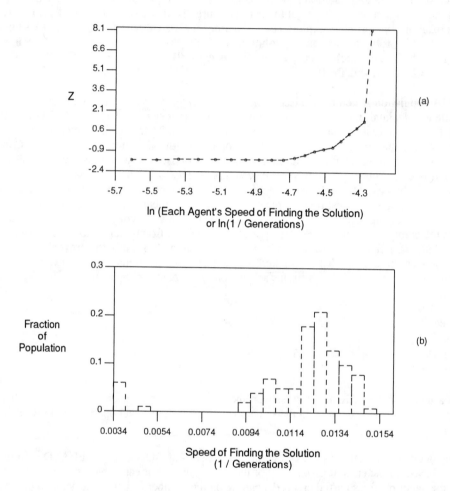

CRYPTARITHMETIC PROBLEM: DONALD + GERALD = ROBERT
Population = 100, P_M = 0.1, P_C = 0.18, Scaling Window = 5

Figure 4: Test Results for the Cryptarithmetic Problem

Figure 4 shows the result of one of the runs. In Figure 4(a), the piecewise linear appearance of the plot suggests that for pieces having a non-zero slope, the distribution is a mixture of lognormal distributions. Slightly curved shapes represent transitions between the distributions. The disruptive effect of mutation prevents a more continuous growth in the solutions.

Figure 5 shows the result of having a varying number of crossover points, ranging from one to nine. There are essentially two separate intervals where the agents discover the solution. The additional disruptive effect of the crossover prevents the continued growth of

CRYPTARITHMETIC PROBLEM: DONALD + GERALD = ROBERT
Population = 100; P_M = 0.1, P_C = 0.1, Scaling Window = 5

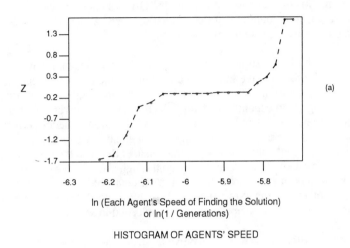

Z

(a)

In (Each Agent's Speed of Finding the Solution)
or ln(1 / Generations)

HISTOGRAM OF AGENTS' SPEED

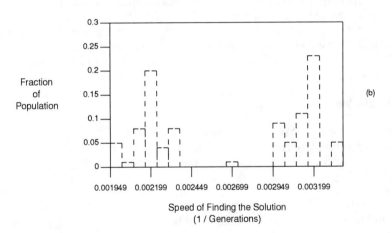

Fraction
of
Population

(b)

Speed of Finding the Solution
(1 / Generations)

Figure 5: Additional Test Results for the Cryptarithmetic Problem

the solution starting around generation 310 so that the remaining agents do not discover
the solution until around generation 450.

5 CONCLUSIONS

Viewing the GA as a cooperative learner provides additional insight into the function of the algorithm. This paper begins to explore the GA as a cooperative process. Due to information sharing among members of the population, the GA can be described metaphorically using a cooperative learning framework.

The actual distribution of solution speeds for the GA depends on many factors. The shape of the distribution is sensitive to parameter settings which affect the balance between cooperation and exploration in the GA. In addition, the propagation of solutions through the population using the mechanisms of reproduction and crossover affect the shape of the speed distribution. The GA increases its memory allocation to better performing schemata in order to increase the rate of access to those schemata; however, the same memory space is used to test potential solutions in the search space. Because the GA uses its memory for both purposes, a relatively fit schema may go through short cycles of growth and decay toward the beginning of the search. In effect, a schema needs to time-share the population memory with other schema. Nevertheless, the overall speed distribution among the agents for finding a solution may be approximately lognormal in some cases.

Future research will address the cooperative aspect of the GA by examining: (1) more test problems; (2) different goodness-of-fit tests; (3) other heuristics used by the GA (e.g., rank based selection), and (4) modifications of the GA. For example, it would be interesting to test what effect, if any, increasing the storage capacity of each agent in the population has on the speed distribution of agents. It might also be interesting to study the effect of having crossover among more than two parents as a way of increasing the communication in the population.

Acknowledgments

I would like to thank the FAW Institute for inviting me to their 1991 "Adaptive Learning and Neural Networks" workshop in Ulm, Germany. It was at this workshop that I first heard Bernardo Huberman speak about his research on cooperative learning.

References

David H. Ackley (1991). Re: GA's and Hill-Climbers. *GA-List, Vol 5, Issue 23* (A moderated bulletin board).

J. Aitchison and J. A. C. Brown (1976). *The Lognormal Distribution*. London, Eng.: Cambridge University Press.

Alan H. Bond and Les Gasser (Eds) (1988). *Readings in Distributed Artificial Intelligence*. San Mateo, CA: Morgan Kaufmann.

Scott H. Clearwater, Bernardo A. Huberman, and Tad Hogg (1991). Problem Solving by Committee. Presented at the FAW workshop on Adaptive Learning and NNs, July 2 - 7, in Ulm, Germany.

Jeffery A. Clouse and Paul E. Utgoff (1992) A Teaching Method for Reinforcement Learning. *Machine Learning: Proceeds of the Ninth International Workshop (ML92)*, pp. 92 - 101, Derek Sleeman and Peter Edwards, eds., San Mateo, CA: Morgan Kaufman.

Kenneth A. De Jong (1975). *An Analysis of the Behavior of a Class of Genetic Adaptive Systems*. Doctoral Dissertation, Univ. of Michigan.

Kenneth A. De Jong (1992). Genetic Algorithms are NOT Function Optimizers. In *Foundations of Genetic Algorithms 2*, D. Whitley (Ed.), San Mateo: Morgan Kaufmann.

Richard D. Fennell and Victor R. Lesser (1989). Parallelism in Artificial Intelligence Problem Solving: A Case Study of Hearsay II. *Readings in Distributed Artificial Intelligence*, pp. 106 - 119, Alan H. Bond and Less Gasser (eds.), San Mateo, CA: Morgan Kaufman.

John J. Grefenstette (1983). *A User's Guide to Genesis*. Technical Report CS-83-11, Computer Science Department, Vanderbuilt University.

David E. Goldberg (1989). *Genetic Algorithms in Search, Optimization & Machine Learning. Reading*, MA: Addison-Wesley.

Gerald J. Hahn and Samuel S. Shapiro (1967). *Statistical Models in Engineering*. New York: John Wiley and Sons, Inc.

John Holland (1975). *Adaptation in Natural and Artificial Systems*. Ann Arbor, MI: The University of Michigan Press.

B. A. Huberman and T. Hogg (1987). Phase transitions in artificial intelligence systems. *Artificial Intelligence*, 33:155-171.

Bernardo A. Huberman (1990). The Performance of Cooperative Processes. *Physica D* 42, 38-37.

William A. Kornfeld (1981). The use of parallelism to implement heuristic search. Technical report, Massachusetts Institute of Technology Artificial Intelligence Laboratory, 1981. AI. Memo No. 627.

James E. Mosimann and Gregory Campbell (1988) Applications in Biology: Simple Growth Models. In Edwin L. Crow and Kunio Shimizu (Eds.). *Lognormal Distributions: Theory and Applications*, NY: Marcel Dekker.

J. David Schaffer, Richard A. Caruana, Larry J. Eshelman, Rajarshi Das (1989). A Study of Control Parameters Affecting Online Performance of Genetic Algorithms for Function Optimization. *Proceedings of the Third Inter. Conf on Genetic Algorithms*. San Mateo, CA: Morgan Kaufman.

Kunio Shimizu (1988). History, Genesis, and Properties. In Edwin L. Crow and Kunio Shimizu (Eds.), *Lognormal Distributions: Theory and Applications*, NY: Marcel Dekker.

Reid G. Smith and Randall Davis (1989). Frameworks for Cooperation in Distributed Problem Solving. *Readings in Distributed Artificial Intelligence,* pp. 61 -70. Alan H. Bond and Les Gasser (Eds). San Mateo, CA: Morgan Kaufmann.

Gilbert Syswerda (1991). Schedule Optimization Using Genetic Algorithms. In *Handbook of Genetic Algorithms.* Lawrence Davis (Ed), NY, NY: Van Nostrand Reinhold.

Hierarchical Automatic Function Definition in Genetic Programming

John R. Koza

Computer Science Department
Stanford University
Stanford, CA 94305 USA
E-MAIL: Koza@Sunburn.Stanford.Edu
PHONE: 415-941-0336   FAX: 415-941-9430

Abstract

A key goal in machine learning and artificial intelligence is to automatically and dynamically decompose problems into simpler problems in order to facilitate their solution. This paper describes two extensions to genetic programming, called "automatic" function definition and "hierarchical automatic" function definition, wherein functions that might be useful in solving a problem are automatically and dynamically defined during a run in terms of dummy variables. The defined functions are then repeatedly called from the automatically discovered "main" result-producing part of the program with different instantiations of the dummy variables. In the "hierarchical" version of automatic function definition, automatically defined functions may call other automatically defined functions, thus creating a hierarchy of dependencies among the automatically defined functions. The even-11-parity problem was solved using using hierarchical automatic function definition.

1 INTRODUCTION AND OVERVIEW

A key goal in machine learning and artificial intelligence is to automatically and dynamically decompose problems into simpler subproblems in order to facilitate their solution.

When a human programmer writes a computer program to solve a problem, he often creates a sub-routine defined in terms of dummy variables (formal parameters) enabling a common calculation to be performed for different instantiations of the dummy variables. For example, suppose a programmer needed the value of the exponential of x, y and $3z^2$ at different points in a main program. In the LISP programming language, the programmer might write a function definition such as

```
(defun our-exp (dv)
    (+ 1.0  dv  (* 0.5 dv dv)  (* 0.1667 dv dv dv))).
```

This function definition (`defun`) assigns a name (`our-exp`) to the function being defined; it identifies the list (`dv`) containing the single dummy variable (formal parameter) `dv` as the argument list to the function; and it contains a body which performs the work (which here happens to be an approximation to the exponential function consisting of the first four terms of the Taylor series). This particular example of a function definition (and all those in this paper) returns a single numerical value, does not refer to any of the actual variables of the overall problem (i.e., contains only a dummy variable), and has no side effects; however, in general, functions may return multiple values, refer to the actual variables of the overall problem, and perform side effects.

Once the function `our-exp` is defined, it can then be called an arbitrary number of times with different instantiations of its dummy variable `dv`.

Function definitions exploit the underlying regularities and symmetries of a problem by obviating the need to tediously rewrite lines of essentially similar code. The process of defining and calling a function, in effect, decomposes the problem into a hierarchy of subproblems.

Genetic programming (Koza 1991, 1992) provides a way to genetically breed a population of computer programs in order to solve a problem. The new processes of "automatic" function definition and "hierarchical automatic" function definition described in this paper provide a way to automatically define and call potentially useful functions on-the-fly during a run of genetic programming.

In section 2 of this paper, we review how the even-parity function of increasing numbers of arguments can be solved with genetic programming. In section 3, we describe the new process of "automatic" function definition and show how it is helpful in facilitating the solution of the Boolean even-parity problem. In section 4, we describe the new process of "hierarchical automatic" function definition and show how it is helpful in facilitating the solution of the even-parity function. Specifically, we use hierarchical automatic function definition to solve the even-parity function for up to eleven arguments. In section 5, we indicate that the technique of hierarchical automatic function definition can be successfully applied to other Boolean functions.

2 LEARNING THE EVEN-PARITY-FUNCTION WITH GENETIC PROGRAMMING

The Boolean even-parity function of k Boolean arguments returns T (True) if an even number of its arguments are T, and otherwise returns NIL (False).

In applying genetic programming to the even-parity function of k arguments, the terminal set T consists of the k Boolean arguments D0, D1, D2, ... involved in the problem, so that

$$T = \{D0, D1, D2, \ldots\}.$$

The function set F for all the examples herein consists of the following computationally complete set of four two-argument primitive Boolean functions:

$$F = \{AND, OR, NAND, NOR\}.$$

Compositions of functions from the function set F and terminals from the terminal set T are called symbolic expressions (S-expressions) in the LISP programming language. An S-expression can be represented as a rooted, point-labeled tree with ordered branches in which the root and other internal points of the tree are labeled with functions and in which the external points of the tree are labeled with terminals. These trees correspond to the parse trees found in other programming languages.

The Boolean even-parity functions appear to be the most difficult Boolean functions to find via a blind random generative search of S-expressions using the above function set F and the terminal set T. For example, even though there are only 256 different Boolean functions with three arguments and one output, the Boolean even-3-parity function is so difficult to find via a blind random generative search that we did not encounter it at all after randomly generating 10,000,000 S-expressions using this function set F and terminal set T. In addition, the even-parity function appears to be the most difficult to learn using genetic programming using this function set F and terminal set T (Koza 1992).

In applying genetic programming to the problem of learning the Boolean even-parity function of k arguments, the 2^k combinations of the k Boolean arguments constitute an exhaustive set of fitness cases for learning this function. The standardized fitness of an S-expression is the sum, over these 2^k fitness cases, of the Hamming distance (error) between the value returned by the S-expression and the correct value of the Boolean function. Standardized fitness ranges between 0 and 2^k; a value closer to zero is better. The raw fitness is equal to the number of fitness cases for which the S-expression is correct (i.e., 2^k minus standardized fitness); a higher value is better.

To establish a baseline for demonstrating the effectiveness of the new processes of automatic function definition and hierarchical automatic function definition, we first consider how genetic programming would solve the problems of learning the even-3-parity function (three-argument Boolean rule 105), the even-4-parity function (four-argument Boolean rule 38,505), and the even-5-parity function (five-argument Boolean rule 1,771,476,585).

Throughout this paper, we employ a numbering scheme for identifying a k-argument Boolean function wherein the value of the function for the 2^k combinations of its k Boolean arguments are concatenated into a 2^k-bit binary number and then converted to the equivalent decimal number. For example, the $2^3 = 8$ values of the even-3-parity function are 0, 1, 1, 0, 1, 0, 0, and 1 (going from the fitness case consisting of three true arguments to the fitness case consisting of three false arguments). Since $01101001_2 = 105_{10}$, the even-3-parity function is referred to as three-argument Boolean rule 105.

The terminal set T for the even-3-parity problem consists of

```
T = {D0, D1, D2}.
```

In one run of genetic programming using a population size of 4,000 (the value of M used consistently herein, except as otherwise noted), genetic programming discovered the following S-expression containing 45 points (i.e., 22 functions and 23 terminals) with a perfect value of raw fitness of 8 (out of a possible value of $2^3 = 8$) in generation 5:

```
(AND (OR (OR D0 (NOR D2 D1)) D2) (AND (NAND (NOR (NOR D0 D2) (AND
(AND D1 D1) D1)) (NAND (OR (AND D0 D1) D2) D0)) (OR (NAND (AND D0
D2) (OR (NOR D0 (OR D2 D0)) D1)) (NAND (NAND D1 (NAND D0 D1))
D2)))).
```

We then considered the even-4-parity function. In one run, genetic programming discovered the following S-expression containing 149 points with a perfect value of raw fitness of 16 (out of $2^4 = 16$) in generation 24:

```
(AND (OR (OR (OR (NOR D0 (NOR D2 D1)) (NAND (OR (NOR (AND D3 D0) D2)
(NAND D0 (NOR D2 (AND D1 (OR D3 D2))))) D3)) (AND (AND D1 D2) D0))
(NAND (NAND (NAND D3 (OR (NOR D0 (NOR (OR D3 D2) D2)) (NAND (AND
(AND (AND D3 D2) D3) D2) D3))) (NAND (OR (NAND (OR D0 (OR D0 D1))
(NAND D0 D1)) D3) (NAND D1 D3))) D3)) (OR (OR (NOR (NOR (AND (OR
(NOR D3 D0) (NOR (NOR D3 (NAND (OR (NAND D2 D2) D2) D2)) (AND D3
D2))) D1) (AND D3 D0)) (NOR D3 (OR D0 D2))) (NOR D1 (AND (OR (NOR
(AND D3 D3) D2) (NAND D0 (NOR D2 (AND D1 D0)))) (OR (OR D0 D3) (NOR
D0 (NAND (OR (NAND D2 D2) D2) D2)))))) (AND (AND D2 (NAND D1 (NAND
(AND D3 (NAND D1 D3)) (AND D1 D1)))) (OR D3 (OR D0 (OR D0 D1)))))).
```

Figure 1 presents two curves, called the performance curves, relating to the even-3-parity function over a series of runs. The curves are based on 66 runs with a population size M of 4,000 and a maximum number of generations to be run G of 51 (the value of G used consistently herein).

The rising curve in figure 1 shows, by generation, the experimentally observed cumulative probability of success, $P(M,i)$, of solving the problem by generation i (i.e., finding at least one S-expression in the population which produces the correct value for all $2^3 = 8$ fitness cases). As can be seen, the experimentally observed value of the cumulative probability of success, $P(M,i)$, is 91% by generation 9 and 100% by generation 21 over the 66 runs.

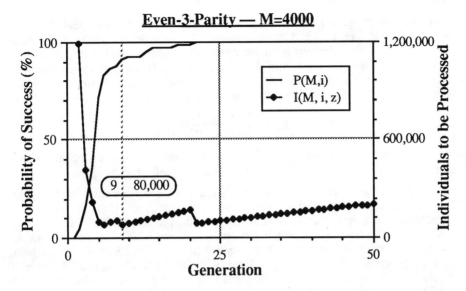

Figure 1 Performance curves for even-3-parity function showing that it is sufficient to process 80,000 individuals to yield a solution with 99% probability with genetic programming.

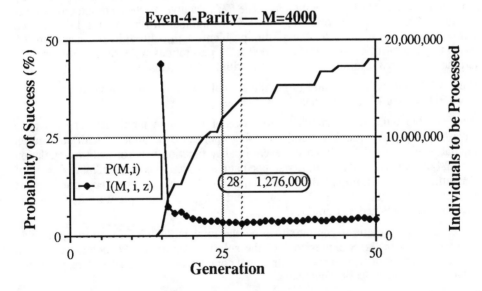

Figure 2 Performance curves for even-4-parity function showing that it is sufficient to process 1,276,000 individuals to yield a solution with 99% probability with genetic programming.

The second curve in figure 1 shows, by generation, the number of individuals that must be processed, $I(M,i,z)$, to yield, with probability z, a solution to the problem by generation i. $I(M,i,z)$ is derived from the experimentally observed values of $P(M,i)$. Specifically, $I(M,i,z)$ is the product of the population size M, the generation number i, and the number of independent runs $R(z)$ necessary to yield a solution to the problem with probability z by generation i. In turn, the number of runs $R(z)$ is given by

$$R(z) = \left[\frac{log(1-z)}{log(1-P(M,i))}\right],$$

where the square brackets indicates the ceiling function for rounding up to the next highest integer. Throughout this paper, the probability z will be 99%.

As can be seen, the $I(M,i,z)$ curve reaches a minimum value at generation 9 (highlighted by the light dotted vertical line). For a value of $P(M,i)$ of 91%, the number of independent runs $R(z)$ necessary to yield a solution to the problem with a 99% probability by generation i is 2. The two summary numbers (i.e., 9 and 80,000) in the oval indicate that if this problem is run through to generation 9 (the initial random generation being counted as generation 0), processing a total of 80,000 individuals (i.e., 4,000 × 10 generations × 2 runs) is sufficient to yield a solution to this problem with 99% probability. This number 80,000 is a measure of the computational effort necessary to yield a solution to this problem with 99% probability.

Figure 2 shows similar performance curves for the even-4-parity function based on 60 runs. The experimentally observed cumulative probability of success, $P(M,i)$, is 35% by generation 28 and 45% by generation 50. The $I(M,i,z)$ curve reaches a minimum value at generation 28. For a value of $P(M,i)$ of 35%, the number of runs $R(z)$ is 11. The two numbers in the oval indicate that if this problem is run through to generation 28, processing a total of 1,276,000 (i.e., 4,000 × 29 generations × 11 runs) individuals is sufficient to yield a solution to this problem with 99% probability.

Thus, according to this measure of computational effort, the even-4-parity problem is about 16 times harder to solve than the even-3-parity problem.

We are unable to directly extend this comparison of the computational effort necessary to solve the even-parity problem with increasing numbers of arguments with our chosen population size of 4,000. When the even-5-parity function was run with a population size of 4,000 and each run arbitrarily stopped at our chosen maximum number $G = 51$ of generations to be run, no solution was found after 20 runs. Even after increasing the population size to 8,000 (with $G = 51$), we did not get a solution until our eighth run. This solution contained 347 points.

Notice that the structural complexity (i.e., the total number of function points and terminal points in the S-expression) of the solutions dramatically increased with an increasing number of arguments (i.e. structural complexity was 45 and 149, respectively, above for the 3- and 4-parity functions).

The population size of 4,000 is undoubtedly not optimal for any particular parity problem and is certainly not optimal for all sizes of parity problems. Nonetheless, it is clear that learning the even-parity functions with increasing numbers of arguments requires dramatically increasing computational effort and that the structural complexity of the solutions become increasingly large.

3   AUTOMATIC FUNCTION DEFINITION

The inevitable increase in computational effort and structural complexity for solving parity problems of order greater than four could be controlled if we could discover the underlying regularities and symmetries of this problem and then hierarchically decompose the problem into more tractable sub-problems. Specifically, we need to discover a function parameterized by dummy variables that would be helpful in decomposing and solving the problem.

If a human programmer were writing code for the even-3-parity or even-4-parity functions, he would probably choose to call upon either the odd-2-parity function (also known as the exclusive-or function XOR, the inequality function, and two-argument Boolean rule 6) or the even-2-parity function (also known as the equivalence function EQV, the not-exclusive-or function, and two-argument Boolean rule 9). If a human programmer were writing code for the even-5-parity function and parity functions with additional arguments, he would probably also want to call upon either the even-3-parity (three-argument Boolean rule 105) or the odd-3-parity (three-argument Boolean rule 150). These lower-order parity functions would greatly facilitate writing code for the higher-order parity functions. None of these low-order parity functions are, of course, in the original set F of available primitive Boolean functions.

The potentially helpful role of dynamically evolving useful building blocks in genetic programming has been recognized for some time [Koza 1990]. When we talk about "function definition" in this paper, we are not contemplating merely defining a function in terms of a sub-expression composed of particular fixed terminals (i.e., actual variables) of the problem. Instead, we are contemplating defining functions *parameterized* by dummy variables (formal parameters). Specifically, if the exclusive-or function XOR were to be dynamically defined during a run, it would be a version of XOR parameterized by two dummy variables (called ARG0 and ARG1), not a mere call to XOR with particular fixed terminals (e.g., D0 and D1). When this parameterized version of the XOR function is called, its two dummy variables ARG0 and ARG1 would be instantiated with two specific values, which would either be the values of two terminals (i.e., actual variables of the problem) or the values of two expressions (each composed ultimately of terminals). For example, the exclusive-or function XOR might be called via (XOR D0 D1) on one occasion and via (XOR D2 D3) on another occasion. On yet another occasion, XOR might be called via

    (XOR (AND D1 D2) (OR D0 D2)),

where the two arguments to XOR are the values returned by the expressions (AND D1 D2) and (OR D0 D2), respectively. Each of these expressions is ultimately composed of the actual variables (i.e., terminals) of the problem.

Moreover, when we talk about "automatic" and "dynamic" function definition, the goal is to dynamically evolve a dual structure containing both function-defining branches and result-producing (i.e., value-returning) branches by means of natural selection and genetic operations. We expect that genetic programming will dynamically evolve potentially useful function definitions during the run and also dynamically evolve an appropriate result-producing "main" program that calls these automatically defined functions.

Note that many existing paradigms for machine learning and artificial intelligence automatically and dynamically define functional subunits during runs (the specific terminology, of course, being specific to the particular paradigm). For example, when a set of weights are discovered enabling a particular neuron in a neural network to perform some subtask, that learning process can be viewed as a process of defining a function (i.e., a function taking the values of the specific inputs to that neuron as arguments and returning an output signal, perhaps a zero or one). Note, however, that the function thus defined is called only once from within the neural network. It is called only in the specific part of the neural net (i.e., the neuron) where it was created and it is called only with the original, fixed set of inputs to that specific neuron. Note also that existing paradigms for neural networks do not provide a way to re-use the set of weights discovered in that part of the network in other parts of the network where a similar subtask must be performed on a different set of inputs. The recent work of Gruau [1992] on recursive solutions to Boolean functions is a notable exception.

3.1   EVEN-4-PARITY FUNCTION

Automatic function definition can be implemented within the context of genetic programming by establishing a constrained syntactic structure [Koza 1992, Chapter 19] for the individual S-expressions in the population in which each individual contains one or more function-defining branches and one or more "main" result-producing branches which may call the defined functions.

The number of result-producing branches is determined by the nature of the problem. Since Boolean parity functions return only a single Boolean value, there would be only one "main" result-producing branch to the S-expression in the constrained syntactic structure required.

We usually do not know *a priori* the optimal number of functions that will be useful for a given problem or the optimal number of arguments for each such function; however, considerations of computer resources (time, virtual memory usage, CONSing, garbage collection, and memory fragmentation) necessitate that choices be made. Additional computer resources are required for each additional function definition. There is a considerable increase in the computer resources required to support the ever-larger S-expressions associated with each larger number of arguments. There will usually be no advantage to having defined functions that take more arguments than there are terminals in the problem. When Boolean functions are involved, there is no advantage to evolving one-argument function definitions (since there are only four rather impotent one-argument Boolean functions).

Thus, for the Boolean even-4-parity problem, it would seem reasonable to permit one two-argument function definition and one three-argument function definition within each S-expression. Thus, each individual S-expression in the population would have three branches. The first (leftmost) branch permits a two-argument function definition (defining a function called ADF0); the second (middle) branch permits a three-argument function definition (defining a function called ADF1); and the third (rightmost) branch is the result-producing branch. The first two branches are function-defining branches which may or may not be called upon by the result-producing branch.

Figure 3 shows an abstraction of the overall structure of an S-expression with two function-defining branches and one result-producing branch. There are 11 "types" (as defined in Koza 1992, Chapter 19) of points in each individual S-expression in the population for this problem. The first eight types are an invariant part of each individual S-expression. The 11 types are as follows: (1) the root (which will always be the placeholding PROGN function), (2) the top point DEFUN of the function-defining branch for ADF0, (3) the name ADF0 of the function defined by this first function-defining branch, (4) the argument list (ARG0 ARG1) of ADF0, (5) the top point DEFUN of the function-defining branch for ADF1, (6) the name ADF1 of the function defined by this second function-defining branch, (7) the argument list (ARG0 ARG1 ARG2) of ADF1, (8) the top point VALUES of the result-producing branch for the individual S-expression as a whole, (9) the body of ADF0, (10) the body of ADF1, and (11) the body of the "main" result-producing branch.

Three syntactic rules of construction govern points of types 9, 10, and 11.

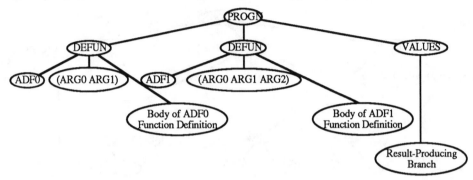

Figure 3 Abstraction of the overall structure of an S-expression with two function-defining branches and the one result-producing branch.

First, the body of ADF0 (i.e., the points of type 9) is a composition of functions from the given function set F and terminals from the terminal set A_2 of two dummy variables, namely A_2 = {ARG0, ARG1}.

Second, the body of ADF1 (i.e., the points of type 10) is a composition of functions from the given function set F and terminals from the set A_3 of three dummy variables, namely A_3 = {ARG0, ARG1, ARG2}.

Third, the body of the result-producing branch (i.e., the points of type 11) is a composition of terminals (i.e., actual variables of the problem) from the terminal set T, namely T = {D0, D1, D2, D3}, as well as functions from the set F_3. F_3 contains the four original functions from the function set F as well as the two-argument function ADF0 defined by the first branch and the three-argument function ADF1 defined by the second branch. That is, the function set F_3 is

$$F_3 = \{AND, OR, NAND, NOR, ADF0, ADF1\},$$

taking two, two, two, two, two, and three arguments, respectively. Thus, the result-producing branch is capable of calling the two defined functions ADF0 and ADF1.

When the overall S-expression is evaluated, the PROGN evaluates each branch; however, the value(s) returned by the PROGN consists only of the value(s) returned by the VALUES function in the final result-producing branch.

Note that one might consider including the terminals from the terminal set T (i.e., the actual variables of the problem) in the function-defining branches; however, we do not do so here.

Figure 4 shows an expansion of figure 3 representing a hypothetical program for the even-4-parity function containing two defined functions, ADF0 and ADF1. The first branch of the illustrative program in figure 4 contains a function definition for the two-argument defined function ADF0. This function is expressed in terms of the two dummy variables ARG0 and ARG1 and happens to be the even-2-parity function (i.e., the equivalence function EQV).

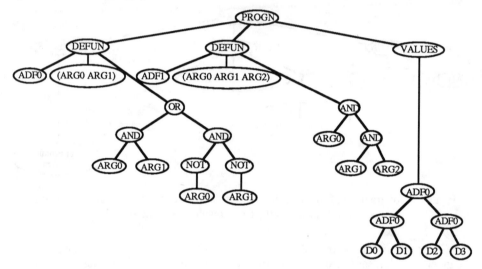

Figure 4 Illustrative program for the even-4-parity function containing two function definitions.

The second branch contains a function definition for the three-argument defined function ADF1. This function is expressed in terms of the three dummy variables ARG0, ARG1, and ARG2 and happens to be the ordinary three-argument conjunction function. The result-producing branch contains the actual variables of the problem (i.e., terminals) D0, D1, D2, and D3 and three calls to the defined function ADF0 (i.e., the equivalence function EQV) and no calls to the defined function ADF1. The result-producing branch computes

```
(EQV (EQV D0 D1) (EQV D2 D3)).
```

When the entire program is evaluated, the value returned is the value of the even-4-parity function.

In what follows, genetic programming will be allowed to evolve two function definitions in the function-defining branches of each S-expression and then, at its discretion, to call one, two, or none of these defined functions in the result-producing branch. We do not specify what functions will be defined in the two function-defining branches. We do not specify whether the defined functions will actually be used (it being, of course, possible to solve this problem without any function definition by evolving the correct program in the result-producing branch). We do not favor one function-defining branch over the other. We do not require that a function-defining branch use all of its available dummy variables. The structure of all three branches is determined by the combined effect, over many generations, by the selective pressure exerted by the fitness measure and by the effects of the operations of Darwinian fitness proportionate reproduction and crossover.

Since a constrained syntactic structure is involved, we must create the initial random generation so that every individual S-expression in the population has the syntactic structure specified by the syntactic rules of construction presented above. Specifically, every individual S-expression must have the invariant structure represented by the eight points of types 1 through 8.

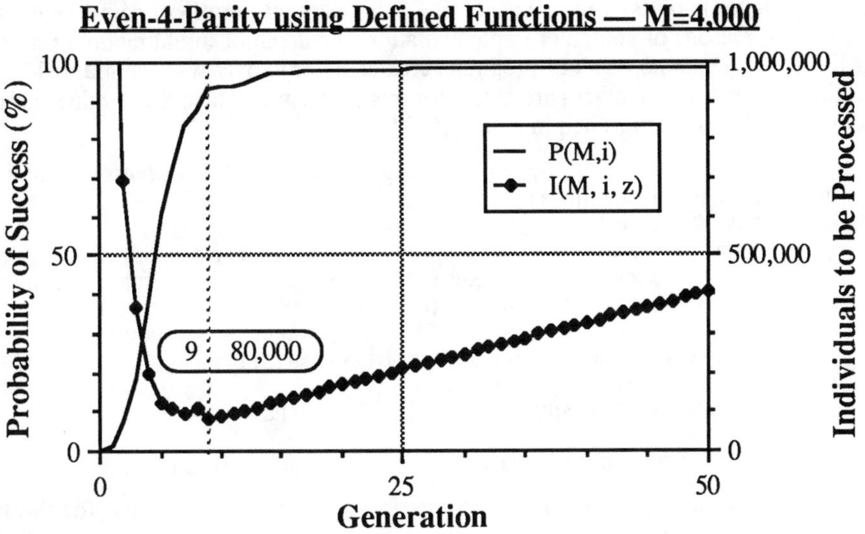

Even-4-Parity using Defined Functions — M=4,000

Figure 5 Performance curves for the even-4-parity problem showing that it is sufficient to process 80,000 individuals to yield a solution with automatic function definition.

In addition, the bodies of ADF0 (type 9), ADF1 (type 10), and the result-producing branch (type 11) must be composed of the functions and terminals specified by the above syntactic rules of construction.

Moreover, since a constrained syntactic structure is involved, we must perform structure-preserving crossover [Koza 1992, Chapter 19] so as to ensure the syntactic validity of all offspring as the run proceeds from generation to generation. Structure-preserving crossover is implemented by first allowing the selection of the crossover point in the first parent to be any point from the body of ADF0 (type 9), ADF1 (type 10), or the result-producing branch (type 11). However, once the crossover point in the first parent has been selected, the crossover point of the second parent must be of the same type (i.e., types 9, 10, or 11). This restriction on the selection of the crossover point of the second parent assures syntactic validity of the offspring. For example, if the crossover point of the first parent comes from the body of ADF0 (type 9), a crossover fragment from the body of ADF1 (type 10) might contain the dummy variable ARG2, which is not an allowable point in the body of ADF0.

In one run of the even-4-parity problem, the best-of-generation individual from the initial random generation (i.e., generation 0) contained 23 points and had a raw fitness of 10 (out of a possible 16). As one would expect, generation 0 contains individuals with useless function definitions and equally useless result-producing branches. For example, ADF0 in the above best-of-generation individual was defined as the ordinary two-argument disjunction function (which is already in the original function set F).

Each new generation of the population is created from the preceding generation by applying the fitness proportionate reproduction operation to 10% of the population and by applying the crossover operation to 90% of the population (with both parents selected with a probability proportionate to fitness). In selecting crossover points, 90% were internal (function) points of the tree and 10% were external (terminal) points of the tree. For practical reasons of computer implementation, the depth of initial random programs was limited to 6 and the depth of programs created by crossover was limited to 17. The selection of values of the other parameters for this problem are the default values used on most of the other problems cited in Koza [1992].

In generation 12, the following S-expression appeared containing 74 points and attaining a perfect value of 16 for raw fitness:

```
(PROGN (DEFUN ADF0 (ARG0 ARG1)
         (NAND (OR (AND (NOR ARG0 ARG1) (NOR (AND ARG1 ARG1) ARG1)) (NOR
         (NAND ARG0 ARG0) (NAND ARG1 ARG1))) (NAND (NOR (NOR ARG1 ARG1)
         (AND (OR (NAND ARG0 ARG0) (NOR ARG1 ARG0)) ARG0)) (AND (OR ARG0
         ARG0) (NOR (OR (AND (NOR ARG0 ARG1) (NAND ARG1 ARG1)) (NOR
         (NAND ARG0 ARG0) (NAND ARG1 ARG1))) ARG1)))))
       (DEFUN ADF1 (ARG0 ARG1 ARG2)
         (OR (AND ARG2 (NAND ARG0 ARG2)) (NOR ARG1 ARG1)))
       (VALUES
         (ADF0 (ADF0 D0 D2) (NAND (OR D3 D1) (NAND D1 D3))))).
```

The first branch of this best-of-run S-expression is a function definition for the two-argument defined function ADF0, which, when simplified, is equivalent to the exclusive-or (XOR) function (i.e. the odd-2-parity function).

The second branch defines the three-argument defined function ADF1, but this defined function is not called by the result-producing branch.

The result-producing branch of this best-of-run individual contains two references to ADF0. Upon substitution of XOR for ADF0, it becomes

```
(XOR (XOR D0 D2) (NAND (OR D3 D1) (NAND D1 D3))),
```

which, when simplified, is equivalent to

```
(XOR (XOR D0 D2) (EQV D3 D1)),
```

which is a correct solution to the even-4-parity problem.

Note that we did not specify that the exclusive-or function would be defined in ADF0, as opposed to, say, the equivalence function, the if-then function, or some other function. We did not specify that the exclusive-or function would be defined in the first branch as opposed to the second branch. We did not specify that the second branch would be ignored. Genetic programming created the two-argument defined function ADF0 on its own in the first branch to help solve this problem. Having done this, genetic programming then used ADF0 in an appropriate way in the result-producing branch to solve the problem. Notice that the 41 points above are considerably fewer than the 149 points contained in the S-expression cited earlier for the even-4-parity problem.

Figure 5 presents the performance curves based on 168 runs for the even-4-parity function with automatic function definition. The cumulative probability of success $P(M,i)$ was 93% by generation 9 and was 99% by generation 50. The two numbers in the oval indicate that if this problem is run through to generation 9, processing a total of 80,000 individuals (i.e., $4,000 \times 10$ generations $\times 2$ runs) is sufficient to yield a solution to this problem with 99% probability.

The 80,000 individuals that had to be processed for the even-4-parity problem using automatic function definition is one sixteenth of the 1,276,000 individuals needed when automatic function definition was not used (as shown in figure 2).

3.2 EVEN-5-PARITY FUNCTION

If automatic function definition is used, solutions to the even-5-parity and other higher order parity functions are readily found with a population size of 4,000.

For the even-5-parity problem, each S-expression has four branches with automatic function definition. The first three branches permit creation of function definitions with two, three, and four dummy variables. The result-producing (i.e., fourth) branch is an S-expression incorporating the four Boolean functions from the function set F; the three defined functions ADF0, ADF1, and ADF2; and the five terminals D0, D1, D2, D3, and D4.

In one run of the even-5-parity problem, genetic programming created a function definition for the even-3-parity function and created the necessary combination of operations in its result-producing branch to perform the behavior of the odd-2-parity function XOR.

Figure 6 graphically depicts the result-producing branch of this best-of-run individual.

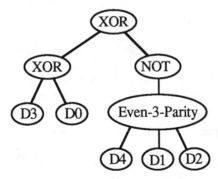

Figure 6 Result-producing branch of the best-of-run individual for the even-5-parity problem with automatic function definition.

Figure 7 shows the performance curves based on 7 runs for the even-5-parity function with automatic function definition. The cumulative probability of success $P(M,i)$ is

100% by generation 37. The two numbers in the oval indicate that if this problem is run through to generation 37, processing a total of 152,000 individuals (i.e., 4,000 × 38 generations × 1 run) is sufficient to yield a solution to this problem with 99% probability.

The 152,000 individuals that must be processed for the even-5-parity problem using automatic function definition is less than an eighth of the 1,276,000 individuals shown in figure 2 for the even-parity function of *only four* arguments.

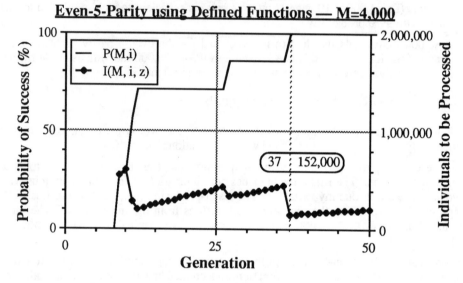

Figure 7 Performance curves for the even-5-parity problem show that it is sufficient to process 152,000 individuals to yield a solution with automatic function definition.

3.3   EVEN-6-PARITY FUNCTION

Similarly, the even-6-parity problem can be solved using automatic function definition. Processing a total of 812,000 individuals is sufficient to yield a solution to this problem with 99% probability. This 812,000 is less than the 1,276,000 individuals shown in figure 2 for the even-parity function of *only four* arguments.

In summary, automatic function definition considerably improved performance for the 4-, 5-, and 6-parity problems.

4   HIERARCHICAL AUTOMATIC FUNCTION DEFINITION

In the previous section on automatic function definition, the definition of a particular defined function never included a call to another defined function. However, it is common in ordinary programming to define one function in terms of other already-defined functions.

In the *hierarchical* form of automatic function definition, any function definition can call upon any other already-defined function. That is, there is a hierarchy (lattice) of function definitions wherein any function can be defined in terms of any combination of already-defined functions.

The number of function-defining branches and the number of dummy variables to appear in each function-defining branch is a matter of computer resources. We decided to use two function-defining branches and one result-producing branch. We also decided that, if the problem involves n terminals, then all function-defining branches will have $n - 1$ dummy variables; however, after the even-6-parity function, we reverted to four dummy variables to reduce the bushiness of the S-expression and thereby save computer resources.

4.1   EVEN-4-PARITY FUNCTION

Each S-expression in the population for solving the even-4-parity function has one result-producing branch and two function-defining branches, each permitting the definition of one function of three dummy variables.

In one run of the even-4-parity function, the following 100%-correct solution containing 45 points with a perfect value of 16 for raw fitness appeared on generation 4:

```
(PROGN (DEFUN ADF0 (ARG0 ARG1 ARG2)
          (NOR (NOR ARG2 ARG0) (AND ARG0 ARG2)))
       (DEFUN ADF1 (ARG0 ARG1 ARG2)
         (NAND (ADF0 ARG2 ARG2 ARG0)
               (NAND (ADF0 ARG2 ARG1 ARG2)
                     (ADF0 (OR ARG2 ARG1)
                           (NOR ARG0 ARG1)
                           (ADF0 ARG1 ARG0 ARG2)))))
       (VALUES
         (ADF0 (ADF1 D1 D3 D0)
               (NOR (OR D2 D3) (AND D3 D3))
               (ADF0 D3 D3 D2))))).
```

The first branch of this best-of-run S-expression is a function definition establishing the defined function ADF0 as the two-argument exclusive-or (XOR) function. The definition of ADF0 ignores one of the available dummy variables, namely ARG1.

The second branch of the above S-expression calls upon the defined function ADF0 (i.e., XOR) to define ADF1. This second branch appears to use all three available dummy variables; however, it reduces to the two-argument equivalence function EQV.

The result-producing (i.e., third) branch of this S-expression uses all four terminals and both ADF0 and ADF1 to solve the even-4-parity problem. This branch reduces to

```
(ADF0 (ADF1 D1 D0) (ADF0 D3 D2)).
```

Then, this result-producing branch is equivalent to

```
(XOR (EQV D1 D0) (XOR D3 D2)).
```

That is, genetic programming decomposed the even-4-parity problem into two different parity problems of lower order (i.e., XOR and EQV).

Figure 8 shows the hierarchy (lattice) of function definitions used in this solution to the even-4-parity problem. Note also that the second of the two functions in this decomposition (i.e., EQV) was defined in terms of the first (i.e., XOR).

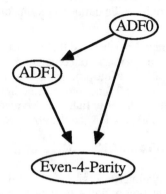

Figure 8 Hierarchy (lattice) of function definitions.

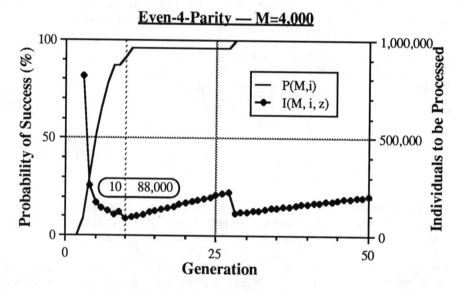

Figure 9 Performance curves for the even-4-parity problem show that it is sufficient to process 88,000 individuals to yield a solution with hierarchical automatic function definition.

Figure 9 presents the performance curves based on 23 runs for the even-4-parity with hierarchical automatic function definition. The cumulative probability of success $P(M,i)$ is 91% by generation 10 and 100% by generation 50. The two numbers in the oval indicate that if this problem is run through to generation 10, processing a total of 88,000 individuals (i.e., 4,000 × 11 generations × 2 runs) is sufficient to yield a solution to this problem with 99% probability.

4.2   EVEN-5-PARITY FUNCTION

In one run of the even-5-parity problem, the following 100%-correct solution containing 160 points with a perfect value of raw fitness of 64 emerged on generation 12:

```
(PROGN (DEFUN ADF0 (ARG0 ARG1 ARG2 ARG3)
         (OR (OR (NOR (NOR ARG3 ARG1) (OR ARG1 ARG3)) (AND (NAND ARG1
         ARG3) (NOR ARG1 ARG2))) (NAND (AND (OR ARG1 ARG2) (NAND ARG1
         ARG2)) (NAND ARG1 (AND (NOR ARG3 ARG1) ARG0)))))
       (DEFUN ADF1 (ARG0 ARG1 ARG2 ARG3)
         (NAND (NAND (AND (NAND ARG1 ARG2) (ADF0 ARG0 ARG3 ARG0 ARG2))
         (NOR (NAND ARG3 ARG1) (AND ARG1 ARG1))) (AND (ADF0 ARG0 (NAND
         ARG1 ARG2) (ADF0 ARG3 ARG0 ARG3 ARG0) (AND ARG1 ARG1)) (ADF0
         (ADF0 ARG3 ARG2 ARG3 ARG0) (ADF0 ARG0 ARG2 ARG2 ARG1) (ADF0
         ARG3 ARG3 ARG3 ARG0) (NOR ARG3 ARG0)))))
       (VALUES
         (OR (OR (NOR (ADF0 D3 D1 D1 D3) (OR D0 D1)) (NOR (NAND D1 D2)
         (OR (OR D3 D2) (NOR D4 D4)))) (ADF1 (ADF1 D4 D0 D4 D1) (OR (OR
         (NOR (OR (NAND D1 D0) (ADF1 D1 D2 D3 D1)) (AND D4 D0)) D2) (NOR
         (OR (NAND D1 D0) (ADF1 D1 D2 D3 D1)) (AND D4 D0))) (NAND (ADF1
         D1 D0 D0 D1) (NAND D0 D2)) (NAND (ADF1 D3 D4 D0 D0) (ADF0 D3 D1
         D1 D3)))))).
```

The first branch is equivalent to the four-argument Boolean rule 50,115, which is

```
(EQV ARG2 ARG1),
```

and which is an even-2-parity function that ignores two of the four available dummy variables.

The second branch is equivalent to the four-argument Boolean rule 38,250, which is

```
(OR (AND (NOT ARG2) (XOR ARG3 ARG0))
    (AND ARG2      (XOR ARG3 (XOR ARG1 ARG0)))).
```

Notice that this rule is not a parity function of any kind.

The result-producing (i.e., third) branch calls on defined functions ADF0 and ADF1 and solves the problem.

Figure 10 presents the performance curves based on 11 runs for the even-5-parity function with hierarchical automatic function definition. The cumulative probability of success $P(M,i)$ is 100% by generation 35. The two numbers in the oval indicate that if this problem is run through to generation 35, processing a total of 144,000 individuals (i.e., $4,000 \times 36$ generations $\times$ 1 run) is sufficient to yield a solution to this problem with 99% probability.

The number of individuals that must be processed to solve this problem with hierarchical automatic function definition (i.e., 144,000) is smaller than the 152,000 shown in figure 7, where only the non-hierarchical version of automatic function definition was employed.

4.3   EVEN 6-, 7-, 8-, 9-, AND 10-PARITY FUNCTIONS

The even 6- and 7-parity problems can be similarly solved using hierarchical automatic function definition. Processing a total of 864,000 individuals is sufficient to yield a solution to the even-6-parity problem with 99% probability. 1,440,000 individuals is sufficient to yield a solution to the even-7-parity problem with 99% probability.

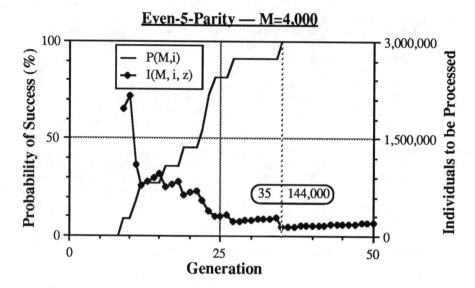

Figure 10 Performance curves for the even-5-parity problem using hierarchical automatic function definition show that it is sufficient to process 144,000 individuals.

The 8-, 9-, and 10-parity problems can be similarly solved using hierarchical automatic function definition. Each problem was solved within the first four runs.

For example, in one run of the even-8-parity function, the best-of-generation individual containing 186 points and attaining a perfect value of raw fitness of 256 appeared in generation 24. The first branch of this S-expression defined a four-argument defined function ADF0 (four-argument Boolean rule 10,280). The second branch of this S-expression defined a four-argument defined function ADF1 (four-argument Boolean rule 26,214) which ignored two of its four arguments and is equivalent to

```
(XOR D0 D1).
```

In one run of the even-9-parity function, the best-of-generation individual containing 224 points and attaining a perfect value of raw fitness of 512 appeared in generation 40. The first branch of this S-expression defined a four-argument defined function ADF0 (four-argument Boolean rule 1,872). The second branch of this S-expression defined a four-argument defined function ADF1 (four-argument Boolean rule 27,030) which is equivalent to the odd-4-parity function.

In one run of the even-10-parity function, the best-of-generation individual containing 200 points and attaining a perfect value of raw fitness of 1,024 appeared in generation 40. The first branch of this S-expression defined a four-argument defined function ADF0 (four-argument Boolean rule 38,791). The second branch of this S-expression defined a four-argument defined function ADF1 (four-argument Boolean rule 23,205) which ignored one of its four arguments. This rule is equivalent to

```
(EVEN-3-PARITY D3 D2 D0).
```

4.4   **EVEN-11-PARITY FUNCTION**

In one run of the even-11-parity function, the following best-of-generation individual containing 220 points and attaining a perfect value of raw fitness of 2,048 appeared in generation 21:

```
(PROGN (DEFUN ADF0 (ARG0 ARG1 ARG2 ARG3)
         (NAND (NOR (NAND (OR ARG2 ARG1) (NAND ARG1 ARG2)) (NOR (OR ARG1
         ARG0) (NAND ARG3 ARG1))) (NAND (NAND (NAND (NAND ARG1 ARG2)
         ARG1) (OR ARG3 ARG2)) (NOR (NAND ARG2 ARG3) (OR ARG1 ARG3)))))
       (DEFUN ADF1 (ARG0 ARG1 ARG2 ARG3)
         (ADF0 (NAND (OR ARG3 (OR ARG0 ARG0)) (AND (NOR ARG1 ARG1) (ADF0
         ARG1 ARG1 ARG3 ARG3))) (NAND (NAND (ADF0 ARG2 ARG1 ARG0 ARG3)
         (ADF0 ARG2 ARG3 ARG3 ARG2)) (ADF0 (NAND ARG3 ARG0) (NOR ARG0
         ARG1) (AND ARG3 ARG3) (NAND ARG3 ARG0))) (ADF0 (NAND (OR ARG0
         ARG0) (ADF0 ARG3 ARG1 ARG2 ARG0)) (ADF0 (NOR ARG0 ARG0) (NAND
         ARG0 ARG3) (OR ARG3 ARG2) (ADF0 ARG1 ARG3 ARG0 ARG0)) (NOR
         (ADF0 ARG2 ARG1 ARG2 ARG0) (NAND ARG3 ARG3)) (AND (AND ARG2
         ARG1) (NOR ARG1 ARG2))) (AND (NAND (OR ARG3 ARG2) (NAND ARG3
         ARG3)) (OR (NAND ARG3 ARG3) (AND ARG0 ARG0)))))
       (VALUES
         (OR (ADF1 D1 D0 (ADF0 (ADF1 (OR (NAND D1 D7) D1) (ADF0 D1 D6 D2
         D6) (ADF1 D6 D6 D4 D7) (NAND D6 D4)) (ADF1 (ADF0 D9 D3 D2 D6)
         (OR D10 D1) (ADF1 D3 D4 D6 D7) (ADF0 D10 D8 D9 D5)) (ADF0 (NOR
         D6 D9) (NAND D1 D10) (ADF0 D10 D5 D3 D5) (NOR D8 D2)) (OR D6
         (NOR D1 D6))) D1) (NOR (NAND D1 D10) (ADF0 (OR (ADF0 D6 D2 D8
         D4) (OR D4 D7)) (NOR D10 D6) (NOR D1 D2) (ADF1 D3 D7 D7
         D6)))))))  .
```

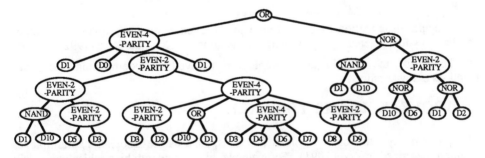

Figure 11 The best-of-run individual from generation 21 of one run of the even-11-parity problem is a composition of even-2-parity and even-4-parity functions.

The first branch of this S-expression defined the four-argument defined function ADF0 (four-argument Boolean rule 50,115) which ignored two of its four arguments. ADF0 is equivalent to the even-2-parity function, namely

```
(EQV ARG1 ARG2).
```

The second branch defined a four-argument defined function ADF1 which is equivalent to the even-4-parity function.

Substituting the definitions of the defined functions ADF0 and ADF1, the result-producing (i.e., third) branch becomes the program shown below.

```
(OR (EVEN-4-PARITY
        D1
        D0
        (EVEN-2-PARITY (EVEN-4-PARITY
                            (EVEN-2-PARITY D3 D2)
                            (OR D10 D1)
                            (EVEN-4-PARITY D3 D4 D6 D7)
                            (EVEN-2-PARITY D8 D9))
                 (EVEN-2-PARITY (NAND D1 D10)
                            (EVEN-2-PARITY D5 D3)))
        D1)
    (NOR (NAND D1 D10)
        (EVEN-2-PARITY (NOR D10 D6) (NOR D1 D2))))
```

which is equivalent to the target even-11-parity function. Note that the even-2-parity function (ADF0) appears six times in this solution and that the even-4-parity function (ADF1) appears three times. Note that this entire solution contains only 220 points (compared to 347 points for the solution to the even-5-parity using genetic programming without any function definitions).

Figure 11 shows the simplified version of the result-producing branch of this best-of-run individual for the even-11-parity problem. As can be seen, the even-11-parity problem was decomposed into a composition of even-2-parity functions and even-4-parity functions.

We found the above solution to the even-11-parity problem on our first completed run. The search space of 11-argument Boolean functions returning one value is of size $2^{2,048} \approx 10^{616}$.

A videotape visualization of the solution to the even-11-parity problem (and other problems involving genetic programming) can be found in Koza and Rice [1992].

In other words, the even-11-parity problem was solved by decomposing into parity functions of lower orders.

5 BOOLEAN 11-MULTIPLEXER

We can similarly learn other Boolean functions, such as the Boolean 11-multiplexer function (described in Koza 1991). For example, in one run with hierarchical automatic function definition, the program shown below appeared in generation 18.

```
(PROGN (DEFUN ADF0 (ARG0 ARG1 ARG2 ARG3)
           (NOR (OR ARG1 (NOR (NOR ARG2 ARG2)
                   (NAND (NAND ARG3 ARG3) (OR ARG0 ARG2))))
               (AND ARG3 ARG0)))
       (DEFUN ADF1 (ARG0 ARG1 ARG2 ARG3)
           (NOR (AND ARG0 ARG3)
               (AND (OR ARG1 ARG2) (NAND ARG1 ARG3))))
       (VALUES
           (ADF0 (ADF1 D2 D0 A0 A1)
               (AND D3 A2)
               (NAND (OR (NOR (AND D3 A2) A1) D3) (OR D1 A1))
               (ADF1 D2 A0 (AND D3 A2) A2)))).
```

The first branch of this 56-point best-of-run individual defines the four-argument defined function ADF0 which is equivalent to Boolean 4-argument rule 4,355. The second branch defines function ADF1 which is equivalent to 4-argument Boolean rule 17,667. The third branch calls on ADF0 once and calls ADF1 twice in order to create the Boolean 11-multiplexer.

6 CONCLUSIONS

The first conclusion of this paper is that the problems of learning the even-parity functions and the Boolean multiplexer function can be solved with the newly developed technique of hierarchical automatic function definition in the context of genetic programming.

The second conclusion is that the technique of hierarchical automatic function definition facilitates the solution of these problems. That is, when problems are decomposed into a hierarchy of function definitions and calls, many fewer individuals must be processed in order to yield a solution to the problem. Moreover, the solutions discovered are comparatively smaller. As can be seen in table 1, automatic function definition and hierarchical automatic function definition are helpful in solving the even-parity problems.

Table 1 Number of individuals $I(M,i,z)$ required to be processed to yield a solution to various even-parity problems with 99% probability.

Even-parity function	Genetic Programming	Genetic Programming with automatic function definition	Genetic Programming with hierarchical automatic function definition
3	80,000		
4	1,276,000	80,000	88,000
5		152,000	144,000
6		812,000	864,000
7			1,440,000

6.1   Acknowledgments

James P. Rice of the Knowledge Systems Laboratory at Stanford University made numerous contributions in connection with the computer programming of the above.

6.1.1   References

Gruau, Frederic. Genetic synthesis of Boolean neural networks with a cell rewriting developmental process. In Schaffer, J. D. and Whitley, Darrell (editors). *Proceedings of the Workshop on Combinations of Genetic Algorithms and Neural Networks 1992*. The IEEE Computer Society Press. 1992.

Koza, John R. *Genetic Programming: A Paradigm for Genetically Breeding Populations of Computer Programs to Solve Problems*. Stanford University Computer Science Department technical report STAN-CS-90-1314. June 1990.

Koza, John R. A hierarchical approach to learning the Boolean multiplexer function. In Rawlins, Gregory (editor). *Foundations of Genetic Algorithms*. San Mateo, CA: Morgan Kaufmann Publishers Inc. 1991. Pages 171-192.

Koza, John R. *Genetic Programming: On the Programming of Computers by Means of Natural Selection and Genetics*. Cambridge, MA: The MIT Press 1992.

Koza, John R. and Rice, James P. *Genetic Programming: The Movie*. Cambridge, MA: The MIT Press 1992.

Author Index

Key Word Index